C. Bittner, A. S. Busemann, U. Griesbach, F. Haunert,
W.-R. Krahnert, A. Modi, J. Olschimke, P. L. Steck

Organic Synthesis Workbook II

Foreword by Stuart Warren

WILEY-VCH

Further Reading from Wiley-VCH

Gewert, J.A./Görlitzer, J./
Götze, S./Looft, J./Menningen, P./
Nöbel, T./Schirok, H./Wulff, C.
Organic Synthesis Workbook
2000. ISBN 3-527-30187-9

Constable, E.C.
Metals and Ligand Reactivity
An Introduction to the Organic Chemistry of Metal Complexes
1996. 308 pp. ISBN 3-527-29278-0

Ansari, F. L./Qureshi, R./Qureshi, M.L.
Electrocyclic Reactions
From Fundamentals to Research
1998. 288 pp. ISBN 3-527-29755-3

Lehn, J.-M.
Supramolecular Chemistry
Concepts and Perspectives
1995. 288 pp. ISBN 3-527-29311-6

Waldmann, H./Mulzer, J. (eds.)
Organic Synthesis Highlights III
1998. ISBN 3-527-29500-3

Nicolaou, K.C./Sorensen, E.J.
Classics in Total Synthesis
1996. ISBN 3-527-29231-4

Hopf, H.
Classics in Hydrocarbon Chemistry
Syntheses, Concepts, Perspectives
2000. ISBN 3-527-29606-9

Lindhorst, T.K.
Essentials of Carbohydrate Chemistry and Biochemistry
2000. ISBN 3-527-29543-7

C. Bittner, A. S. Busemann, U. Griesbach, F. Haunert,
W.-R. Krahnert, A. Modi, J. Olschimke, P. L. Steck

Organic Synthesis Workbook II

Foreword by Stuart Warren

WILEY-VCH

Weinheim · New York · Chichester · Brisbane · Singapore · Toronto

C. Bittner
A. S. Busemann
U. Griesbach
F. Haunert
W.-R. Krahnert
A. Modi
J. Olschimke
P. L. Steck
Institut für Organische Chemie der
Universität Göttingen
Tammannstraße 2
D-37077 Göttingen

> This book was carefully produced. Nevertheless, authors and publisher do not warrant the information contained therein to be free of errors. Readers are advised to keep in mind that statements, data, illustrations, procedural details or other items may inadvertently be inaccurate.

Library of Congress Card No. applied for.
A cataloque record for this book is available from the British Libary.

Die Deutsche Bibliothek – Cataloguing-in-Publication Data
A catalogue record for this book is available from Die Deutsche Bibliothek

ISBN 3-527-30415-0

© WILEY-VCH Verlag GmbH. D-69469 Weinheim (Federal Republic of Germany), 2001

Printed on acid-free paper.

All rights reserved (including those of translation in other languages). No part of this book may be reproduced in any form – by photoprinting, microfilm, or any other means – nor transmitted or translated into machine language without written permission from the publishers. Registered names, trademarks, etc. used in this book, even when not specifically marked as such, are not to be considered unprotected by law.

Printing: betz-druck GmbH, D-69291 Darmstadt
Bookbinding: Buchbinderei J. Schäfer, D-67269 Grünstadt

Printed in the Federal Republic of Germany.

*Dedicated to our PhD adviser Prof. Dr. Dr. h. c. L. F. Tietze
on the occasion of his 60th birthday*

Foreword

Organic chemistry is easy to teach but difficult to learn. Students often complain that they understand the lectures or the book but 'can't do the exam questions'. This is largely because of the unique nature of the subject – at once more unified than any other branch of chemistry (or of science?) and more diverse in its applications. Research workers similarly often feel they understand the basic principles of the subject but fail to find a solution to a problem even though they understand their molecules very well. All organic chemists need to match intellectual learning with the skill to deal with the difficulty of the moment.

The answer to these dilemmas is problem solving. Or more exactly solving invented problems on paper at the same time as mastering the intellectual understanding. Now a new difficulty arises. Where is one to find a carefully graded set of problems arranged around a comprehensible framework that gives significance to the answers by showing that solving these problems is practical and useful? It is not easy to compile such a set of problems. I know, as I wrote both the problems in our recent textbook and the solutions manual.[1]

Organic Synthesis Workbook II will be the answer to many young organic chemists' prayers. It is a set of problems of extraordinary diversity set within the framework of large syntheses. This gives the young authors (all members of Professor Lutz Tietze's research group at Göttingen) the freedom to reveal details or to conceal them. The reader might be asked simply to furnish a reagent for a given step, or more challenging questions like explaining a mechanism or a stereoselectivity. Even prediction appears as some of the intermediates in the big syntheses are blank spaces to be filled in. The layout is intriguing - one wants to read on, as in the best novels, first to find out what happens and then to find out how it was done. Needless to say, just turn the page and the answers appear. And just because you couldn't do that problem, you're not handicapped when it comes to the next.

You should not suppose that this book is simply about organic synthesis. It has a lot to offer to the general student of organic chemistry at the advanced undergraduate and graduate level. The problems vary in difficulty but there is something to suit us all. The rewards of tackling the problems seriously will be great. I am very enthusiastic about this book and I know a lot of readers will share my enthusiasm.

[1] J. Clayden, N. Greeves, S. Warren, P. Wothers, Organic Chemistry.

Stuart Warren
Cambridge 2001

Preface

Thank you for purchasing this book; we hope you will enjoy it.
Based on a seminar in the research group of *Prof. Dr. Dr. h. c. L. F. Tietze* at the University of Göttingen, Germany, eight members of the group contributed to a collection of synthesis problems in 1998, and this was published by Wiley-VCH under the title "Organic Synthesis Workbook". Encouraged by the success of this approach toward understanding organic synthesis we decided to write a sequel containing more recent chemistry. In addition we have included carbohydrate and industrial scale chemistry.
We have not changed the proved original concept, and therefore we hope that those who already know *Organic Synthesis Workbook* will feel at home.
This book contains 16 independent chapters, based on publications of well known scientists.
Each chapter is divided into five parts. First, the **Introduction** will give you a brief view of the target molecule and its background. The **Overview** shows the complete synthetic problem on two pages. In the **Synthesis** section the reaction sequence is divided into individual *Problems*. Afterwards *Hints* are given to assist you in solving the problem. Each further hint will reveal more and more of the solution; therefore it might be useful to cover the remaining page with a piece of paper. The *Solution* will show if your answer is correct. In the *Discussion* section the problem is explained in detail. However this book cannot serve as a substitute for an organic textbook. After the last problem, the **Conclusion** briefly comments on the synthesis, highlighting the key steps. The original references can be found in the **Literature** section for further reading.

We are very grateful for the support we received while writing this book, in particular to our PhD adviser Prof. Lutz F. Tietze and the members of his research group. We would also like to thank H. Bell, H. Braun, G. Brasche, S. Hellkamp, and S. Hölsken for proof reading. J. A. Gewert, J. Görlitzer, S. Götze, J. Looft, P. Menningen, T. Nöbel, H. Schirock and C. Wulff are the authors of the first problems workbook which made this sequel possible.

Christian Bittner Göttingen, 2001
Anke S. Busemann
Ulrich Griesbach
Frank Haunert
Wolf-Rüdiger Krahnert
Andrea Modi
Jens Olschimke
and Peter L. Steck

Contents

Chapter 1:   (+)-Asteriscanolide (Paquette 2000)   1

Chapter 2:   (−)-Bafilomycin A_1 (Roush 1999) 15

Chapter 3:   Curacin A (Wipf 1996) 35

Chapter 4:   Dysidiolide (Corey 1997) 55

Chapter 5:   Efavirenz (Merck, DuPont 1999) 71

Chapter 6:   (+)-Himbacine (Chackalamannil 1999) 85

Chapter 7:   Hirsutine (Tietze 1999) 101

Chapter 8:   (+)-Irinotecan® (Curran 1998) 121

Chapter 9:   (+)-Laurallene (Crimmins 2000) 137

Chapter 10: Myxalamide A (Heathcock 1999) 157

Chapter 11: (+)-Paniculatine (Sha 1999) 177

Chapter 12: (+)-Polyoxin J (Gosh 1999) 193

Chapter 13: (−)-Scopadulcic Acid (Overman 1999) 209

Chapter 14: Sildenafil (VIAGRA™) (Pfizer 1998) 231

Chapter 15: GM2 (Schmidt 1997) 245

Chapter 16: H-Type II Tetrasaccharide Glycal (Danishefsky 1995) . . 265

Abbreviations 281

Index 285

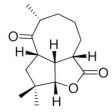

1

(+)-Asteriscanolide (Paquette 2000)

1.1 Introduction

The sesquiterpene (+)-asteriscanolide **1** was first isolated from *Asteriscus aquaticus* L and characterized by *San Feliciano* in 1985.[1] It has captured the attention of organic chemists mainly because of its uncommon bicyclo[6.3.0]undecane ring system bridged by a butyrolactone fragment. The only prior enantioselective synthesis of **1** has been described by *Wender* in 1988 featuring an Ni(0)-promoted [4 + 4]-cycloaddition.[2] *Booker-Milburn* and co-workers described the sequential application of intramolecular [2 + 2]-photocycloaddition, *Curtius* rearrangement, and oxidative fragmentation to produce the 7-desmethyl derivative in 1997.[3]
This problem is based on the work of *Paquette* published in 2000.[4]

1
(+)-asteriscanolide

1.2 Overview

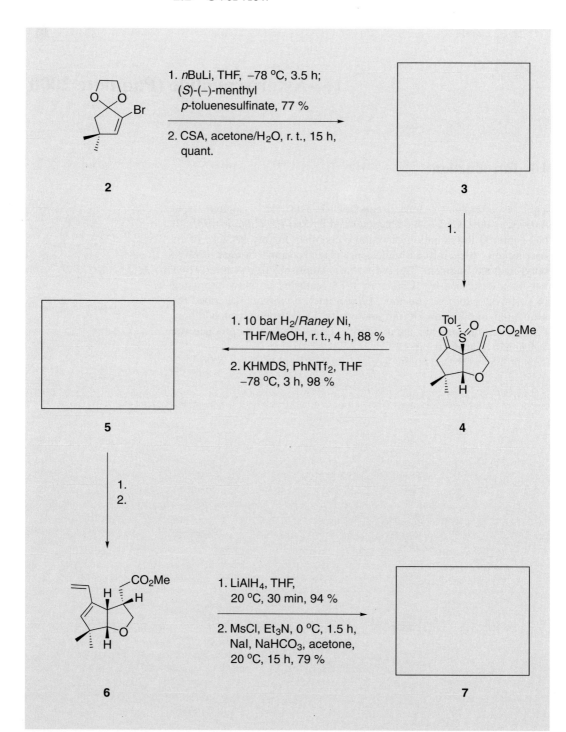

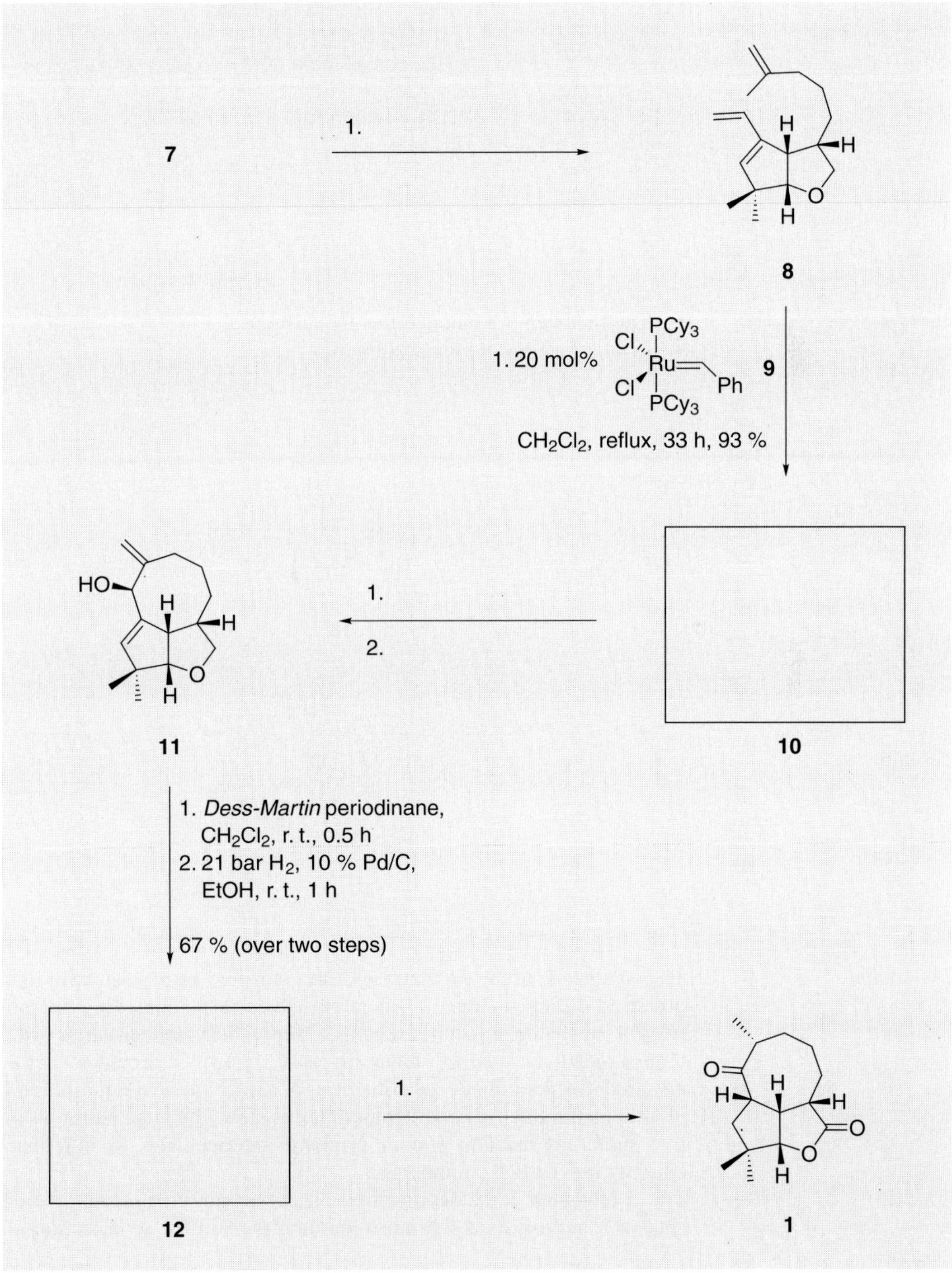

1 (+)-Asteriscanolide

7 →(1.) **8**

8 → 1. 20 mol% **9** (Cl₂(PCy₃)₂Ru=CHPh), CH₂Cl₂, reflux, 33 h, 93 % → **10**

10 → 1. / 2. → **11**

11:
1. *Dess-Martin* periodinane, CH₂Cl₂, r. t., 0.5 h
2. 21 bar H₂, 10 % Pd/C, EtOH, r. t., 1 h

67 % (over two steps)

12 →(1.) **1**

1.3 Synthesis

Problem

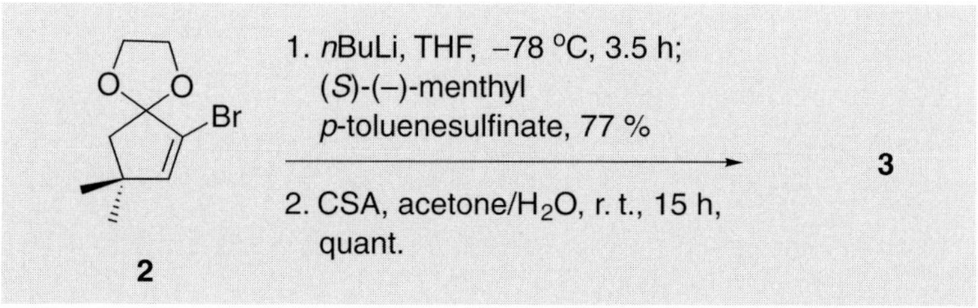

[Scheme: compound **2** (bromocyclopentenone with dioxolane protection and methyl group) → **3** via: 1. nBuLi, THF, −78 °C, 3.5 h; (S)-(−)-menthyl p-toluenesulfinate, 77 %; 2. CSA, acetone/H₂O, r. t., 15 h, quant.]

Hints

- The first step is a halogen-metal exchange.
- (S)-(−)-menthyl p-toluenesulfinate:

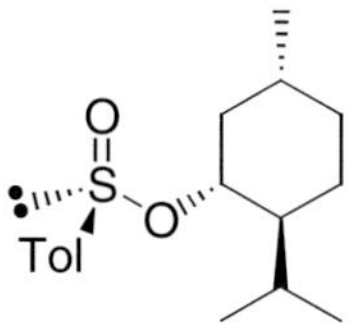

[Structure **13**: (S)-(−)-menthyl p-toluenesulfinate]

Solution

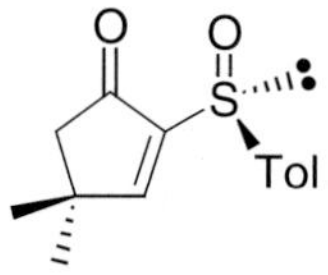

[Structure **3**: cyclopentenone bearing methyl group and (S)-tolyl sulfoxide]

Discussion

The synthesis of **3** was initiated by reaction of nBuLi with the protected cyclopentenone **2** generating the corresponding vinyllithium reagent by halogen-metal exchange. Subsequent condensation with (S)-(−)-menthyl *para*-toluenesulfinate (**13**) provides the enantiodefined sulfoxide substituent in **3**.[5] Since thermal equilibration of chiral sulfoxides at room temperature is slow, the large sulfur atom is a preferred reaction site in synthetic intermediates to introduce chirality into carbon compounds.

The second step is the deprotection of the ketone functionality. Acid-catalyzed hydrolysis is the most common method for deprotection of acetals or ketals. However, *Lewis* acids can also be used to effect deacetalization.

Problem

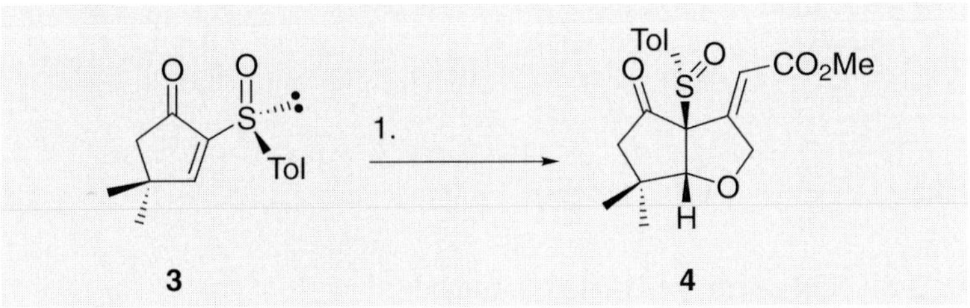

3 → 4 (via step 1.)

Hints

- **3** is a *Michael*-system.
- **3** reacts with methyl-4-hydroxy-2-butynoate (**14**).
- The ester **14** reacts as an oxygen-centered hetero-nucleophile with the *Michael*-system.
- The reaction is a twofold *Michael* reaction with a second stage intramolecular conjugate addition.

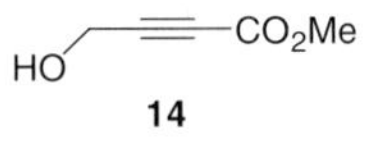

Solution

1. Methyl-4-hydroxy-2-butynoate (**14**), K$_2$CO$_3$, THF, r. t., 5 h, 38 %

Discussion

This domino[6] *Michael-Michael* reaction sequence is one of the key steps in this synthesis and proceeds with complete asymmetric induction, which could be confirmed by X-ray crystallographic analysis. The two five-membered rings in the target molecule are thereby generated in a single step.

The butynoate **14** adds to the chiral enone **3** from the α–surface with asymmetric induction probably rationalized by the chelate model **15**. The initial product **16** of the 1,4-conjugate addition is capable of another intramolecular *Michael* addition to the triple bond resulting in conversion of **16** to **4**.

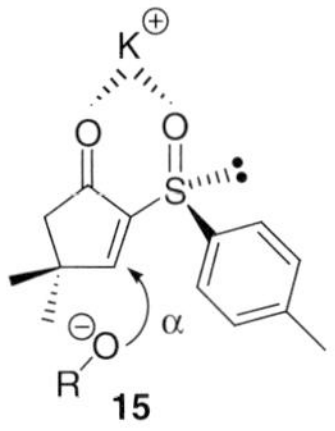

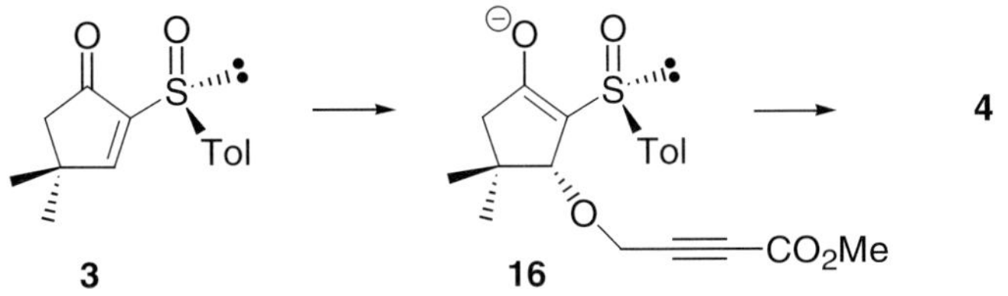

1 (+)-Asteriscanolide

Problem

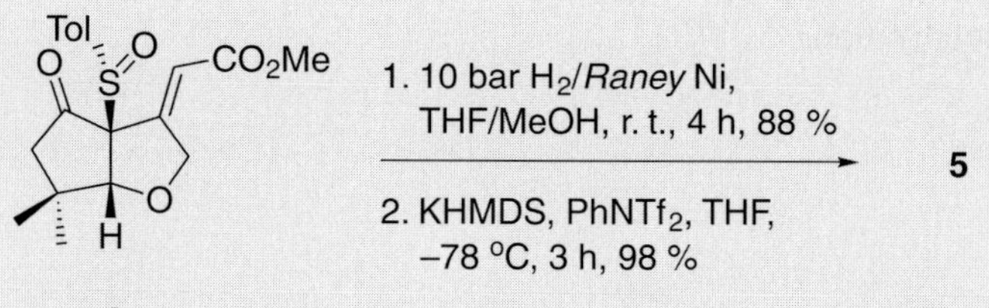

4 → 5

1. 10 bar H_2/*Raney* Ni, THF/MeOH, r.t., 4 h, 88 %
2. KHMDS, PhNTf$_2$, THF, −78 °C, 3 h, 98 %

Hints

- Two equivalents of H_2 are consumed.
- Reductive desulfurization as well as reduction of the double bond take place.
- KHMDS (potassium hexamethyldisilazide) is a strong non-nucleophilic base.
- An enol triflate is generated.

Solution

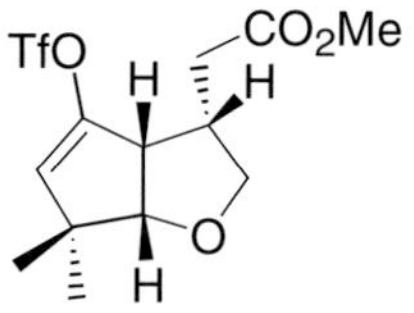

5

Discussion

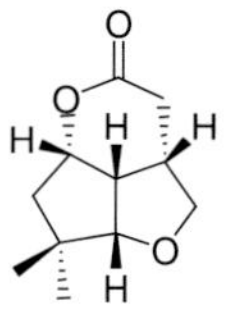

The most common application of *Raney* nickel is the desulfurization of a wide range of compounds including thioacetals, thiols, sulfides, disulfides, sulfoxides, and sulfur-containing heterocycles. In addition it can be used to reduce benzylic nitrogen and oxygen atoms. Hydrogenation of **4** in the presence of *Raney* nickel results in carbon-sulfur bond cleavage concomitant with saturation of the olefinic bond in 88 % yield. The configuration of the newly generated stereogenic centers was proved by facile overreduction. An increase of hydrogen pressure up to 70 bar was sufficient to reduce the ketone as well. The following intramolecular cyclization gives lactone **17** which could not take place if a different diastereomer had been initially produced.

Potassium hexamethyldisilazide is a strong non-nucleophilic base which deprotonates in α-position to the ketone. The resulting enolate can be captured as the enol triflate **5** by reaction with *N*-phenyl triflimide (PhNTf$_2$) and is directly used in the next reaction step.[7]

1 (+)-Asteriscanolide

Problem

Hints

- This reaction involves a palladium (0) catalyst.
- **5** is a coupling partner for tributylvinylstannane.
- What is the name of this reaction?

Solution

1. Tributylvinylstannane, LiCl, 10 mol% $Pd_2(dba)_3 \cdot CHCl_3$, THF, 20 °C, 15 h, 95 %

Discussion

One general reaction of organostannanes is the cross coupling with organic halides or triflates promoted by catalytic amounts of palladium, known as the *Stille* reaction.[8] The nature of such transformations involves a transfer of a carbon ligand from tin to palladium. The carbon-carbon bond formation proceeds *via* a reductive elimination. The reaction has proven to be very general with respect to both the halides (or triflates) and the types of stannanes that can be used. The groups that can be theoretically transferred from tin include alkyl, alkenyl, aryl and alkynyl. The approximate effectiveness of group-transfer is alkynyl > alkenyl > aryl > benzyl > methyl > alkyl. Unsaturated groups are normally transferred selectively. The reaction tolerates a broad range of functionalities both in the halide (or triflate) and in the tin reagent, such as ester, nitrile, nitro, and formyl groups.

The catalytic cycle in the *Stille* coupling reaction is accepted to involve formation of an active palladium(0) species **20**. The next step is the oxidative addition of the organic moiety ROTf (or RX) to palladium to give **21**. The subsequent transmetalation with $R'SnR_3''$ forms a species with an R-Pd-R' linkage (**24**). The catalytic cycle is completed by *cis/trans* isomerization (**25**) and reductive elimination to give **6** and the regenerated palladium species. The role of the often used additive lithium chloride is not certain. It had been demonstrated that the success of the intermolecular palladium(0)-catalyzed coupling of enol triflates with vinylstannanes depends upon the presence of LiCl in the reaction mixture.

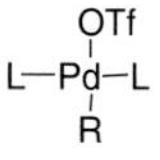

8   1   (+)-Asteriscanolide

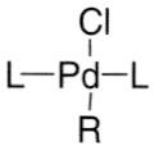

However, there are examples in intramolecular reactions where it is not necessary. An explanation is the formation of the unstable complex **18**. The palladium catalyst may be transformed into an uncharacterized, catalytically ineffective complex. This problem is overcome by addition of LiCl, which results in the production of the stable palladium complex **19**.[9]

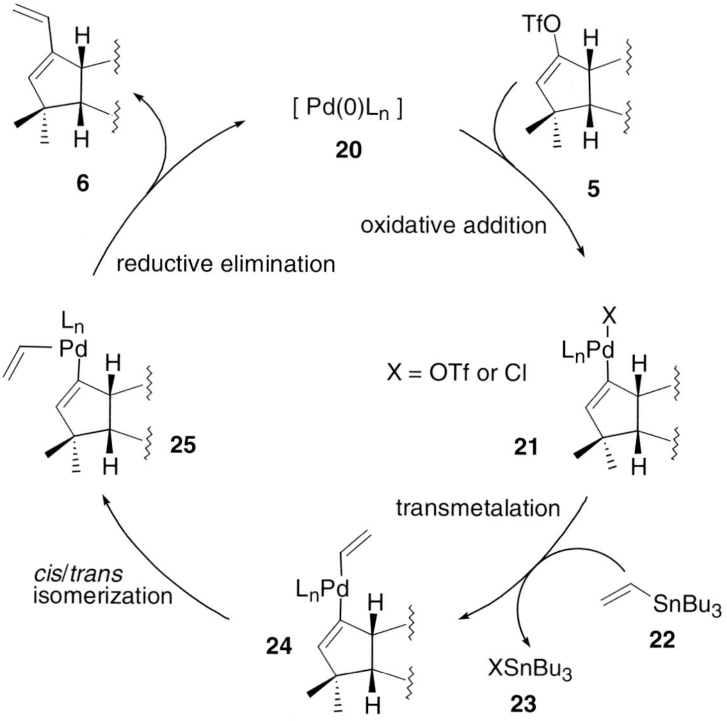

Problem

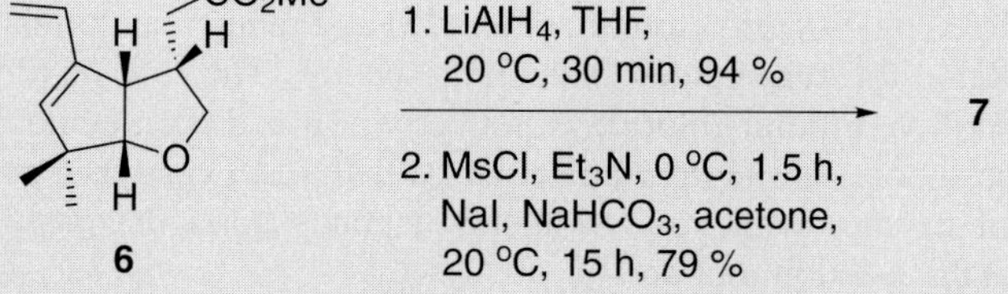

- LiAlH₄ reduces the ester to the primary alcohol. *Hints*
- Mesylates are good leaving groups.
- The second step is a substitution.

Solution

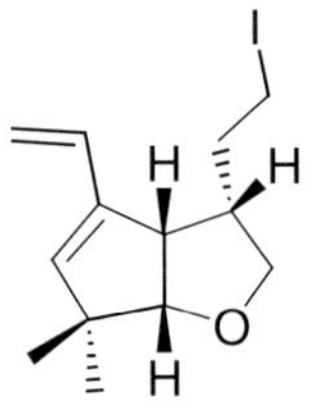

7

These two steps involve fairly standard procedures. LiAlH₄ is a widespread reagent for the reduction of esters to alcohols.
The transformation of an alcohol into a halide can be done either by substitution of a good leaving group such as mesylate by I⁻ (as in this case) or alternatively for example by *Appel* analog reactions involving PPh₃.

Discussion

Problem

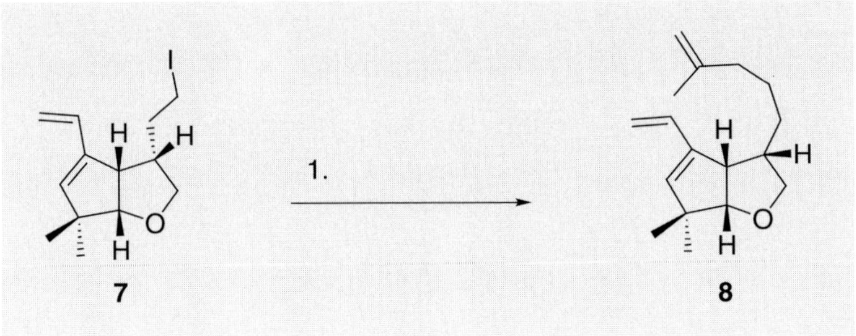

7 → **1.** → **8**

- The iodide is substituted by a *Grignard* reagent. *Hints*

1. Methallylmagnesium chloride, CuI, THF, 0 °C, 4 h, 98 % *Solution*

Discussion

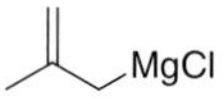

Coupling of alkyllithium and *Grignard* reagents with alkyl halides gives poor yields and if possible tends to produce mixtures of regio- and stereoisomers. However, effective procedures have been developed involving stoichiometric or catalytic use of Cu(I) salts.[10] Thus, the copper-catalyzed substitution of iodide **7** by methallylmagnesium chloride **26** proceeds smoothly in 98 % yield.

Problem

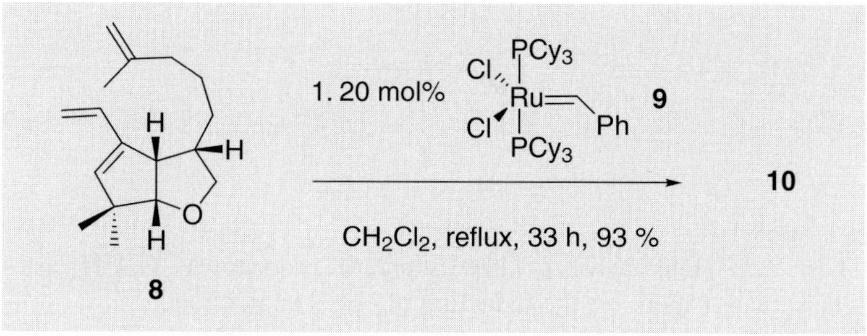

Hints

- What is the name of the Ru-catalyst **9**?
- What type of reaction does it catalyze?
- An eight-membered ring is formed in a ring-closing metathesis.

Solution

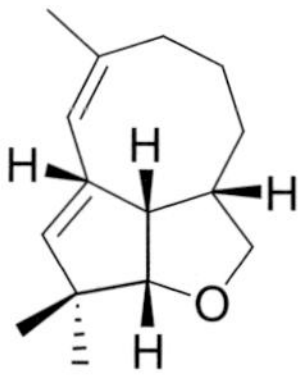

Discussion

The RCM[11] (ring-closing metathesis, see also chapter 9) of **8** using the *Grubbs* catalyst **9**[12] provides an eight-membered ring in which a conjugate 1,3-diene unit resides. The excellent yield of this metathesis reaction is remarkable because of entropic and enthalpic factors that impede the preparation of eight-membered rings.[13] It has become apparent that polar functions such as ethers, amides, urethanes, sulfonamides and esters greatly facilitate the assembly of cyclooctyl derivatives. In the absence of these internal ligands, the formation of eight-membered rings has been documented much less frequently. Evidently, the limited conformational flexing available to the side chains in **8** serves to facilitate their conjoining *via* the ruthenium carbenoid.

Problem

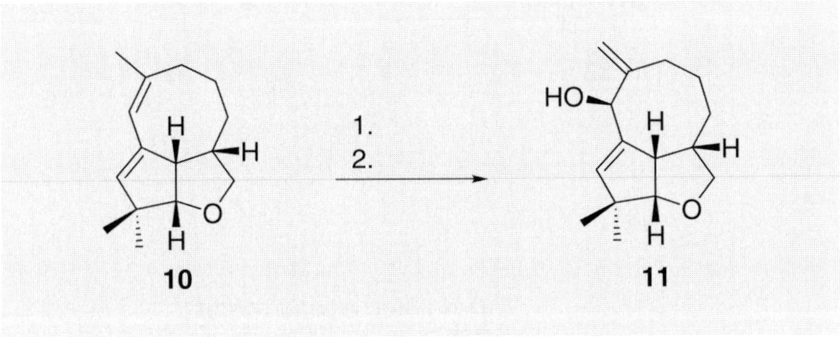

Hints

- The double bond is oxygenated selectively.
- Singlet oxygen is used to effect the photooxidation. What is the mechanism of the reaction?
- Final reduction leads to **11**.

Solution

1. O_2, TPP, CH_2Cl_2, r. t., 40 min
2. $LiAlH_4$, THF, 20 °C, 30 min
61 % (over two steps)

Discussion

The critical step in this synthesis was to achieve suitable oxygenation of the double bond internal and not external to the eight-membered ring. Experiments involving epoxidation or hydroboration were not successful. The reagent of choice turned out to be singlet oxygen in CH_2Cl_2.[14] The most common method for generating $^1O_2$ in solution is the dye sensitized photochemical excitation of triplet oxygen. In this case 5,10,15,20-tetraphenyl-21*H*,23*H*-porphine (TPP) (**27**) was used as sensitizer. Other common dyes are for example methylene blue, Rose Bengal, chlorophyll or riboflavin. For other reactions involving singlet oxygen see Chapter 4.

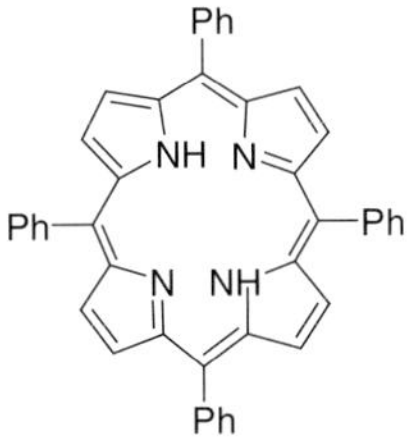

Mechanistically, the reaction is explained as an ene-type reaction involving a concerted electron shift (see **28**) forming an allylic hydroperoxide and direct hydride reduction of **29** gives rise to the diallylic alcohol **11**.

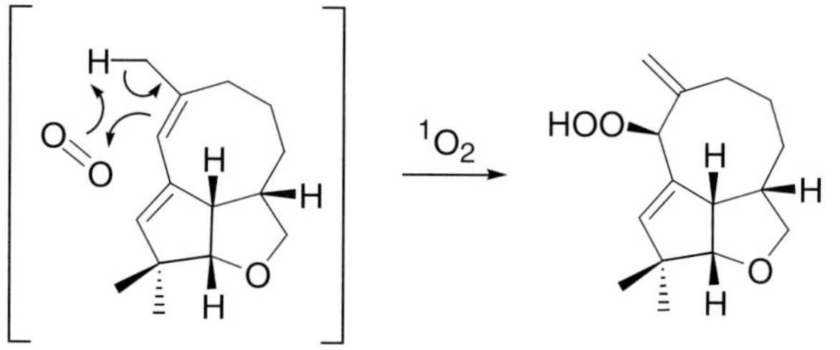

1 (+)-Asteriscanolide

Problem

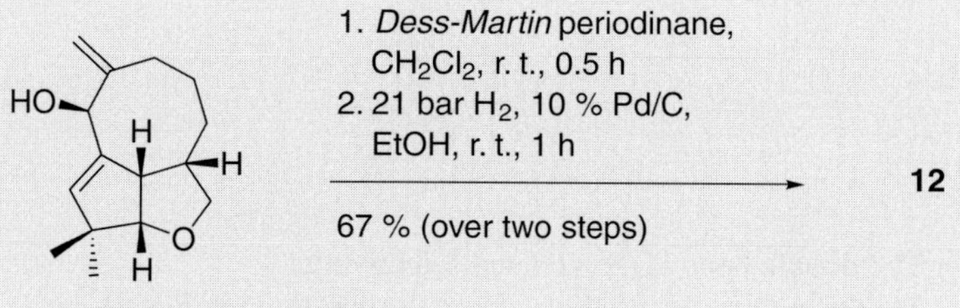

Hints

- The *Dess-Martin* periodinane is an oxidating reagent.
- All olefinic double bonds are reduced in the second step.

Solution

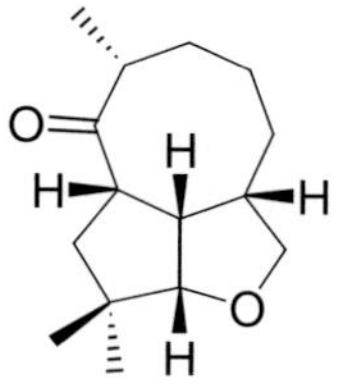

Discussion

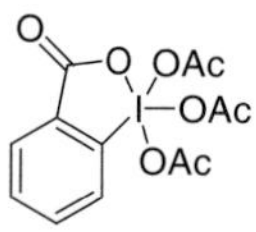

The *Dess-Martin* periodinane **30** (1,1,1-triacetoxy-1,1-dihydro-1,2-benziodoxol-3(*1H*)-one) was originally described in 1983 and has become a widespread reagent for the oxidation of complex, sensitive and multifunctional alcohols.[15] The periodinane is a hypervalent iodine species and a number of related compounds also serve as oxidizing agents.

Problem

Hints
- A regioselective oxidation takes place.
- The cyclic ether is oxidized to a lactone.

Solution

1. RuCl$_3$, NaIO$_4$, MeCN/CCl$_4$/H$_2$O, r. t., 5 h, 63 %

Discussion

The reactive species in this last step for the synthesis of (+)-asteriscanolide is RuO$_4$ prepared *in situ* from RuCl$_3$ and NaIO$_4$. Other common co-oxidants are for example sodium bromate, peracetic acid, oxygen or potassium permanganate. RuO$_4$ is a strong oxidant, however, conditions for ruthenium mediated reactions are very mild (usually a few hours at room temperature) and often only catalytic amounts are sufficient. Water is important for the reaction; thus many ruthenium mediated reactions have been performed in the CCl$_4$-H$_2$O solvent system. The addition of MeCN improves yields and reaction times. The configuration of stereocenters close to the reaction site normally remains unaffected. The most common synthetic use of ruthenium is the reaction with alcohols. Cyclic ethers as in this case, are oxidized, yielding lactones, but also a lot of other functional groups are converted: RuO$_4$ usually reacts with unsaturated systems, cleaving the C-C bonds; alkylamines are oxidized to mixtures of nitriles and amides, cyclic amines to lactams and amides to imides. The perruthenate ion RuO$_4^-$ for example in TPAP (see Chapter 10) is also useful for the oxidation of several functional groups especially primary alcohols.

1.4 Conclusion

The preceding synthesis realized by *Paquette* provides (+)-asteriscanolide in 13 steps starting from protected 2-bromo-4,4-dimethylcyclopentenone (**2**) in an overall yield of 4 %. The key steps are the convergent merging of the readily available enantiopure cyclopentanone sulfoxide **3** and the methyl 4-hydroxybutynoate **14** This domino *Michael-Michael* addition with a heteronucleophile has not been previously described. The use of the *Stille* coupling protocol followed by a few more steps furnishes the substrate for a ring-closing metathesis demonstrating that a conjugated diene typified by **10** can be produced by RCM with exceptional efficiency.

1.5 References

1   A. San Feliciano, A. F. Barrero, M. Medarde, J. M. Miguel del Corral, A. Aramburu, A. Perales, J. Fayos, *Tetrahedron Lett.* **1985**, *26*, 2369-2372.
2   P. A. Wenders, N. C. Ihle, C. R. D. Correia, *J. Am. Chem. Soc.* **1988**, *110*, 5904-5906.
3   K. I. Booker-Milburn, J. K. Cowell, L. J. Harris, *Tetrahedron.* **1997**, *53*, 12319-12338.
4   L. A. Paquette, J. Tae, M. P. Arrington, A. H. Sadoun, *J. Am. Chem. Soc.* **2000**, *122*, 2742-2748.
5   G. H. Posner, J. P. Mallamo, M. Hulce, L. L. Frye, *J. Am. Chem. Soc.* **1982**, *104*, 4180-4185.
6   L. F. Tietze, *Chem. Rev.* **1996**, *96*, 115-136.
7   K. Ritter, *Synthesis*, **1993**, 735-762.
8   J. Stille, *Angew. Chem.* **1986**, *98*, 504-519; *Angew. Chem. Int. Ed. Engl.* **1986**, *25*, 508-524.
9   E. Piers, R. W. Friesen, B. A. Keay, *Tetrahedron* **1991**, *47*, 4555-4570.
10   B. H. Lipschutz, S. Sengupta, *Org. React.* **1992**, *41*, 135.
11   R. H. Grubbs, S. J. Miller, G. Fu, *Acc. Chem. Res.* **1995**, *28*, 446-452.
12   P. Schwab, R. H. Grubbs, J. W. Ziller, *J. Am. Chem Soc.* **1996**, *118*, 100-110.
13   S. K. Amstrong, *J. Chem Soc., Perkin Trans. 1* **1998**, 371-388.
14   N. M. Hasty, D. R. Kearns, *J. Am. Chem. Soc.* **1973**, *95*, 3380-3381.
15   a) D. B. Dess, J. C. Martin, *J. Org. Chem.* **1983**, *48*, 4155-4156; b) D. B. Dess, J. C. Martin, *J. Am. Chem. Soc.* **1991**, *113*, 7277-7287.

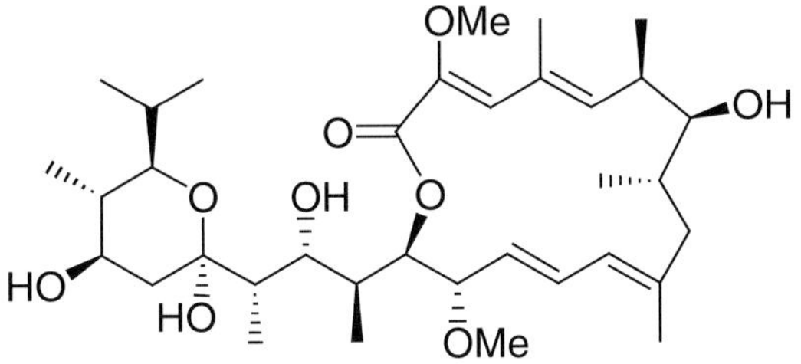

2

(−)-Bafilomycin A$_1$
(Roush 1999)

2.1 Introduction

(−)-Bafilomycin A$_1$ was first isolated in 1983 by *Werner* and *Hagenmaier* from a culture of *Streptomyces griseus* sp. *Sulphuru*.[1] Bafilomycin belongs to a family of macrolide antibiotics. It was found to exhibit activity against Gram-positive bacterial and fungi;[2] it also showed immunosuppressive activity and proved to be the first specific potent inhibitor of vacuolar H$^+$-ATPase.[3] Structurally, bafilomycin A$_1$ is constructed from a 16-membered tetraenic lactone ring and a β-hydroxyl-hemiacetal side chain. The intramolecular hemiacetal ring and the macrolactone are linked by a C$_3$ spacer and a hydrogen-bonding system.

The biological activity and the interesting structure stimulated efforts towards its total synthesis. *Evans* and *Calter* reported the first synthesis by an efficient aldol method.[4] *Toshima* and co-workers also succeeded in the total synthesis of bafilomycin A$_1$.[5]

This chapter is based on the enantioselective total synthesis by *William R. Roush* and co-workers, which was published in 1999.[6]

2.2 Overview

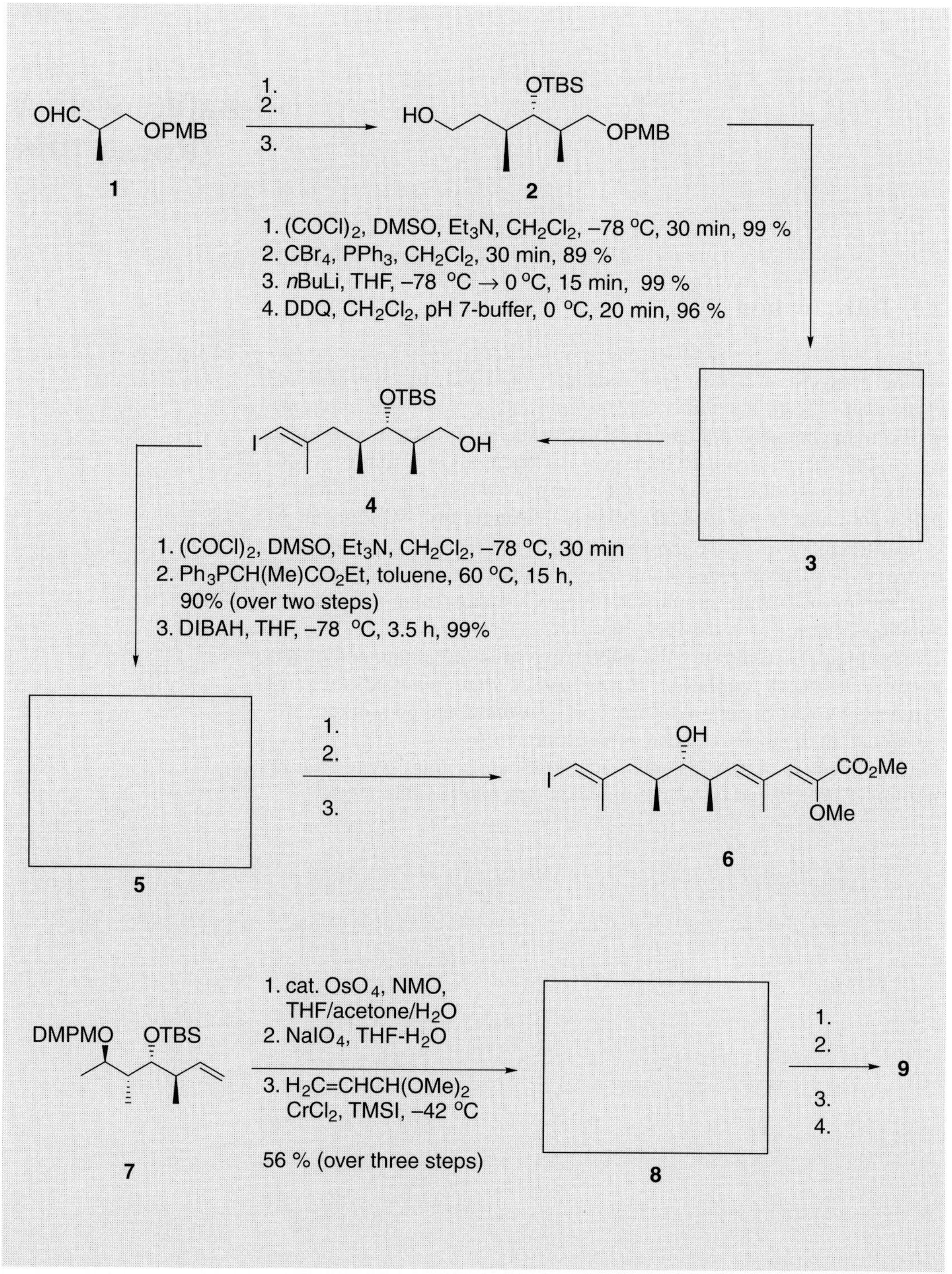

1 OHC-CH(Me)-CH₂-OPMB

1.
2.
3.

2 HO-CH₂-CH₂-CH(Me)-CH(OTBS)-CH(Me)-CH₂-OPMB

1. (COCl)$_2$, DMSO, Et$_3$N, CH$_2$Cl$_2$, −78 °C, 30 min, 99 %
2. CBr$_4$, PPh$_3$, CH$_2$Cl$_2$, 30 min, 89 %
3. nBuLi, THF, −78 °C → 0 °C, 15 min, 99 %
4. DDQ, CH$_2$Cl$_2$, pH 7-buffer, 0 °C, 20 min, 96 %

4 I-C(Me)=CH-CH$_2$-CH(Me)-CH(OTBS)-CH(Me)-CH$_2$-OH

1. (COCl)$_2$, DMSO, Et$_3$N, CH$_2$Cl$_2$, −78 °C, 30 min
2. Ph$_3$PCH(Me)CO$_2$Et, toluene, 60 °C, 15 h, 90% (over two steps)
3. DIBAH, THF, −78 °C, 3.5 h, 99%

5

1.
2.
3.

6 I-C(Me)=CH-CH$_2$-CH(Me)-CH(OH)-CH(Me)-CH=C(Me)-C(OMe)=CH-CO$_2$Me

7 DMPMO-CH(Me)-CH(Me)-CH(OTBS)-CH(Me)-CH=CH$_2$

1. cat. OsO$_4$, NMO, THF/acetone/H$_2$O
2. NaIO$_4$, THF-H$_2$O
3. H$_2$C=CHCH(OMe)$_2$, CrCl$_2$, TMSI, −42 °C

56 % (over three steps)

8

1.
2.
3.
4.

9

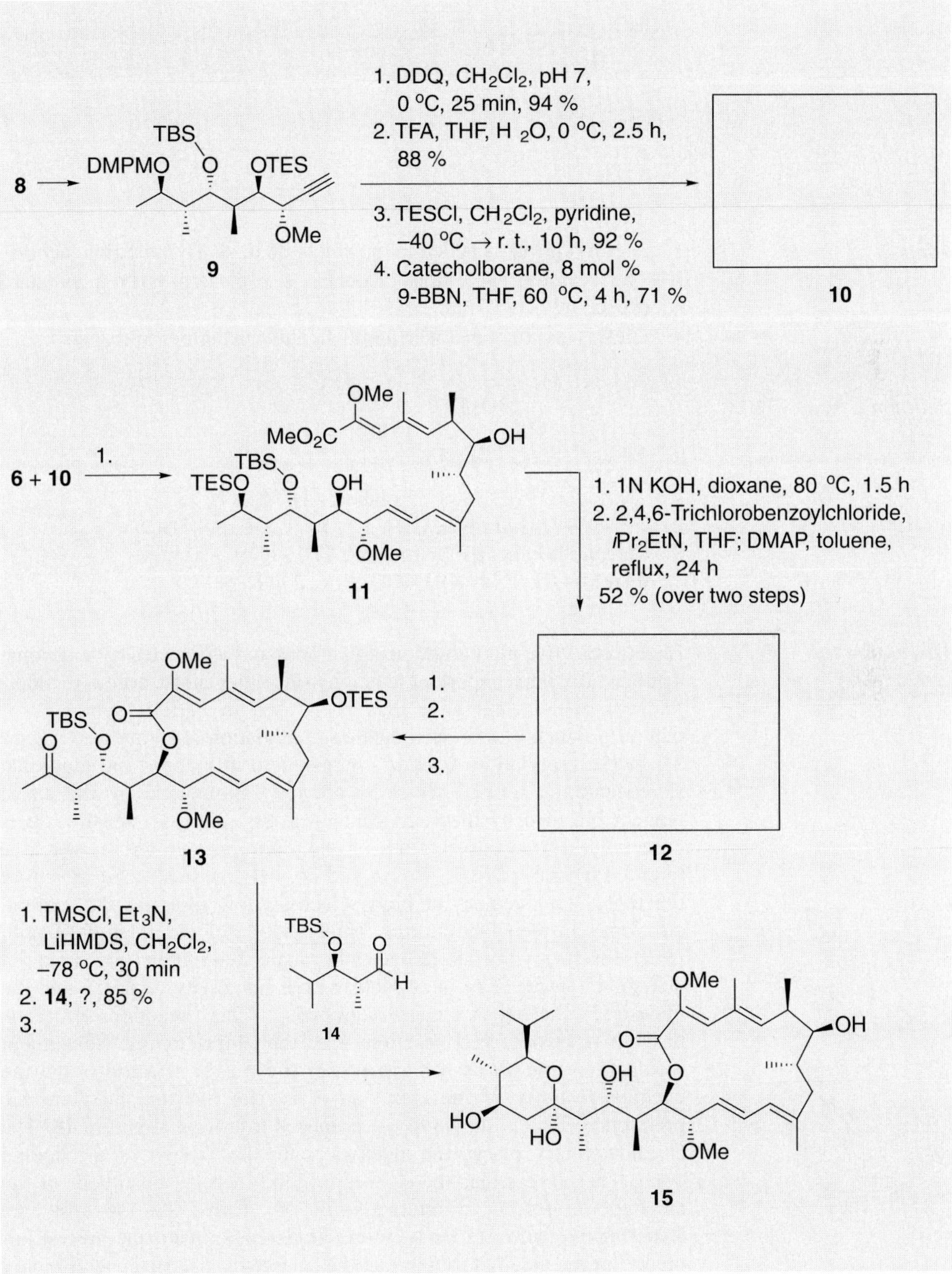

2 (−)-Bafilomycin A₁

9: DMPMO, TBS, OTES, OMe (from **8**)

Reagents:
1. DDQ, CH₂Cl₂, pH 7, 0 °C, 25 min, 94 %
2. TFA, THF, H₂O, 0 °C, 2.5 h, 88 %
3. TESCl, CH₂Cl₂, pyridine, −40 °C → r. t., 10 h, 92 %
4. Catecholborane, 8 mol % 9-BBN, THF, 60 °C, 4 h, 71 %

→ **10**

6 + 10 → 1. → **11**

1. 1N KOH, dioxane, 80 °C, 1.5 h
2. 2,4,6-Trichlorobenzoylchloride, iPr₂EtN, THF; DMAP, toluene, reflux, 24 h
52 % (over two steps)

→ **12**

1.
2.
3.
→ **13**

1. TMSCl, Et₃N, LiHMDS, CH₂Cl₂, −78 °C, 30 min
2. **14**, ?, 85 %
3.

14: TBS-O-CH(iPr)-CH(Me)-CHO

→ **15**

2.3 Synthesis

Problem

1 → [1., 2., 3.] → 2

Hints

- A crotyl group is added to the aldehyde in an asymmetric reaction.
- The resulting secondary alcohol is protected with a standard procedure.
- Olefins can be transformed into alcohols using boron reagents.

Solution

1. **16**, toluene, –78 °C, 8 h, 78 %
2. TBSOTf, 2,6-lutidine, CH_2Cl_2, –50 °C, 30 min, 99 %
3. Catecholborane, [$(PPh_3)_3RhCl$], THF, –5 °C, 30 min, then MeOH, 1N NaOH, H_2O_2, r. t., 2 h, 87 %

Discussion

Enantioselective allyl additions to ketones and aldehydes have become synthetically very important reactions since they allow access to aldol-like compounds and have thus been used in various syntheses of natural products (for an introduction to allylation reagents see Chapter 3). In the crotylation reaction compared to allylations an additional stereocenter is formed which is not only influenced by the chiral reagent but also by the stereocenters at the aldehyde substrate. It is especially difficult to synthesize the *anti-anti*-stereotriad which is required by bafilomycin. *Roush* and co-workers succeeded in setting the three stereocenters in high selectivity by applying the method developed in their laboratories. Thus reaction of the aldehyde **1** with (*R,R*)-diisopropyltartrate-(*E*)-crotylboronate (**16**) gave the required alcohol **17** in 78 % isolated yield with a selectivity of 85:15 and the undesired 3,4-*anti*-4,5-*syn*-diastereomer. This reaction proceeds through a mismatched reaction, i. e. the chiral methyl substituent would favor the *anti-syn* diastereomer (by *Felkin* selectivity) but the enantioselectivity of the chiral auxiliary can override this intrinsic preference. This is shown in the proposed transition structure **18**: The chiral auxiliary places the aldehyde onto the *Si* side of the double bond. Therefore either the R group or the methyl substituent of the aldehyde is forced to interact with the methyl on the crotylate. Experiments with sterically more hindered substituents instead of methyl at the aldehyde show that the selectivity decreases with higher steric requirements, making the *Felkin* selectivity the prominent factor (i. e. the aldehyde will then be on the *Re* side of the alkene).

Generally the reactivity of alcohols towards protection or deprotection decreases with higher substitution. In order to protect secondary alcohols as TBS ethers it is usually necessary to use the highly reactive silyl trifluoromethanesulfonate (triflate) instead of TBSCl, which is often used to protect primary alcohols selectively in the presence of secondary and tertiary alcohols. 2,6-Lutidine is used as the base and the TBS protection succeeds in almost quantitative yield.

Hydroborations are standard procedures to transform double bonds regioselectively into the less substituted alcohols. Catecholborane (**19**) is a much more stable reagent for hydroboration than diborane and has the advantage that the boronic acid byproducts are more easily hydrolyzed than the corresponding dialkylboranes. Catecholborane reacts with alkenes to form an alkoxyborinate but usually requires elevated temperatures.[7] Hydroborations using catecholborane can be catalyzed by Rhodium(I) complexes:[8] By using 3 % $(PPh_3)_3RhCl$ and one equivalent of **19** the reaction proceeds smoothly at −5 °C over 30 min. Oxidative work-up with hydrogen peroxide in the presence of base gives **2** in 87 % yield.

Problem

Hints

- Oxidation of the free alcohol is achieved in the first step.
- PPh_3 and CBr_4 react to form $Ph_3P=CBr_2$ which then reacts with the aldehyde.
- Steps 2 and 3 transform an aldehyde into an alkyne.
- Step 4 is an oxidative deprotection reaction.

Solution

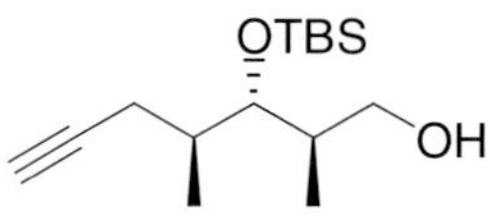

3

Discussion

The primary alcohol is oxidized in the standard *Swern* procedure to give aldehyde **19**.[9] This very popular oxidation method creates a reactive intermediate (**22**) from dimethylsulfoxide and oxalyl chloride. This intermediate is then attacked in S_N2 fashion at the sulfur atom by the substrate alcohol. Upon work-up with triethylamine the desired aldehyde or ketone and dimethylsulfide is formed.

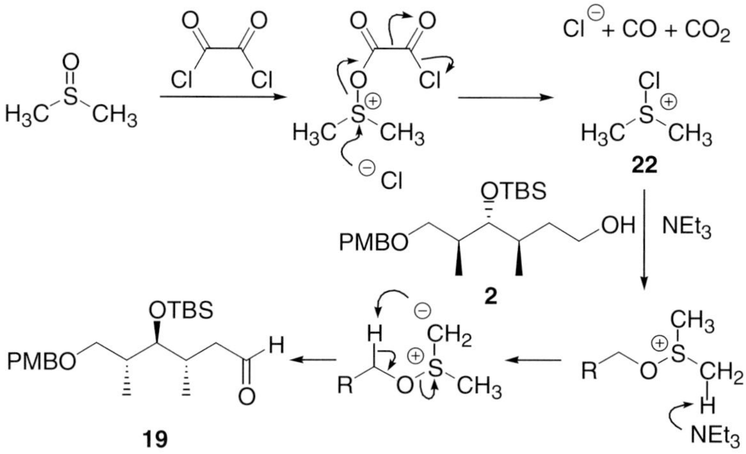

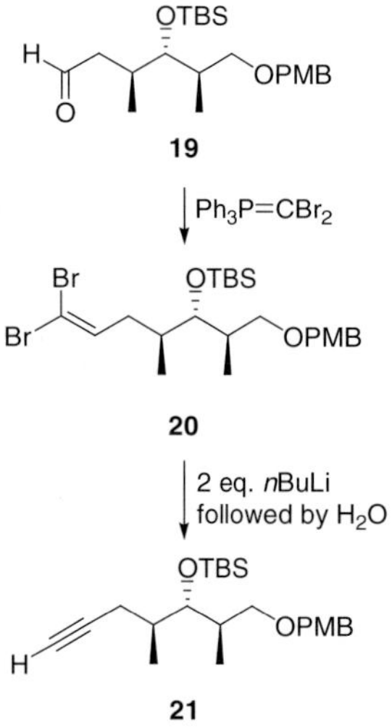

19

22

2

The reaction sequence in steps two and three is known as the *Corey-Fuchs* method to create an alkyne from an aldehyde:[10] Reaction of triphenylphosphane with carbontetrabromide gives phenylphosphane-dibromomethylene. This reagent then transforms aldehyde **19** into the corresponding dibromoalkene **20** thereby extending the chain by one carbon. Reaction of the bromo compound with two equivalents of *n*-butyllithium in THF at –78 °C results in the rapid formation of the acetylenic lithio derivative which forms the terminal acetylene **21** upon aqueous work-up.

The *para*-methoxybenzyl group belongs to a class of alcohol protecting groups that are stable to basic conditions but can be removed by oxidation. Here DDQ (2,3-dichloro-5,6-dicyano-1,4-benzoquinone) is used to yield the free primary alcohol **3**.

Problem

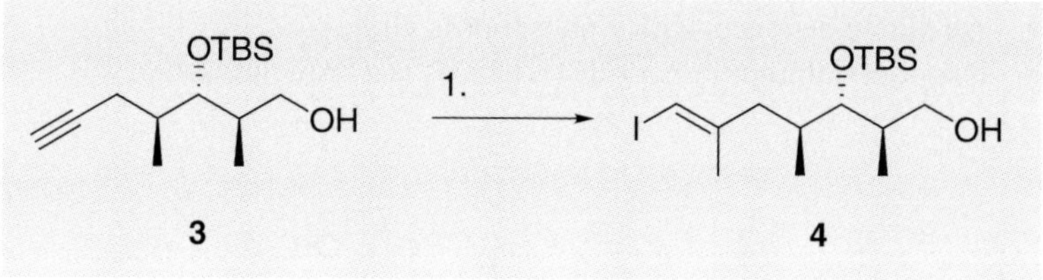

OTBS
OH
1.
OTBS
I
OH
3
4

Hints

- This reaction is a carbometalation.
- Trimethylaluminum is used. Which other metal is necessary?
- The carboalumination intermediate is treated with iodine.

Solution

1. $AlMe_3$, $[Cp_2ZrCl_2]$, $Cl(CH_2)_2Cl$, 60 °C, 14 h, then I_2, –30 °C, 1 h, 65 %

Discussion

Negishi and co-workers developed this carbometalation reaction of alkynes with organoalane-zirconocene derivatives, and it has since turned into an often used route to stereo- and regiodefined trisubstituted olefins.[11]

Applying a methylalane and a zirconocene derivative (*E*)-2-methyl-1-alkenylalanes can thus be synthesized with stereoselectivity generally greater than 98 %, the regioselectivity observed with terminal alkynes being ca. 95 %. This Zr catalyzed carboalumination reaction most likely involves direct Al-C bond addition assisted by zirconium to yield the carboalane **23**. The carboalanes are versatile intermediates, since the aluminum moiety can be easily replaced by hydrogen, iodine and various carbon electrophiles to produce the trisubstituted olefin **4**.

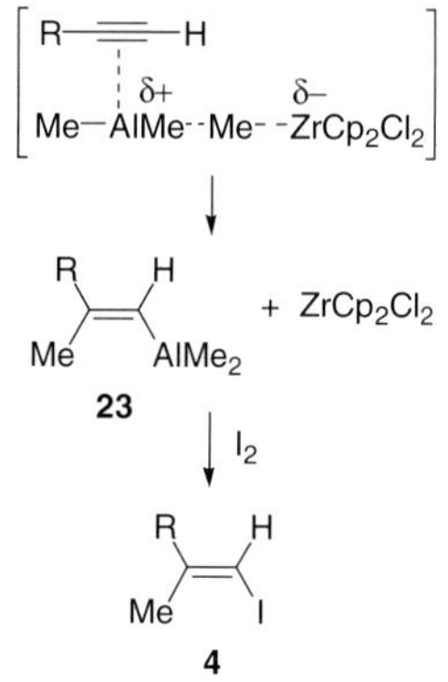

R
H
δ+
δ–
Me—AlMe- -Me- -ZrCp2Cl2
R
H
Me
AlMe2
+ ZrCp2Cl2
23
I2
R
H
Me
I
4

Problem

1. (COCl)2, DMSO, Et3N,
CH2Cl2, –78 °C → 0 °C,
30 min
2. Ph3PCH(Me)CO2Et,
toluene,60 °C, 15 h,
90 % (over two steps)
3. DIBAH, THF, –78 °C
3.5 h, 99 %
OTBS
I
OH
4
5

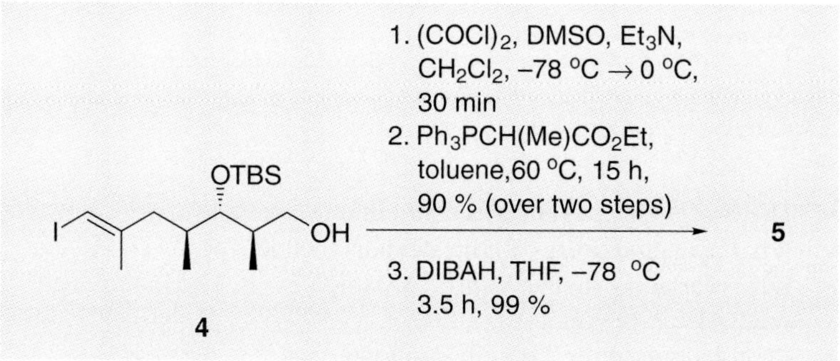

2 (−)-Bafilomycin A₁

Hints
- Another *Swern* reaction is performed.
- An aldehyde reacts with a phosphorus ylide.
- Diisobutylaluminumhydride (DIBAH) is a reducing agent.

Solution

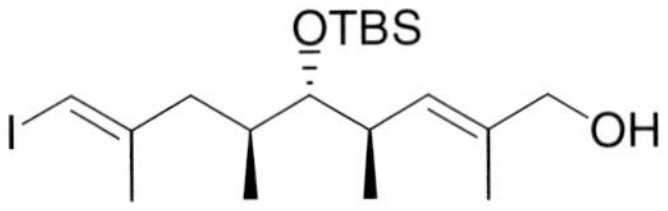

[Structure of compound 5: iodide-substituted diene with OTBS and primary OH]

5

Discussion

The primary alcohol is first oxidized to an aldehyde, which is then the substrate in a *Wittig* olefination reaction. Here a stabilized ylide is employed and therefore the E double bond is formed exclusively. (For a detailed description of the *Wittig* reaction see Chapter 13; the selectivity issues are explained in Chapter 9.)

The resulting ester can then be reduced with diisobutylaluminumhydride (DIBAH) to synthesize the primary alcohol **5**.

Problem

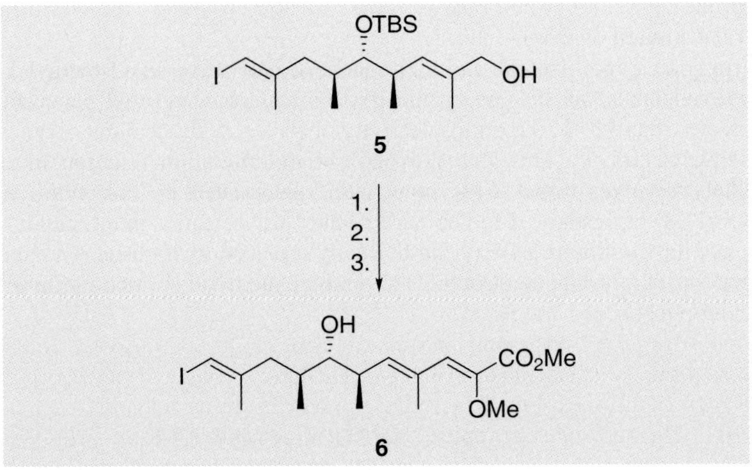

[Transformation from compound 5 to compound 6 via steps 1, 2, 3; compound 6 bears free OH, CO₂Me and OMe groups]

Hints
- The first two steps turn an alcohol into an olefin again.
- What reagent oxidizes allylic alcohols to aldehydes?
- What variation in the *Wittig* methodology is also often used to synthesize E olefins?
- Deprotection of the TBS ether follows.

1. MnO$_2$, CH$_2$Cl$_2$, r. t., 18 h, 99 %
2. KHMDS, THF, (*i*PrO)$_2$P(O)CH(OMe)CO$_2$Me, [18]crown-6, 0 °C → r. t., 8 h, 85 %
3. TBAF, THF, r. t., 2 h, 82 %

Solution

Discussion

Manganese dioxide is an important reagent, since it can oxidize primary or secondary alcohols to the aldehydes or ketones in neutral media. Oxidation of allylic and benzylic alcohols with MnO$_2$ is faster than that of saturated alcohols. The primary synthetic utility of MnO$_2$ is therefore the selectivity of oxidation of allylic over saturated alcohols. There is rather poor selectivity in the oxidation of primary allylic alcohols over secondary allylic alcohols, though. The mechanism is believed to proceed through radical intermediates. The reactivity of manganese dioxide is strongly influenced by the method of preparation. One of the more common methods involves precipitation of MnO$_2$ from a warm aqueous solution of KMnO$_4$ and MnSO$_4$. The reagent is then activated by heating it to ca. 200 °C for several hours.

Thus the primary allylic alcohol **5** is transformed into an aldehyde, which can now be used in a *Horner-Wadsworth-Emmons* reaction. In this reaction the dienoate moiety was obtained in a Z,E:E,E-selectivity of 95:5.

This variation of the *Wittig* reaction uses ylides prepared from phosphonates.[12] The *Horner-Wadsworth-Emmons* method has several advantages over the use of phosphoranes. These ylides are more reactive than the corresponding phosphoranes, especially when substituted with an electron withdrawing group. In addition the phosphorus product is a phosphate ester and soluble in water – unlike the Ph$_3$PO product of the *Wittig* reaction – which makes it easy to separate from the olefin product. Phosphonates are also cheaper than phosphonium salts and can easily be prepared by the *Arbuzov* reaction from phosphanes and halides.

The silyl ether protecting groups are commonly removed by acidic conditions or a fluoride ion source.[13] The high stability of the fluorine-silicon bond is exploited by many standard fluorine reagents such as HF, HF-pyridine complex as acidic and TBAF (tetra-*n*-butylammonium fluoride, as basic deprotecting agents. TBAF is commercially available as trihydrate which is highly hygroscopic, a fact that sometimes limits its use with water sensitive substrates.

Compound **6** is later used in a coupling reaction. First we turn our attention to the synthesis of the other coupling partner.

Arbuzov reaction:

(EtO)$_3$P + RCH$_2$X

↓ -EtX

(EtO)$_2$P(O)–CH$_2$R

2 (−)-Bafilomycin A$_1$

Problem

DMPMO, OTBS substrate **7**

1. cat. OsO$_4$, NMO, THF/acetone/H$_2$O
2. NaIO$_4$, THF-H$_2$O
3. H$_2$C=CHCH(OMe)$_2$, CrCl$_2$, TMS-I, −42 °C

→ **8**

56 % (over three steps)

Hints

- OsO$_4$ oxidizes the double bond.
- A diol is formed by the OsO$_4$ oxidation, NMO is cooxidant.
- NaIO$_4$ cleaves the diol creating an aldehyde.
- The acrolein dimethylacetal is reduced by the CrCl$_2$ generating an active species that adds to the aldehyde.
- An *anti* diol is formed.

Solution

DMPMO, TBS, OH, OMe — compound **8**

Discussion

The substrate **7** had been previously synthesized by *Roush* using the crotylation method described above.[14]

Osmium tetroxide is commonly used to add two OH groups to a double bond.[15] The mechanism gives *syn* addition from the less hindered side of the alkene. Since OsO$_4$ is expensive and highly toxic it is therefore mostly used in a catalytic fashion using stoichiometric cooxidants, like H$_2$O$_2$ or *N*-methylmorpholine-*N*-oxide (NMO).

1,2-Glycols are easily cleaved under mild conditions and in good yield by lead tetraacetate in organic solvents or periodic acid in water solutions. The yields are so good that olefins are often transformed into the diol and then cleaved to form two aldehydes – or ketones depending on the substrate – rather than cleaving the double bond directly with O$_3$. The mechanism was proposed by *Criegee* to involve the intermediate **24**[16] and yields aldehyde **25**.

Takai and co-workers introduced the use of the *in situ* generated γ-methoxyallylchromium reagent to synthesize diol derivatives stereoselectively.[17] Chromium(II) chloride has the ability to afford umpolung, transforming acrolein dialkyl acetate into the γ-alkoxy substituted allylic chromium reagent. This mild nucleophilic species will then add to the aldehyde placing the methoxy group *anti* to the alcohol created from the aldehyde. The major diastereomer is formed with a 10:2:1 selectivity in 67 % yield.

Problem

[Scheme: compound **8** (TBS, DMPMO, OMe, vinyl) → compound **9** (TBS, DMPMO, OTES, OMe, alkyne) via steps 1–4]

Hints

- The reactive alcohol needs protection.
- The double bond is transformed into an aldehyde first.
- Bishydroxylation and oxidative cleavage creates the aldehyde.
- Aldehydes can be transformed into alkynes by a special phosphorane.

Solution

1. TESOTf, 2,6-lutidine, CH_2Cl_2, –50 °C, 1 h, 99 %
2. OsO_4, NMO, THF, pH 7-buffer, 16 h
3. $Pb(OAc)_4$, EtOAc, 0 °C, 10 min
4. $(MeO)_2P(O)CHN_2$, tBuOK, THF, –78 °C → r. t., 15 min
85 % (over three steps)

Discussion

The free alcohol **8** needed to be protected temporarily since it interfered with following three steps.
Generally the stability of the silyl protecting groups increases with increasing steric hindrance of the alkyl substituents. Thus trimethylsilyl (TMS) ethers are only used as intermediates since they are labile to even weak acids. The triethylsilyl (TES) protecting group is more stable and survives column chromatography as well as oxidation, reduction and organometallic reagents, but is much more labile than the TBS group.
We have already seen the bishydroxylation using OsO_4 and NMO; this time lead tetraacetate is used to cleave the diol and yield the aldehyde. The phosphorane $(MeO)_2P(O)CHN_2$ is named the *Gilbert-Seyferth* reagent.[18] It basically behaves like the phosphoranes in the *Horner-Wadsworth-Emmons* reaction described above, except that the olefin subsequently loses nitrogen, creating the desired triple bond (also see Chapter 10).

Problem

9 → [1. DDQ, CH$_2$Cl$_2$, pH 7, 0 °C, 25 min, 94 %; 2. TFA, THF, H$_2$O, 0 °C, 2.5 h, 88 %; 3. TESCl, CH$_2$Cl$_2$, pyridine, −40 °C → r.t., 10 h, 92 %; 4. Catecholborane, 8 % 9-BBN, THF, 60 °C, 4 h, followed by pH 7-buffer; 71 %] → **10**

Compound **9**: DMPMO–CH(CH$_3$)–CH(OTBS)–CH(OMe)–CH(OTES)–C≡CH

Hints

- Three protecting group manipulations are performed.
- 3,4-Dimethoxybenzyl (DMPM) groups are cleaved oxidatively.
- TES groups are acid labile.
- Which alcohol is protected as TES ether now?
- Hydroboration is regioselective with alkynes, too.

Solution

Compound **10**: TESO–CH(CH$_3$)–CH(OTBS)–CH(OMe)–CH(OH)–CH=CH–B(OH)$_2$

Discussion

Previous studies had shown that the 3,4-dimethoxybenzyl (DMPM) group was impossible to cleave at a later stage of the synthesis, since it interfered with the diene moiety at C-10-C-13. Therefore it had to be replaced by a more labile protecting group like the TES group. DMPM is a variation of the *para*-methoxybenzyl (PMB) group, which is cleaved under oxidative conditions using DDQ or cerium ammonium nitrate (CAN). Therefore it can be cleaved while silyl and benzyl ethers (reductive deprotection) are stable. A protecting group strategy that employs different protecting groups which can each be cleaved selectively by using different conditions is called *orthogonal*.

The TES group is then cleaved by trifluoroacetic acid (TFA). The next step again introduces a TES group using the TES-chloride. Here the sterically less encumbered alcohol is more reactive and therefore can be selectively protected. In a *parallel* protecting group strategy one uses the same protecting groups and selectively protects and deprotects depending on the reactivity of the substrate.

With the TES protected alcohol at C-19 and the free alcohol at C-15 the subsequent reactions could be performed. Hydroboration of the triple bond and aqueous work-up creates vinylboronic acid **10**, which is the other coupling partner with **6**.

Problem

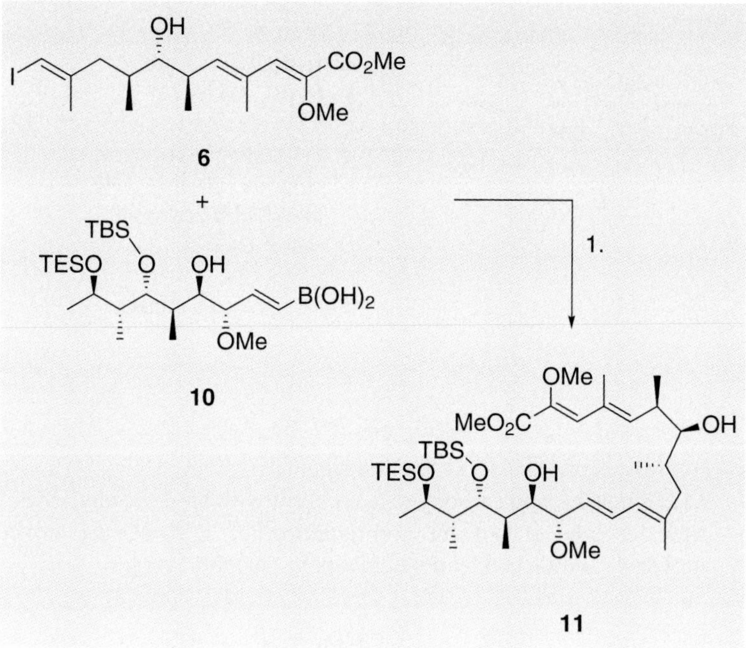

Hints
- What metal catalyzes this reaction?
- Do you know variations in the *Suzuki* reaction?

Solution

1. 20 mol% [Pd(PPh$_3$)$_4$], aq. TlOH, THF, r. t., 30 min, 65 %

Discussion

The diene synthesis using vinyl boranes or vinyl boronic acids with alkene halides catalyzed by palladium(0) has been developed by *Suzuki* and co-workers.[19] This reaction is explained in Chapter 10. The *Suzuki* coupling has not been as widely used in total synthesis as the *Stille* reaction, even though tin reagents are much more toxic. One of the drawbacks of the *Suzuki* reaction is the decreased rate of reaction when substrates of higher molecular weight are used. *Kishi* found that addition of thallium hydroxide increases the rate of reaction by a factor of 1000.[20] Under the modified conditions the reactions now proceed almost instantaneously even at 0 °C and produce less side products, allowing its application to substrates with fragile functional groups and high molecular weight. The only drawback is the high toxicity of thallium compounds.

Problem

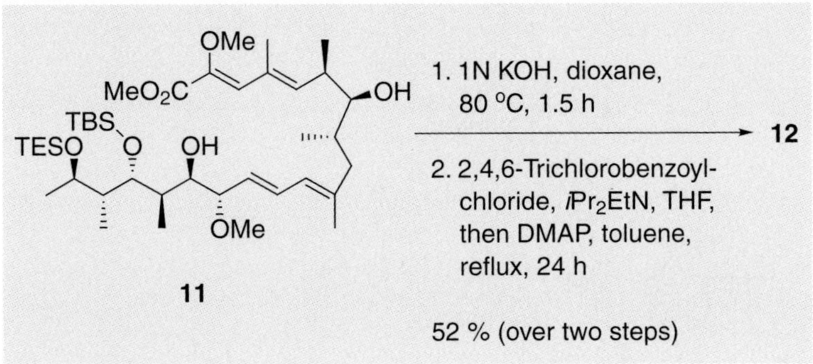

[Structure **11**: MeO₂C-containing polyene with TBS, TESO, OH, OMe substituents]

1. 1N KOH, dioxane, 80 °C, 1.5 h
2. 2,4,6-Trichlorobenzoyl-chloride, iPr₂EtN, THF, then DMAP, toluene, reflux, 24 h

→ **12**

52 % (over two steps)

Hints

- Hydrolysis of the ester is the first step.
- Acid chlorides and carboxylic acids react with base to anhydrides.
- After change of solvent and addition of DMAP, the mixed anhydride is attacked by the nucleophilic alcohol oxygen.

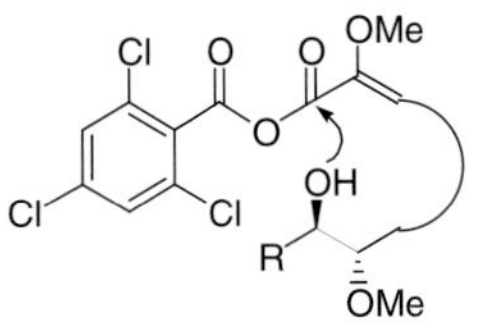

Solution

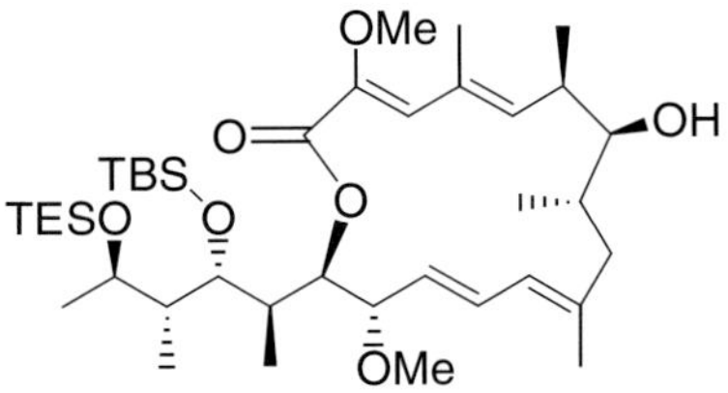

[Macrolactone structure **12**]

Discussion

Hydrolysis of the methyl ester is afforded by saponification with KOH in dioxane and the *seco*-acid was used in the subsequent steps without purification.

Macrocycles are often not easy to form, especially when they include many substituents. The applied procedure to synthesize macrolactones was introduced by *Yamaguchi* and co-workers and has since been used extensively.[21] First the acid is transformed into the mixed anhydride **26**. Refluxing the anhydride in toluene will yield the desired macrocycle. The attack of the alcohol at the trichlorobenzoic acid carbonyl moiety is not favored because of steric hindrance by the chlorine substituents *ortho* to the acid. Therefore the ring is closed selectively. Usually in macrocyclizations high dilution conditions are applied, too, in order to avoid intermolecular reactions.

[Structure **26**: mixed anhydride of 2,4,6-trichlorobenzoic acid]

Problem

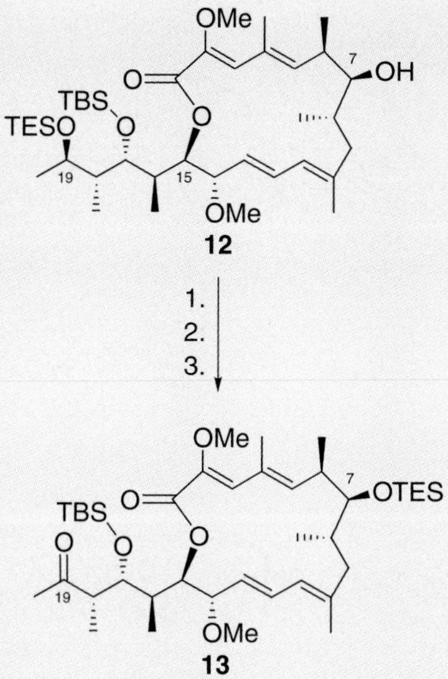

[Structures of compounds **12** and **13** with steps 1, 2, 3 between them]

Hints

- Protecting group operations are performed in the first two steps.
- An alcohol is oxidized to the ketone.

Solution

1. TESOTf, 2,6-lutidine, CH_2Cl_2, –50 °C, 20 min, 85 %
2. TFA, THF, 5 °C, 6 h, 90 %
3. *Dess-Martin* periodinane, pyridine, CH_2Cl_2, 0 °C, 4 h, 98 %

Discussion

The triethylsilyltriflate is used to protect the secondary alcohol **12** at C-7 first. Now the alcohols at C-7 and C-19 are both TES protected. Since the secondary alcohol is less hindered having one methyl substituent the authors succeeded in selectively deprotecting only C-19 with trifluoroacetic acid. This strategy appears to be complicated but proved to be necessary since attempts to use a TBS or TES protected C-7 alcohol in the macrocyclization failed to react probably because of interference of the protecting group with the methyl substituents at C-6 and C-8 in the transition state. The experiment using unprotected C-7, C-15 and C-19 alcohols gave mainly a 20-membered macrolactone. The hydroxy functionality at C-7 also needed to be protected for the following step since the C-19 alcohol could not be selectively oxidized in its presence. The C-19 oxidation was then accomplished using the standard *Dess-Martin* reagent (see Chapter 1)

2 (−)-Bafilomycin A₁

Problem

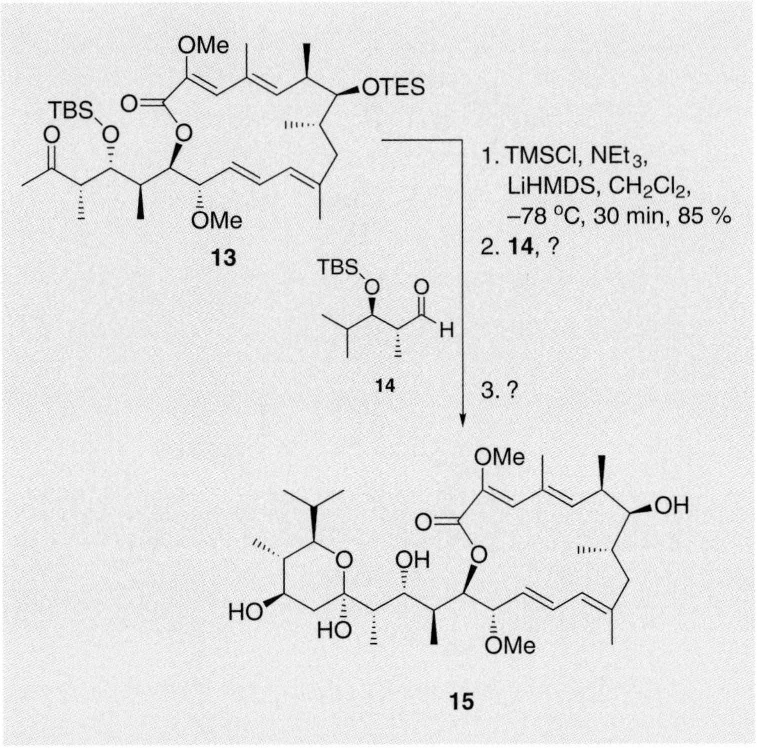

[Scheme: Conversion of methylketone **13** to aldol product **15** via aldehyde **14**.
Reagents: 1. TMSCl, NEt₃, LiHMDS, CH₂Cl₂, −78 °C, 30 min, 85 %; 2. **14**, ?; 3. ?]

Hints

- The methylketone is enolized and transformed into the TMS ether.
- Aldehyde **14** is reacted with the enolate under *Mukaiyama* conditions.
- The aldol product is deprotected.
- Ring closure in the deprotection step creates (−)-bafilomycin A₁.

Solution

2. **14**, BF₃·OEt₂, −78 °C, 30 min, 85 %
3. TAS-F, DMF, H₂O, r. t., 4 h, 93 %

Discussion

Mukaiyama found that *Lewis* acids can induce silyl enol ethers to attack carbonyl compounds, producing aldol-like products.[22] The reaction proceeds usually at −78 °C without selfcondensation and other *Lewis* acids such as TiCl₄ or SnCl₄ are commonly used. The requisite silyl enol ether **27** was prepared by treatment of ketone **13** with lithium hexamethyl disilazide (LiHMDS) and trapping the kinetic enolate with chlorotrimethylsilane. When the silyl enol ether **27** was mixed with aldehyde **14** in the presence of BF₃·OEt₂ a condensation occurred *via* transition state **28** to produce the product **29** with loss of chlorotrimethylsilane. The induced stereochemistry in *Mukaiyama* reactions using methylketones and α,β-chiral aldehydes as substrates was thoroughly investigated by *Evans*[23] and *Roush*. They found that *anti* substituted α-methyl-β-alkoxy aldehydes react highly *Felkin* selective to give the 1,3-*anti* product diastereomer. The *Felkin-Anh* model[24] is used to interpret the contributions of torsional, steric and electronic factors from the stereogenic center α to the reacting carbonyl. The nucleophile attacks in a trajectory coming over the smallest substituent, with the largest substituent pointing away.

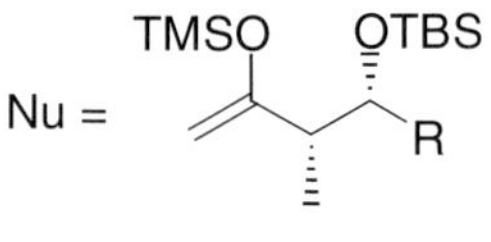

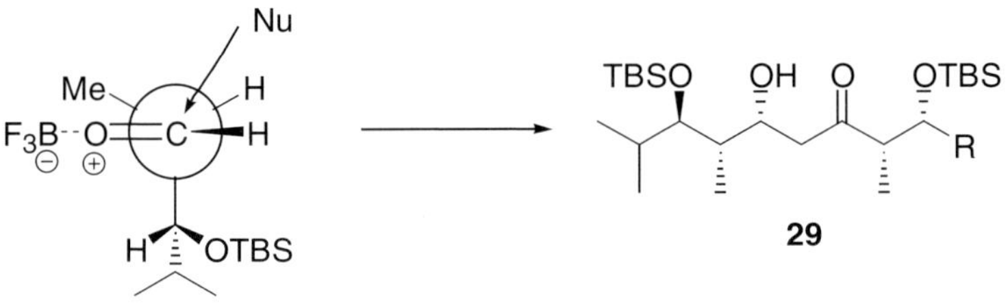

Deprotection of the silyl ether protecting groups had failed in previous attempts with conventional deprotecting agents. TAS-F [Tris-(dimethylamino)sulfonium difluorotrimethylsilicate, $(Me_2N)_3S^+F_2SiMe_3^-$ is one example of new silyl deprotection reagents which are mild and work under neutral conditions.[25] Other reagents are based on hypervalent fluoride complexes of tin[26] and phosphorus. TBAT ($Bu_4N^+Ph_3SiF_2^-$) for example is crystalline and soluble in most organic solvents and deprotection can be afforded under water free neutral conditions.[27]

Even though all the silyl protecting groups are removed in this step, affording **30**, only the alcohol at C-23 will react with the ketone, since it is able to form the stable six-membered heterocycle. The deprotection concludes the total synthesis of (−)-bafilomycin A$_1$.

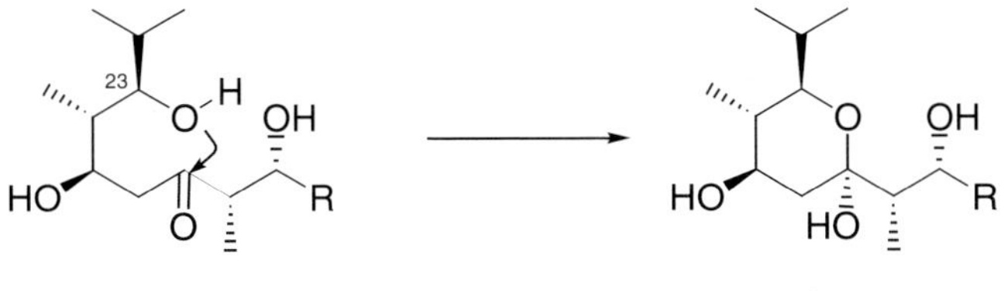

2.4 Conclusion

The total synthesis of (–)-bafilomycin A_1 by *Roush* et al. demonstrates the importance of protecting group selection. The impressive synthesis was only possible because of careful protecting group orchestration. The strategy relies not only on different classes of protecting groups but also on the stabilities and reactivities of functional groups towards protection and deprotection. The challenge of synthesizing Bafilomycin also resulted in the application of TAS-F, a new mild fluoride source, as substitute for the basic TBAF in deprotection of alcohols. *Roush* et al. also demonstrated the power of their asymmetric crotylation protocol using chiral diisopropyltartrate-crotylboronates. The synthesis also inspired extensive work on the selectivities in the *Mukaiyama*-aldol reaction using methyl ketones. Many other important reactions are used during the synthesis showing the broad range of methods that is necessary to achieve a complex synthetic goal.

2.5 References

1. G. Werner, H. Hagenmaier, K. Albert, H. Kohlshorn, H. Drautz, *Tetrahedron Lett.* **1983**, *24*, 5193-5196.
2. J. Haung, G. Albers-Schönberg, R. L. Monaghan, K. Jakubas, S. S. Pong, O. D. Hensens, R. W. Burg, D. A. Ostlind, *J. Antibiot.* **1984**, *37*, 970-975.
3. E. J. Bowman, A. Siebers, K. Altendorf, *Proc. Natl. Acad. Sci. USA* **1988**, *85*, 7972-7976.
4. D. A. Evans, M. A. Calter, *Tetrahedron Lett.* **1993**, *34*, 6871-6874.
5. K. Toshima, T. Jyojima, H. Yamaguchi, Y. Noguchi, T. Yoshida, H. Murase, M. Nakata, S. Matsumura, *J. Org. Chem.* **1997**, *62*, 3271-3284.
6. K. A. Scheidt, A. Tasaka, T. D. Bannister, M. D. Wendt, W. R. Roush, *Angew. Chem. Int. Ed.* **1999**, *38*, 1652-1655; *Angew. Chem.* **1999**, *111*, 1760-1762.
7. H. C. Brown, S. K. Gupta, *J. Am. Chem. Soc.* **1971**, *93*, 1816-1818.
8. D. A. Evans, G. C. Fu, A. H. Hoveyda, *J. Am. Chem. Soc.* **1992**, *114*, 6671-6679.
9. A. J. Mancuso, D. Swern *Synthesis*, **1981**, 165-185.
10. E. J. Corey, P. L. Fuchs, *Tetrahedron Lett.* **1972**, *12*, 3769-3772.
11. E. Negishi, D. E. Van Horn, T. Yoshida, *J. Am. Chem. Soc.* **1985**, *107*, 6639-6647.
12. B. E. Maryanoff, A. B. Reitz, *Chem. Rev.* **1989**, *89*, 863-927.
13. a) A. J. Pearson, W. J. Roush (ed.) *Handbook of Reagents for Organic Synthesis – Activating Agents and Protecting Groups*, John Wiley & Sons, Chichester **1999**; b) T. W. Greene, P. G. W. Wuts, *Protective Groups in Organic Synthesis*, 2nd ed.;

	John Wiley & Sons: New York, **1991**; c) P. J. Kocienski, *Protecting Groups*; Georg Thieme Verlag: Stuttgart, **1994**.
14	W. R. Roush, T. D. Bannister, *Tetrahedron Lett.* **1992**, *33*, 3587-3590.
15	For a review see: M. Schröder, *Chem. Rev.* **1980**, *80*, 187-213.
16	R. Criegee, L. Kraft, B. Rank, *Liebigs Ann. Chem.* **1933**, *507*, 159-167.
17	K. Takai, K. Nitta, K. Utimoto, *Tetrahedron Lett.* **1988**, *29*, 5263-5266.
18	a) J. C. Gilbert, U. Weerasooriya, *J. Org. Chem.* **1979**, *44, 4997-4998*; b) D. Seyferth, R. S. Marmor, P.Hilbert, *J. Org. Chem.* **1971**, *36*, 1379-1386
19	N. Miyaura, K. Yamada, H. Suginome, H. Suzuki, *J. Am. Chem. Soc.* **1985**, *107*, 972-980.
20	J. Uenishi, J.-M. Beau, R. W. Armstrong, Y. Kishi, *J. Am. Chem. Soc.* **1987**, *109*, 4756-4758.
21	J. Inanaga, K. Hirata, H. Saeki, T. Katsuki, M. Yamagushi, *Bull. Chem. Soc. Jpn.* **1979**, *52*, 1989-1993.
22	T. Mukaiyama *Org. React.* **1982**, *28*, 238-248.
23	D. A. Evans, M. J. Dart, J. L. Duffy, M. G. Yang, *J. Am. Chem. Soc.* **1996**, *118*, 4322-4343.
24	a) M. Cherest, H. Felkin, N. Prudent, *Tetrahedron Lett.* **1968**, 2199-2204; b) N. T. Anh, O. Eisenstein, *Nouv. J. Chem.* **1977**, *1*, 61-70.
25	K. A. Scheidt, H. C. Chen, B. C. Follows, S. R. Chemler, D. S. Coffey, W. R. Roush, *J. Org. Chem.* **1998**, *63*, 6436-6437.
26	M. Gingras, *Tetrahedron Lett.* **1991**, *32*, 7381-7384.
27	A. S. Pilcher, P. DeShong, *J. Org. Chem.* **1996**, *61*, 6901-6905.

Curacin A (Wipf 1996)

3.1 Introduction

Marine cyanobacteria offer a wide range of biologically active metabolites, some of which display high potential in treating human diseases.[1] One of these compounds is curacin A (**1**) from *Lyngbya majuscala* collected off the coast of Curaçao and isolated by *Gerwick* et al. in 1994.[2] By inhibiting the microtubule assembly by compound **1** cells are prevented from dividing and growing – the fundamental requirement for an anticancer drug. As well as its promising antiproliferative effects; curacin A (**1**) also inhibits the binding of colchicine (**2**) to tubulin as a consequence of its high binding affinity to the colchicine site of tubulin.[3]

Other remarkable facts are the structural features of compound **1**. It contains a vinyl and cyclopropyl substituent at C-7 and C-5 which makes **1** an unusual example among natural products with a thiazoline ring. Furthermore, three of the four stereogenic centers are included in this moiety.[4] Both biological activity and structural features motivated several research groups to find synthetic routes toward **1**,[5] whose first total synthesis was published by *White* and co-workers in 1995.[6] The same group determined the absolute configuration *via* comparison of products obtained from asymmetric preparation with those acquired by degradation of the natural material. In 1996 *Wipf* et al. reported an interesting synthetic approach to curacin A (**1**) by use of modern organometallic chemistry.[4]

3.2 Overview

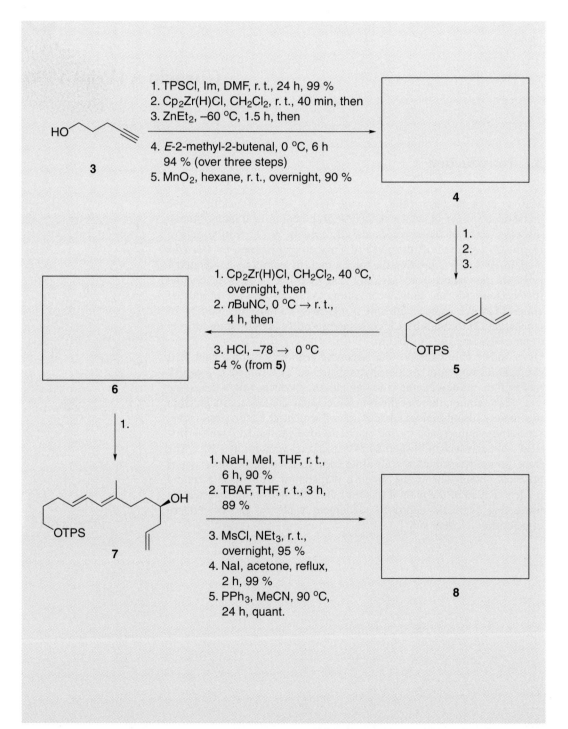

3 Curacin A

38   3   Curacin A

3.3 Synthesis

Problem

HO~~~≡ **3**

1. TPSCl, Im, DMF, r. t., 24 h, 99 %
2. Cp$_2$Zr(H)Cl, CH$_2$Cl$_2$, r. t., 40 min, then
3. ZnEt$_2$, –60 °C, 1.5 h, then
4. *E*-2-methyl-2-butenal **14**, 0 °C, 6 h, 94 % (over three steps)
5. MnO$_2$, hexane, r. t., overnight, 90 %

→ **4**

Hints

- The first step adds a protecting group.
- In the second step the triple bond is attacked.
- An organozirconium species is formed *via* hydrometalation. What are the positions of the zirconium and the hydrogen in the substrate?
- What happens to the organozirconium species on treatment with ZnEt$_2$?
- A transmetalation takes place.
- Compound **14** represents a *Michael* system. Does a 1,2- or a 1,4-addition occur?
- Manganese dioxide is often used for the mild oxidation of special types of alcohol.

14 (E-2-methyl-2-butenal structure)

Solution

TPSO~~~~~=~~C(=O)~C(CH$_3$)=CH-CH$_3$
4

Discussion

First conversion of primary alcohol **3** to the *tert*-butyldiphenylsilyl ether **15** occurs. In the field of silyl ethers the TPS group as well as the triisopropylsilyl (TIPS) group are the most stable protecting groups against a large variety of reaction conditions – consequently they are frequently used in organic synthesis (see Chapter 2).[7]

Generation of organozirconium species **18** follows the hydrozirconation of the triple bond using zirconocene hydrochloride (*Schwartz* reagent) **16**. This 16-electron, d^0 Zr(IV) complex is coordinatively unsaturated, so alkyne **15** coordinates to the electrophilic Zr center followed by insertion of the triple bond into the Zr–H bond. The resulting σ-vinyl-Zr(IV)-species **18** is formed with a high *cis*-selectivity and regioselectivity, such that the bulky zirconocene moiety always adds end-on to the terminal multiple bond.

16
Schwartz reagent (Cp$_2$Zr(H)Cl)

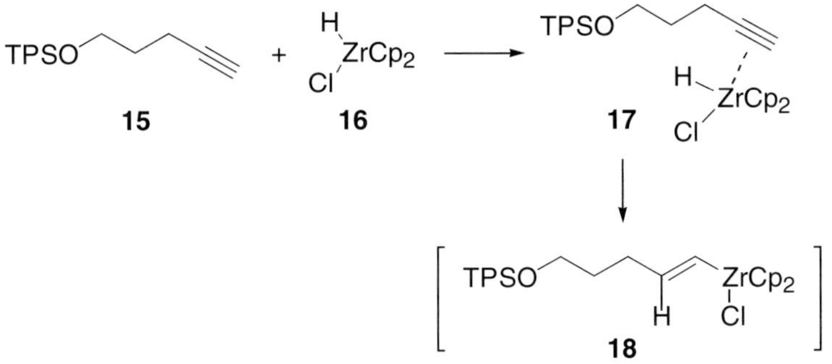

3 Curacin A

Hydrozirconation of terminal triple bonds is an essential method to obtain alkenes with defined stereochemistry. In the case of internal alkynes the zirconocene moiety adds to the sterically less hindered position of the triple bond. *Cis*-selectivity is high, but the regioselectivity is sometimes moderate depending on the nature of the substrate.[8]

Before adding aldehyde **14** a transmetalation from zirconium to zinc is necessary because of low reactivity of the sterically hindered organozirconocene compounds like **18** toward most organic electrophiles.[9] Resulting alkenylethylzinc **19** reacts in a 1,2-addition with the α,β-unsaturated aldehyde **14** transferring exclusively the alkenyl moiety. The formation of *E*-allylic alcohol **20** reveals stereochemical retention of the double bond configuration in the transmetalation and addition steps.

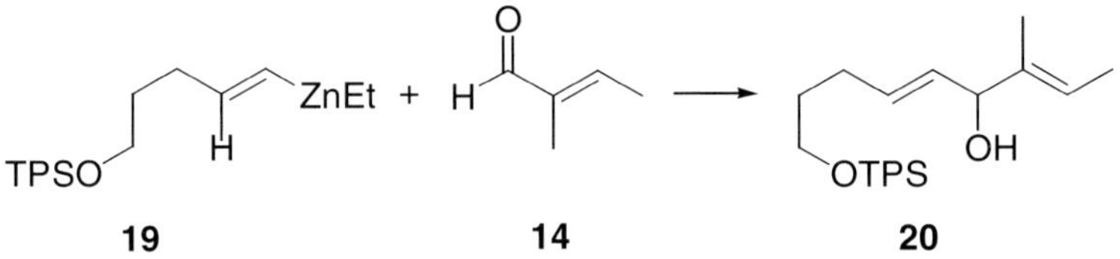

Oxidation of sensitive divinyl alcohol **20** to dienone **4** is achieved by treatment with activated manganese dioxide. Commercially available "active" MnO$_2$ **21** is a synthetic nonstoichiometric hydrated material. This reagent provides mild conditions for oxidation of allylic, propargylic, and benzylic alcohols.[10]

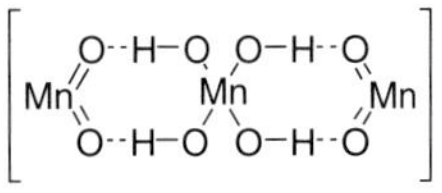

Problem

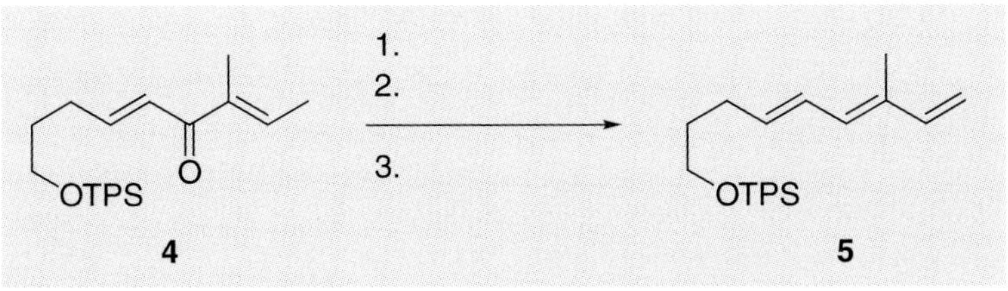

Hints

- Ketone **4** is converted to triene **5** *via* deoxygenation.
- Initially a base deprotonates ketone **4**.
  Which hydrogen atom is abstracted and so which enolate arises?
- Next step is the generation of a good leaving group.
- Reductive elimination of this group yields triene **5**.

Solution

1. KHMDS, THF, –78 °C, 2 h, then
2. Tf$_2$NPh, –78 → 0 °C, overnight
   quant. (over two steps)
3. Bu$_3$SnH, Pd$_2$(dba)$_3$·CHCl$_3$, LiCl, PPh$_3$, THF, reflux, 4 h, 71 %

Discussion

At low temperature (–78 °C) deprotonation with the bulky silyl amide KHMDS runs under kinetic control – enolization takes place regio- and stereoselectively at the terminal methyl group of ketone **4**. The sterically less hindered enolate is transformed to enol triflate **22**, which has all-*trans* configuration.

Vinyl- or aryl triflates have their application in C-C coupling reactions with organometallic reagents to form olefinic compounds. Another possibility is Pd-catalyzed reductive elimination: enol triflate **22** is treated with Bu$_3$SnH in a *Stille*-reaction to give triene **5** (see Chapter 1). Reduction of enol triflates with standard reducing agents such as LiAlH$_4$ or DIBAH yields only the original enolate ion.

Problem

1. Cp$_2$Zr(H)Cl, CH$_2$Cl$_2$, 40 °C, overnight, then
2. *n*BuNC, 0 °C → r. t., 4 h, then
3. HCl, –78 → 0 °C
54 % (from **5**)

→ **6**

Hints

- *Schwartz* reagent Cp$_2$Zr(H)Cl behaves in a similar way to that described above.
- *n*-Butyl isocyanide is an organic electrophile.

Besides hydrozirconation of terminal triple bonds Cp$_2$Zr(H)Cl (**16**) also reacts with double bonds.[8] The mechanism is similar to that described for alkynes. After coordination of alkene **5** to the Zr center giving π-complex **23** the terminal double bond inserts into the Zr-H bond to form the stable σ-alkyl complex **24**. The bulky zirconocene moiety again adds to the end-position of the terminal double bond.

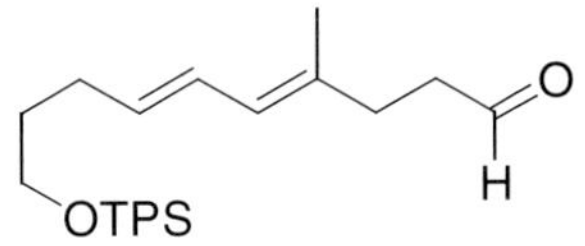

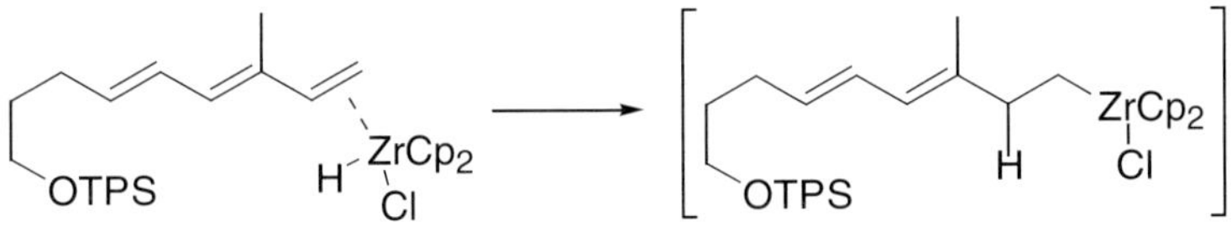

The steric demand of the zirconocene moiety leads to rearrangements in the case of hydrozirconation of internal double bonds. Regardless of position or configuration of the double bond in the substrate, zirconium migrates to the terminal position of the alkyl chain *via* insertion and β-H-elimination steps. Such isomerization does not occur upon hydrozirconization of internal alkynes.

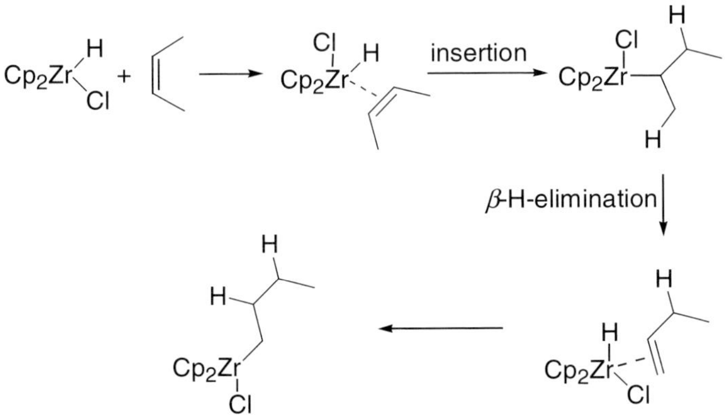

Although reaction of organozirconocene compounds fails with most organic electrophiles, treatment with sterically unhindered isocyanides, such as *n*BuNC, is possible.[11] Organometallic species **24** attacks *n*BuNC **25** to give isocyanide-insertion product **26**, which is finally hydrolyzed to the one-carbon homologated aldehyde **6**.

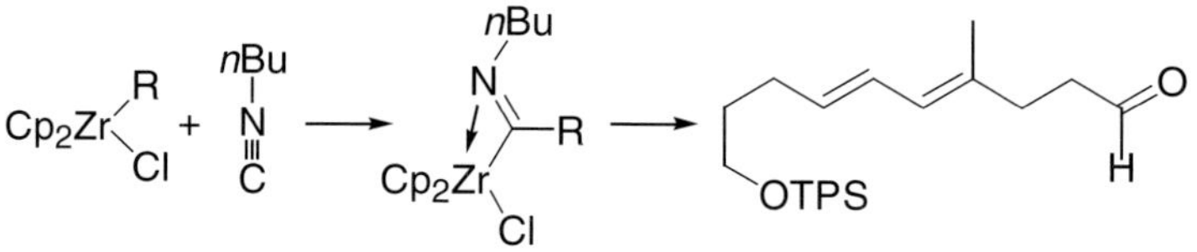

3 Curacin A

Problem

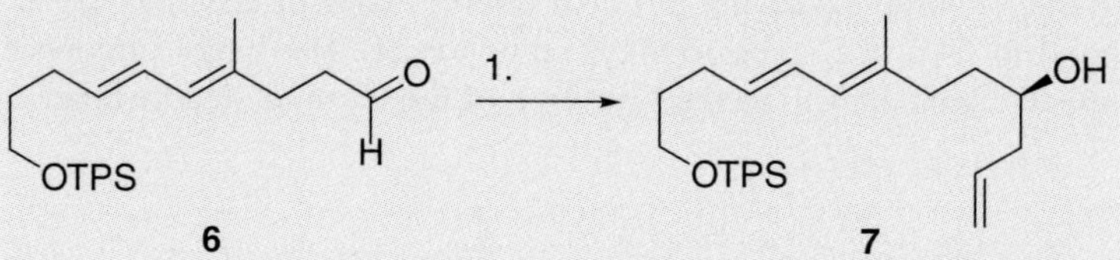

Hints
- A homoallylic alcohol is formed.
- Which reagents for stereoselective allylation of aldehydes do you know?

Solution

1. [(−)IPC]$_2$B-allyl **27**, −78 °C, 4 h, then ethanolamine, −78 °C → r. t., 2 h, 63 %, 93 % ee

Discussion

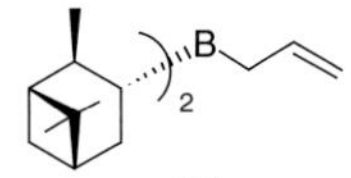

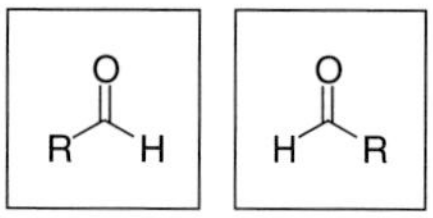

Investigations in the field of organoboron reagents for allylation of aldehydes resulted in a variety of powerful methods, e. g. the allylation by *Brown*.[12] Allylation reaction with [(−)IPC]$_2$B-allyl **27** (allyl diisopinocampheyl borane) proceeds *via* a six-membered chair-like transition state, the so called *Zimmerman-Traxler* transition state **28**. *Lewis* acid boron coordinates to the oxygen of the carbonyl group and activates it for a nucleophilic attack by the allyl group. Chiral ligands at the boron atom are responsible for the facial differentiation of aldehyde **6**, the *si*-face is favored in this case. Because of the closed being transition structure and the relative short B-O bond (147 pm) asymmetric induction is normally very high. Thus, migration of the boron and creation of a terminal double bond furnishes compound **29**.

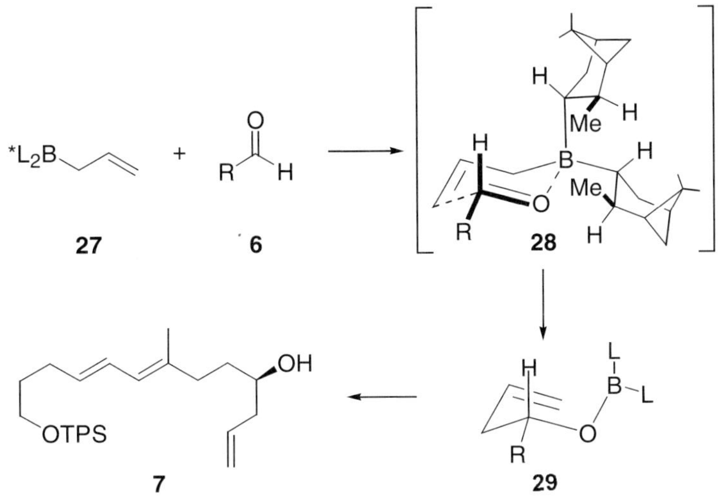

27 6 28 7 29

Dialkylalkoxyborane **29** is treated with ethanolamine to liberate homoallylic alcohol **7** – this work-up allows the recycling of the chiral auxiliary.[13] After precipitation of the $(IPC)_2B$-ethanolamine adduct **30** it can be transformed into the allylating agent **27** *via* compound **31**.

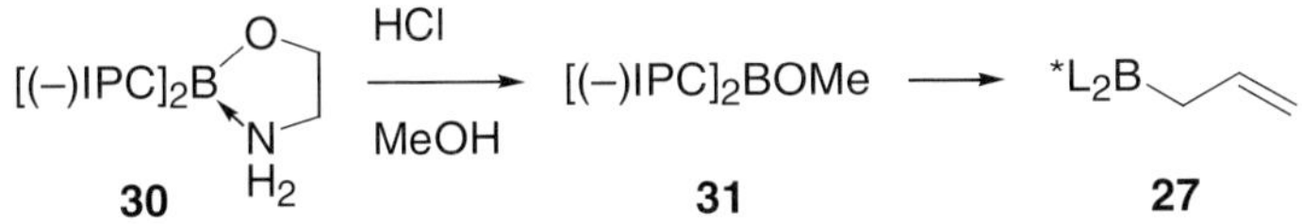

30 31 27

Advantages of *Brown's* chiral allylboranes (isopinocampheyl and later caranyl borane) are the easy access to the ligands (α-pinene is a natural product, chiral pool), the availability of both enantiomers and their low price. Excellent selectivities (96–99% *ee*) can be obtained at reaction temperatures of –100 °C. Other important mediators for enantioselective allylation of aldehydes are shown below.[14]

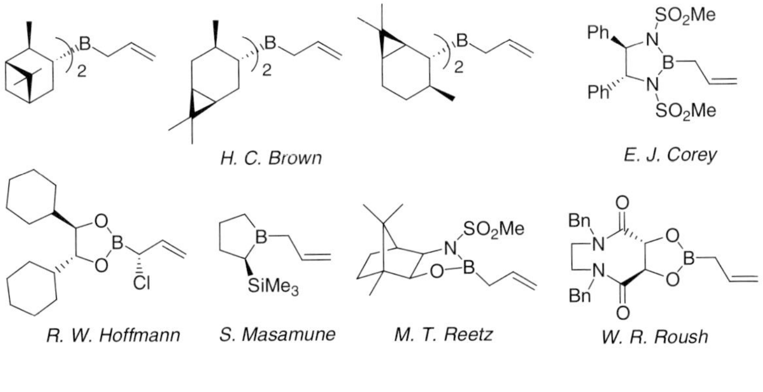

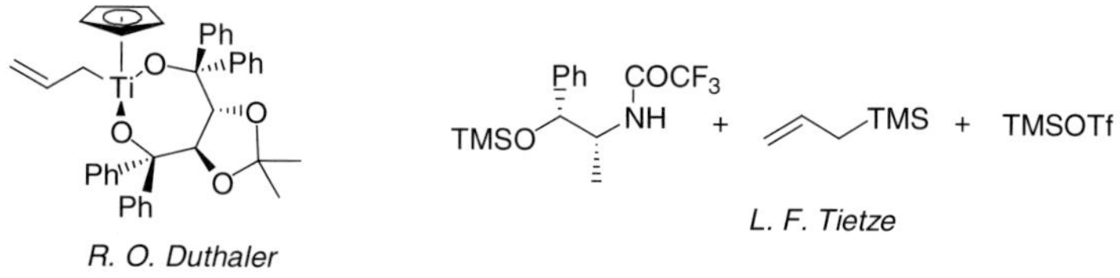

Ph
Ph
Ti
O
O
O
Ph
Ph
O
R. O. Duthaler
TMSO
Ph
COCF3
NH
+
TMS
+
TMSOTf
L. F. Tietze

In recent years methods for the catalytic asymmetric allylation have also been developed.[14]

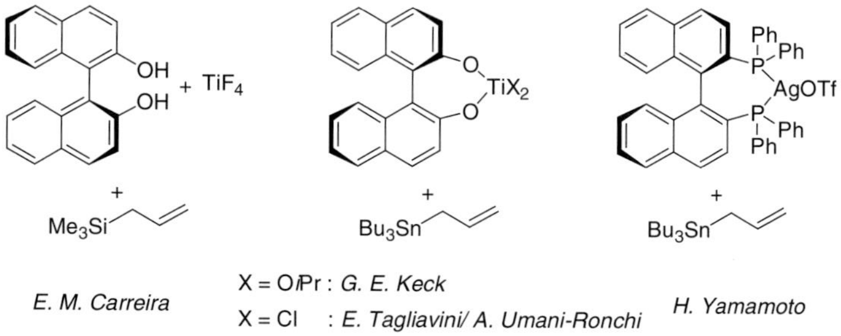

OH
OH
+ TiF4
+
Me3Si
E. M. Carreira
O
O
TiX2
+
Bu3Sn
X = OiPr : G. E. Keck
X = Cl : E. Tagliavini/ A. Umani-Ronchi
Ph
Ph
P
AgOTf
P
Ph
Ph
+
Bu3Sn
H. Yamamoto

Problem

OH
OTPS
7
1. NaH, MeI, THF, r. t., 6 h, 90 %
2. TBAF, THF, r. t., 3 h, 89 %
3. MsCl, NEt3, r. t., overnight, 95 %
4. NaI, acetone, reflux, 2 h, 99 %
5. PPh3, MeCN, 90 °C, 24 h, quant.
8

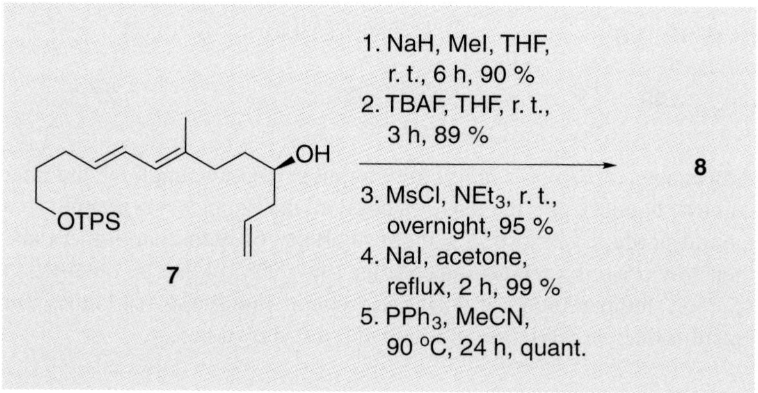

Hints

- An ether is cleaved with TBAF.
- Steps 3 and 4 are an alternative to the *Appel*-reaction.
- Product **8** is a salt.

Solution

OMe
PPh3 I
8

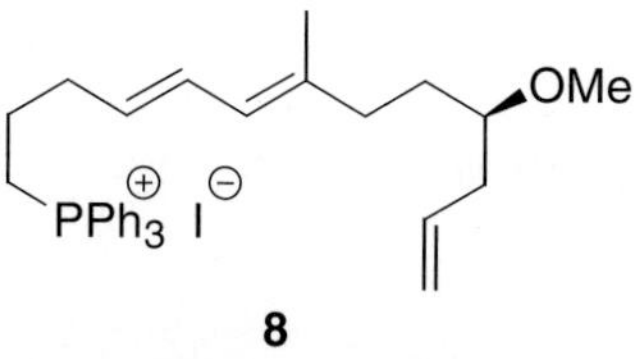

Discussion

Before the TPS ether is cleaved with TBAF[7], secondary alcohol **7** has to be protected as methyl ether. TBAF is a reagent to cleave every silyl ether. Most other functional groups are not affected (see Chapter 2). In the next two steps the conversion of alcohol **33** into mesylate **34**, which is a good leaving group, and then into iodide **35** in a *Finkelstein* type reaction occurs.[15] Acetone is the solvent of choice, because NaI is better soluble in it than NaOMs and consequently reaction equilibrium is forced to the product side. Direct transformation from an alcohol to an iodide is possible with PPh$_3$ and I$_2$ in an *Appel*-like reaction, but in some cases this reaction fails. Final procedure is the generation of phosphonium salt **8**.

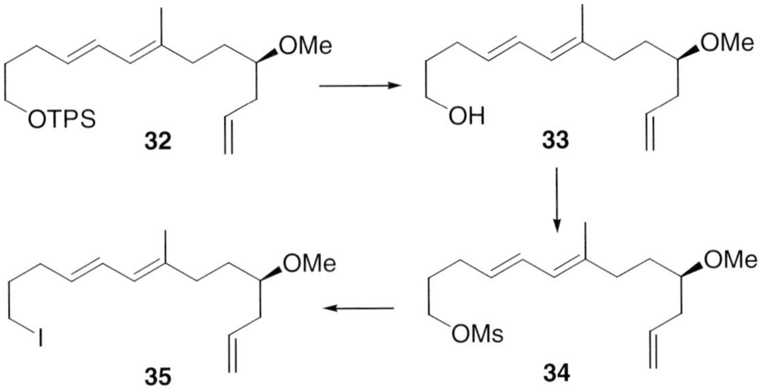

Problem

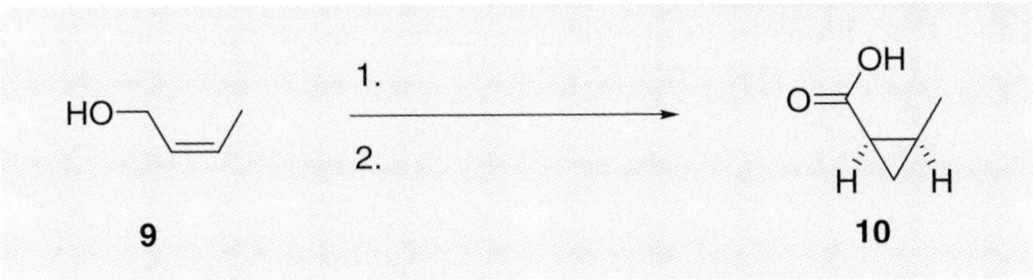

Hints

- First an enantioselective cyclopropanation occurs.
- For the enantioselective cyclopropanation, the presence of an allylic hydroxy group, the use of an organozinc reagent and a chiral ligand are needed.
- How is a primary alcohol converted in one step into a carboxylic acid?

Solution

1. **36**, Zn(CH$_2$I)$_2$·DME, CH$_2$Cl$_2$, −10 °C, 2 h, 75 %, 95 % ee
2. cat. RuCl$_3$·H$_2$O, NaIO$_4$, MeCN, H$_2$O, r. t., 6 h, 71 %

3 Curacin A

Discussion

Cyclopropanes are commonly synthesized with $Zn(CH_2I)_2$ (**37**) using the *Simmons-Smith*-reaction.[16] This reaction proceeds *via* one-step addition of one of the methylene groups of the zinc reagent to olefins. As an advantage of the *Simmons-Smith*-reaction no free carbene is involved.

In the presence of the chiral dioxaborolane **36** the attack of the methylene group proceeds selectively from the *si*-face.[17] This is the asymmetric *Charette* cyclopropanation:[18] The presence of coordinating groups on the dioxaborolane ring in **36** is crucial for obtaining high enantioselectivities. For steric and stereoelectronic reasons it is postulated that the allylic alkoxide in **41** adopts the more stable pseudoaxial configuration and the cyclopropanation occurs *via* the formation of a bidentate chelate **41** between reagent and substrate. The complex **41** is believed to involve the carbonyl oxygen of one of the amide groups and the allylic oxygen.

Catalytic amounts of $RuCl_3$ with $NaIO_4$ as the co-oxidant provides one-pot oxidation of **39**.[10,19] The active agent RuO_4 is formed *in situ* and oxidizes alcohol **39** first to the aldehyde *via* intermediate **42** and then to the carboxylic acid **10** *via* the aldehyde hydrate intermediate **43**.

The use of chrome reagents (*Jones* conditions) is also possible.

Problem

1. *L*-Serine-OMe **44**, DCC, DMAP, CH_2Cl_2, r. t., overnight, 71 %
2. TBSCl, imidazole, DMAP, CH_2Cl_2, r. t., overnight, 89 %
3. LiCl, $NaBH_4$, THF, EtOH, r. t., overnight, 88 %
4. $(COCl)_2$, DMSO, NEt_3, CH_2Cl_2, −60 °C, 20 min, quant.

10 → **11**

Hints

- In the first step an amide bond is formed.
- The second step introduces a protecting group.
- $NaBH_4$ is a reagent to reduce esters.

Solution

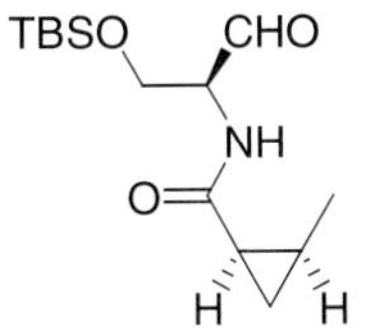

Discussion

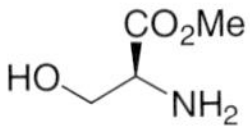

First the carboxylic acid **10** reacts with the amino group of the amino acid *L*-serine methyl ester **44**. This reaction is carried out with DCC **45** and DMAP **46** as activators of the carboxyl group.[20,21]

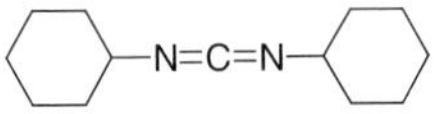

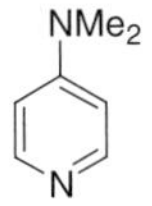

With the basic DMAP **46** as the catalyst, a proton transfer between the carboxylic acid **10** and the diimide **45** yields the carboxylate anion **47** which undergoes nucleophilic addition to the protonated diimide **48**. This activated ester **49** is readily attacked by the amino group of *L*-serine methyl ester **44** as a nucleophile.

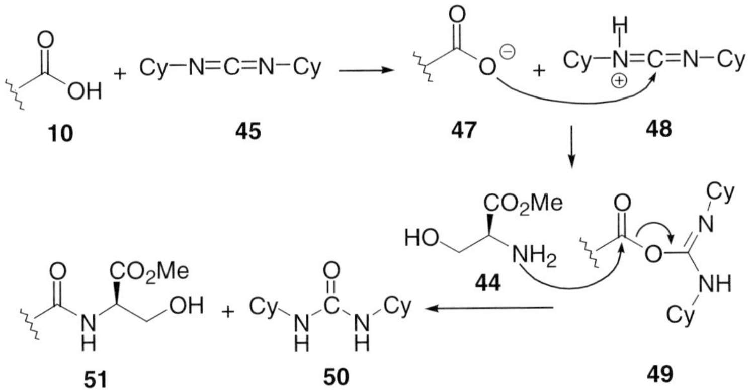

The driving force of the terminal step is the formation of the very stable urea derivative **50**, which is formed stoichiometrically. Further reagents employed in peptide bond forming reactions are diimide EDC **52** and triazole HOBT **53** which react similarly to DCC **45** but give water-soluble by-products.[20]

The hydroxy group of *L*-serine methyl ester does not undergo reaction with the activated ester **49** because of its smaller nucleophilicity.

The next step gives the TBS ether **54** at this position applying standard conditions to protect primary alcohols as silyl ethers.

$NaBH_4$ provides a method to reduce the ester **54** to alcohol **55**. The additive LiCl is used to enhance the reactivity of $NaBH_4$ towards esters.[10] Alcohol **55** is then oxidized to aldehyde **11** using *Swern* conditions with $(COCl)_2$ and DMSO (see Chapter 2).

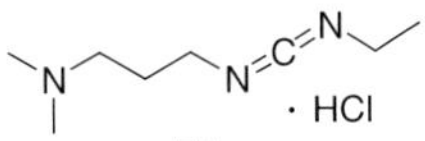

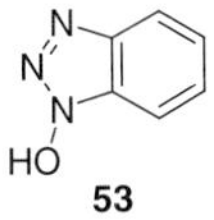

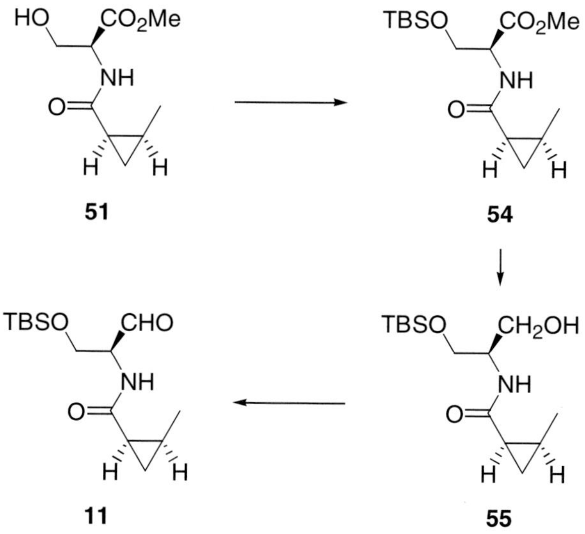

3 Curacin A

51 → **54**

55 → **11**

Problem

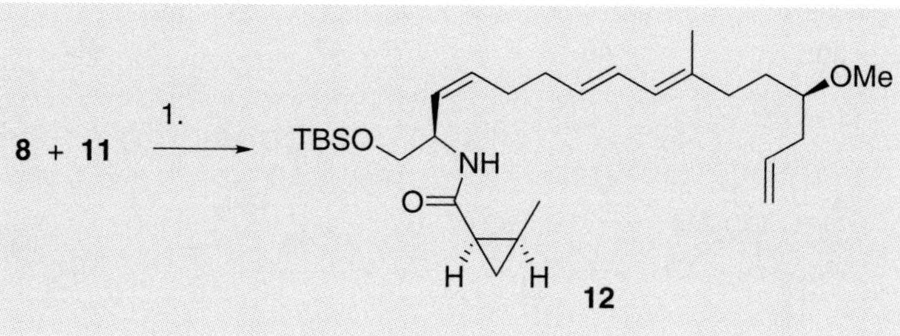

8 + 11 →[1.] **12**

Hints
- How will phosphonium salt **8** react with aldehyde **11** in the presence of base?
- A double bond is formed.

Solution

1. NaHMDS, THF, −78 → 0 °C, 62 %

Discussion

Tetraene **12** is formed following the *Wittig* protocol: Deprotonation of the phosphonium salt **8** yields a phosphorus ylide which is subjected to condensation with aldehyde **11** (see Chapters 9 and 13).

Problem

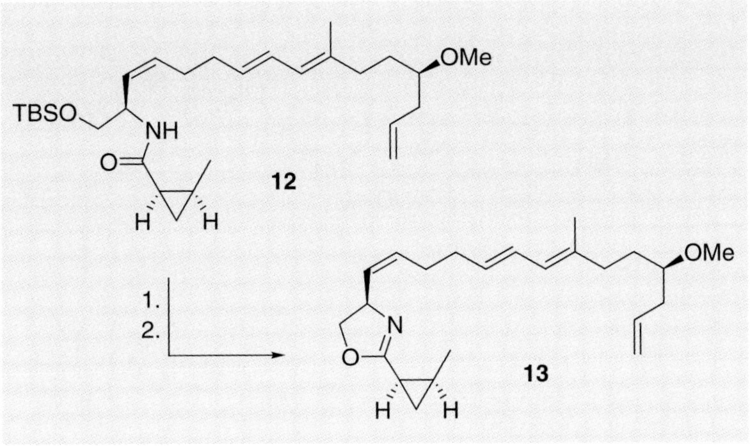

Hints
- The TBS ether is cleaved in the first step.
- Step two is the conversion of an alcohol into a leaving group.
- Then a cyclodehydratization is carried out to form oxazoline **13**.

Solution

1. HF·py **56**, THF, 0 °C → r. t., 3 h, 91 %
2. Et$_3$NSO$_2$NCO$_2$Me **57**, THF, r. t., 9 h, 71 %

Discussion

The HF·pyridine complex **56** is a common reagent to cleave silyl ethers of primary alcohols. Thus, deprotection of TBS ether **12** gives primary alcohol **59**.

To achieve formation of the oxazoline moiety in **13**, *Burgess* reagent[22] **57** is employed as a mild reagent to provide a reactive alcohol derivative and as an intramolecular base to facilitate the cyclization process. Treatment of alcohol **59** with *Burgess* reagent results in nucleophilic attack of the hydroxy group on the sulfur atom and loss of NEt$_3$. The resulting sulfonate **60** is converted to heterocycle **13** by intramolecular attack of the carbonyl group, liberating SO$_3$ and methyl carbamate **61**.

Mitsunobu conditions with DIAD **58** and PPh$_3$ have also been employed for oxazoline formation.[23]

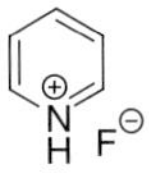

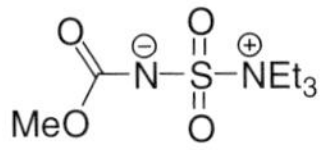

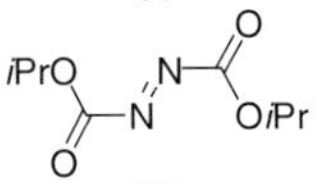

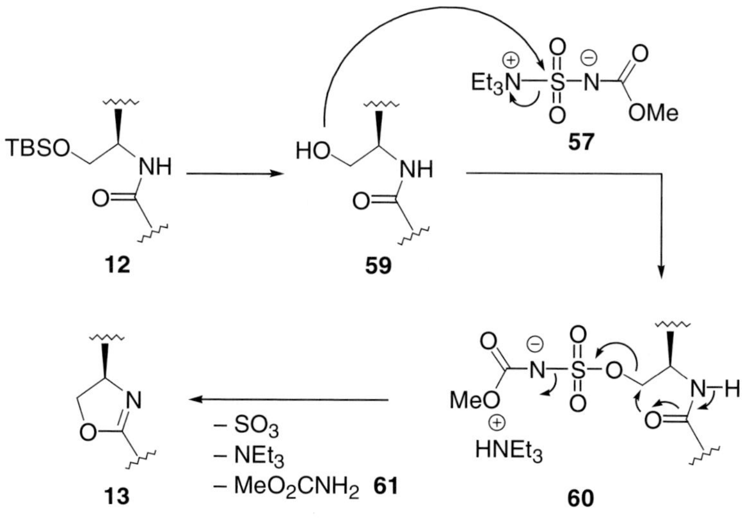

Problem

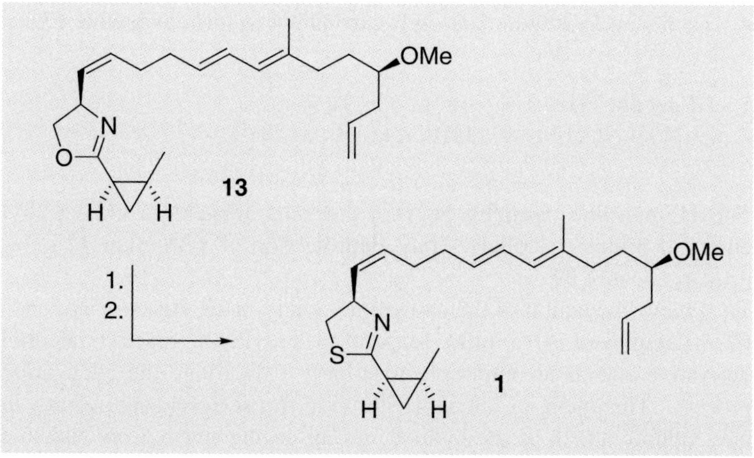

Hints
- Which nucleophile would you use to open the oxazoline ring?
- Another cyclodehydratization is carried out in the final step.

Solution
1. H_2S, MeOH, NEt_3, 35 °C, 20 h, 66 %
2. $Et_3NSO_2NCO_2CH_2CH_2OPEG$ **62**, THF, r. t., 1 h, 63 %

Thiolysis with H₂S as the nucleophile is employed to open the oxazoline ring in **13**.[24] Attack on the sp^2 carbon in the heterocycle leads to tetrahedral intermediate **63** which decomposes preferentially under C-O bond cleavage.

The resulting thioamide **64** is subjected to cyclodehydratization with the PEG supported *Burgess* reagent **62**[25] with the sulfur atom instead of the carbonyl group performing the intramolecular attack on the sulfonate group in a similar way to that described above. The conversion of the oxazoline into a thiazoline ring as the final step yields curacin A (**1**).

Discussion

The use of PEG supported *Burgess* reagent gave superior yields compared to the unsupported reagent because it provides milder reaction conditions for the substrate, which is especially sensitive toward acid or base treatments.

3.4 Conclusion

The structurally novel antimitotic agent curacin A (**1**) was prepared with an overall yield of 2.5 % for the longest linear synthesis. Three of the four stereogenic centers were built up using asymmetric transformations; one was derived from a chiral pool substrate. Key steps of the total synthesis are a hydrozirconation - transmetalation protocol, the stereoselective formation of the acyclic triene segment via enol triflate chemistry and another hydrozirconation followed by an isocyanide insertion. For the preparation of the heterocyclic moiety of curacin A (**1**) the oxazoline – thiazoline conversion provides efficient access to the sensitive marine natural product.

3.5 References

1. G. M. König, A. D. Wright, *Planta Med.* **1996**, *62*, 193-210.
2. W. H. Gerwick, P. J. Proteau, D. G. Nagle, E. Hamel, A. Blokhin, D. Slate, *J. Org. Chem.* **1994**, *59*, 1243-1245.
3. P. Verdier-Pinard, J. Y. Lai, H. D. Yoo, J. Yu, B. Marquez, D. G. Nagle, M. Nambu, J. D. White, J. R. Falck, W. H. Gerwick, B. W. Day, E. Hamel, *Mol. Pharm.* **1998**, *53*, 62-76.
4. a) P. Wipf, W. Xu, *J. Org. Chem.* **1996**, *61*, 6556-6562; b) D. J. Faulkner, *Nat. Prod. Rep.* **1996**, *13*, 75-125 and references therein.
5. a) M. Z. Hoemann, K. A. Agrios, J. Aube, *Tetrahedron Lett.* **1996**, *37*, 953-956; b) H. Ito, N. Imai, S. Tanikawa, S. Kobayashi, *Tetrahedron Lett.* **1996**, *37*, 1795-1798; c) H. Ito, N. Imai, K. Takao, S. Kobayashi, *Tetrahedron Lett.* **1996**, *37*, 1799-1800.
6. a) J. D. White, T. S. Kim, M. Nambu, *J. Am. Chem. Soc.* **1995**, *117*, 5612-5613; b) D. G. Nagle, R. S. Geralds, H. D. Yoo, W. H. Gerwick, T. S. Kim, M. Nambu, J. D. White, *Tetrahedron Lett.* **1995**, *36*, 1189-1192.
7. a) P. J. Kocienski, *Protecting Groups*, Georg Thieme Verlag, New York **1994**; b) T. W. Greene, P. G. M. Wuts, *Protective Groups in Organic Synthesis*, 3rd ed., John Wiley & Sons, New York **1999**.
8. a) J. Schwartz, J. A. Labinger, *Angew. Chem.* **1976**, *88*, 402-409; *Angew. Chem. Int. Ed. Engl.* **1976**, *15*, 333-340; b) L. S. Hegedus, *Transition Metals in the Synthesis of Complex Organic Molecules*, University Science Books, Mill. Valley Calif. **1994**; c) R. B. Grossman, *The Art of Writing Reasonable Organic Reaction Mechanisms*, Springer, New York **1999**.
9. P. Wipf, W. Xu, *Tetrahedron Lett.* **1994**, *35*, 5197-5200.
10. For a review see: S. D. Burke, R. L. Danheiser (ed.) *Handbook of Reagents for Organic Synthesis – Oxidizing and Reducing Agents*, John Wiley & Sons, Chichester **1999**.
11. E. Negishi, D. R. Swanson, S. R. Miller, *Tetrahedron Lett.* **1988**, *29*, 1631-1634.
12. H. C. Brown, K. S. Bhat, *J. Am. Chem. Soc.* **1986**, *108*, 5919-5923.
13. H. C. Brown, U. S. Racherla, Y. Liao, V. V. Khanna, *J. Org. Chem.* **1992**, *57*, 6608-6614.
14. a) H. C. Brown, R. S. Randad, K. S. Bhat, M. Zaidlewicz, U. S. Racherla, *J. Am. Chem. Soc.* **1990**, *112*, 2389-2392; b) R. Stürmer, R. W. Hoffmann, *Synlett* **1990**, 759-761; c) A. Hafner, R. O. Duthaler, R. Marti, G. Rhis, P. Rothe-Streit, F. Schwarzenbach, *J. Am. Chem. Soc.* **1992**, *114*, 2321-2336; d) L. F. Tietze, K. Schiemann, C. Wegner, C. Wulff, *Chem. Eur. J.* **1996**, *2*, 1164-1172; e) L. F. Tietze, C. Wulff, C. Wegner, A.

Schuffenhauer, K. Schiemann, *J. Am. Chem. Soc.* **1998**, *120*, 4276-4280; f) D. R. Gauthier, E. M. Carreira, *Angew. Chem.* **1996**, *108*, 2521-2523; *Angew. Chem. Int. Ed. Engl.* **1996**, *35*, 2363-2365; g) G. E. Keck, K. H. Tarbet, L. S. Geraci, *J. Am. Chem. Soc.* **1993**, *115*, 8467-8468; h) A. L. Costa, M. G. Piazza, E. Tagliavini, C. Trombini, A. Umani-Ronchi, *J. Am. Chem. Soc.* **1993**, *115*, 7001-7002; i) A. Yanagisawa, H. Nakashima, A. Ishiba, H. Yamamoto, *J. Am. Chem. Soc.* **1996**, *118*, 4723-4724.

15 H. Finkelstein, *Ber. Dtsch. Chem. Ges.* **1910**, *43*, 1528-1532.

16 H. E. Simmons, R. D. Smith, *J. Am. Chem. Soc.* **1959**, *81*, 4256-4264.

17 S. L. Eliel, S. H. Wilen, *Stereochemistry of Organic Compounds*, Wiley, New York **1994**.

18 a) A. B. Charette, S. Prescott, C. Brochu, *J. Org. Chem.* **1995**, *60*, 1081-1083; b) A. B. Charette, H. Juteau, *J. Am. Chem. Soc.* **1994**, *116*, 2651-2652.

19 a) H. Niwa, S. Ito, T. Hasegawa, K. Wakamatsu, T. Mori, K. Yamada, *Tetrahedron Lett.* **1991**, *32*, 1329-1330; b) A. K. Singh, R. S. Varma, *Tetrahedron Lett.* **1992**, *33*, 2307-2310.

20 For a review see: A. J. Pearson, W. J. Roush (ed.) *Handbook of Reagents for Organic Synthesis – Activating Agents and Protecting Groups*, John Wiley & Sons, Chichester **1999**.

21 a) B. Neises, W. Steglich, *Angew. Chem.* **1978**, *90*, 556-557; *Angew. Chem. Int. Ed. Engl.* **1978**, *17*, 522-524; b) A. Hassner, V. Alexanian, *Tetrahedron Lett.* **1978**, 4475-4478.

22 a) G. M. Atkins, Jr., E. M. Burgess, *J. Am. Chem. Soc.* **1968**, *90*, 4744-4745; b) P. Wipf, C. P. Miller, *Tetrahedron Lett.* **1992**, *33*, 907-910.

23 P. Wipf, C. P. Miller, *Tetrahedron Lett.* **1992**, *33*, 6627-6630.

24 P. Wipf, S. Venkatraman, *Synlett* **1997**, 1-10.

25 P. Wipf, S. Venkatraman, *Tetrahedron Lett.* **1996**, *37*, 4659-4662.

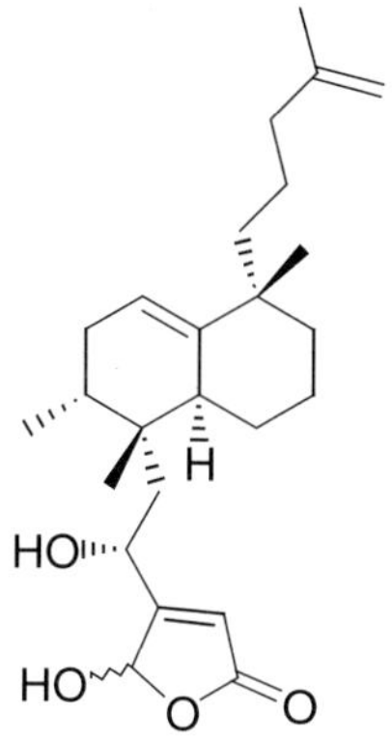

4

Dysidiolide (Corey 1997)

4.1 Introduction

The marine metabolite dysidiolide (**1**) was isolated from the Caribbean sponge *Dysidea etheria* de Laubenfels. Dysidiolide is the first naturally derived inhibitor of the human cdc25A protein phosphatase.[1,2] Since this class of enzymes (cdc25A, B and C) is involved in dephosphorylation of cyclin-dependent kinases, dysidiolide causes growth arrest accompanied by massive apoptosis in several human cancer cell lines.[3] The structure was elucidated by *Gunasekera*, *Clardy* et al. in 1996.[4] Dysidiolide is a γ-hydroxy-butenolide of the rearranged sesterterpene class whose [4.4.0] bicyclic nucleus with an array of stereocenters including two quaternary carbons and two axially disposed appendages defines a structurally attractive synthetic target with several total syntheses reported to date.[5] This problem is based on the first enantioselective total synthesis of dysidiolide realized by *Corey* in 1997, which also proved the absolute stereochemistry of natural **1**.[6]

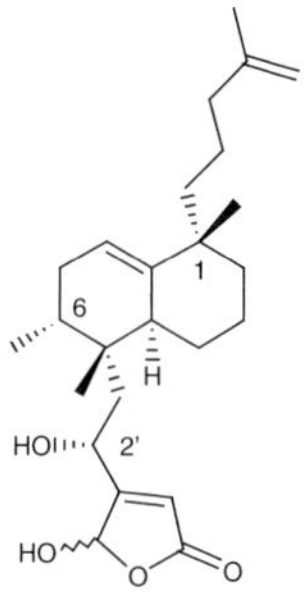

4.2 Overview

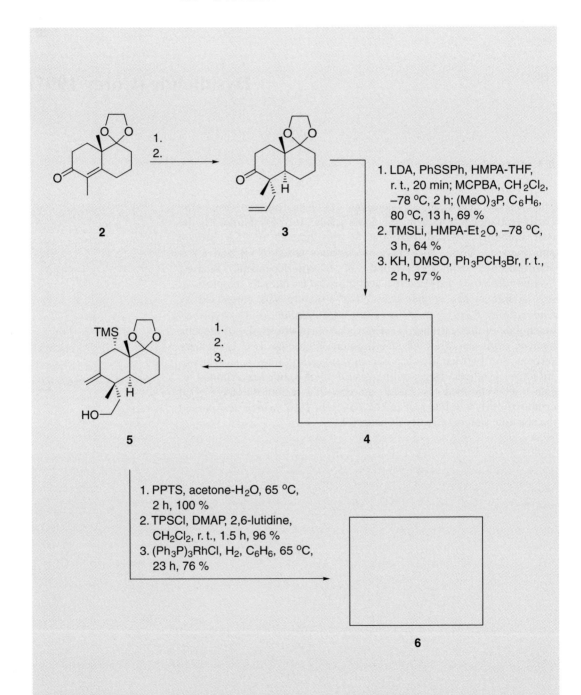

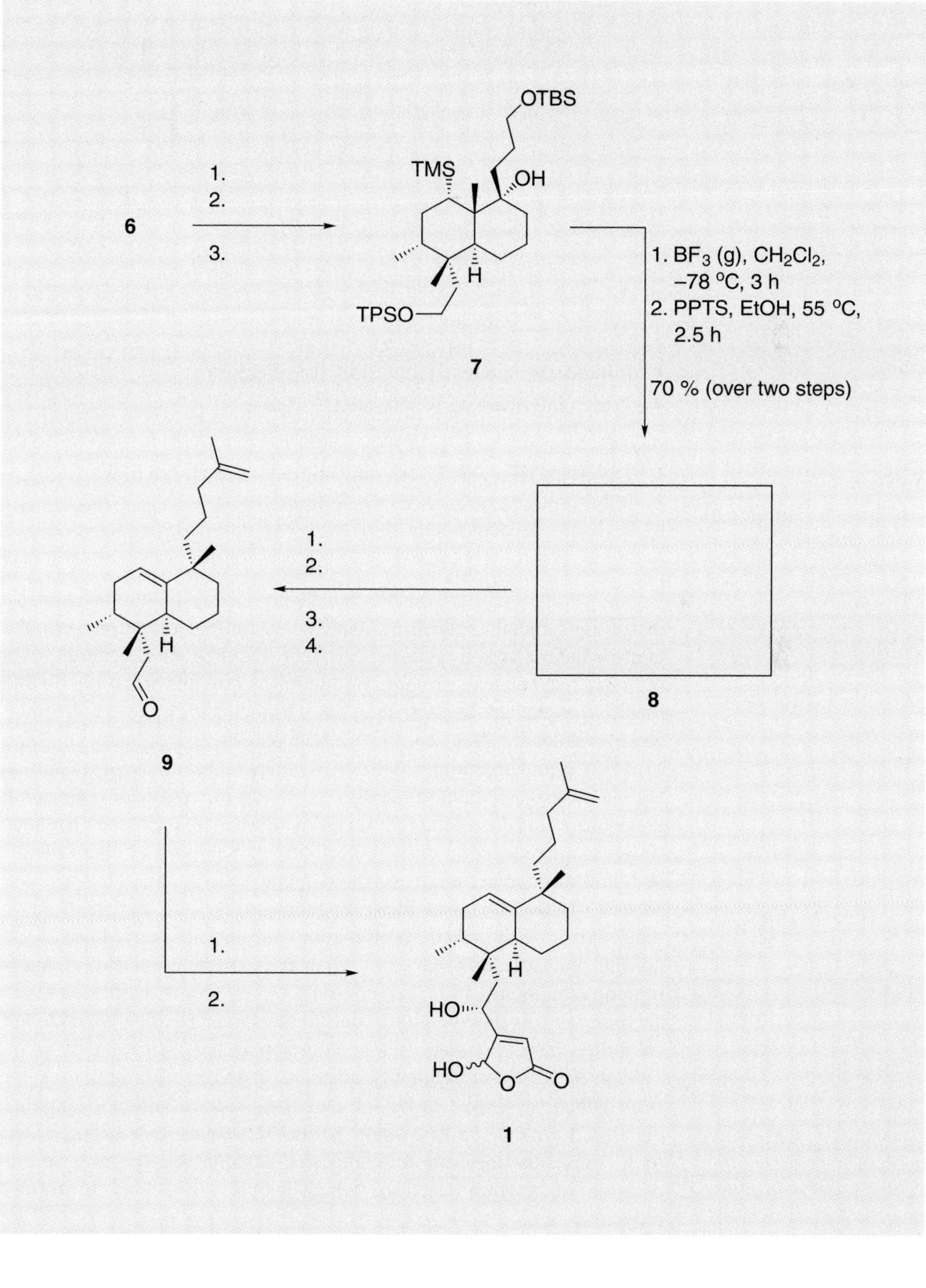

6
1.
2.
3.
OTBS
TMS
OH
H
TPSO
7
1. BF3 (g), CH2Cl2, –78 °C, 3 h
2. PPTS, EtOH, 55 °C, 2.5 h
70 % (over two steps)
8
1.
2.
3.
4.
H
O
9
1.
2.
H
HO
HO
O
O
1

4 Dysidiolide

4.3 Synthesis

Problem

Hints
- Initially the enone is reduced by free electrons.
- Which intermediate is alkylated?

Solution

1. Li-NH$_3$, THF, –40 °C, 10 min; isoprene, –78 °C, 15 min
2. Allyl bromide, –78 → 35 °C, 30 min
82 % (over two steps)

$$M \xrightarrow{NH_3\ (l)} M^+(NH_3)...e^-(NH_3)$$

Discussion

Isolated carbon-carbon double bonds are normally not reduced by dissolving metal reducing agents, because formation of the intermediate electron addition product requires more energy than ordinary reagents can provide. Reduction is possible when the double bond is conjugated, because then the intermediate can be stabilized by electron delocalization. By far the best reagent is a solution of alkali metal in liquid ammonia, with or without addition of an alcohol – known as *Birch* reduction conditions. Using this protocol conjugated dienes, α,β-unsaturated ketones, styrene derivatives and even benzene rings can be reduced to dihydro derivatives. Solutions of alkali metals in liquid ammonia contain solvated metal cations and electrons, and part of the usefulness of these reagents arises from the small steric requirement of the electrons. Reactions are usually carried out at the boiling point of ammonia (–33 °C) and since the solubility of many organic compounds in liquid ammonia is low at this temperature, co-solvents such as diethyl ether or tetrahydrofuran are often added to aid solubility. The first step of the reduction is the formation of anion radical **10** which subsequently abstracts a proton from ammonia (or from added alcohol) to give **11** and then, after addition of another electron, forms the enolate anion **13**. In the absence of a stronger acid (e. g. alcohols) this anion retains its negative charge and resists addition of another electron, which would correspond to further reduction.

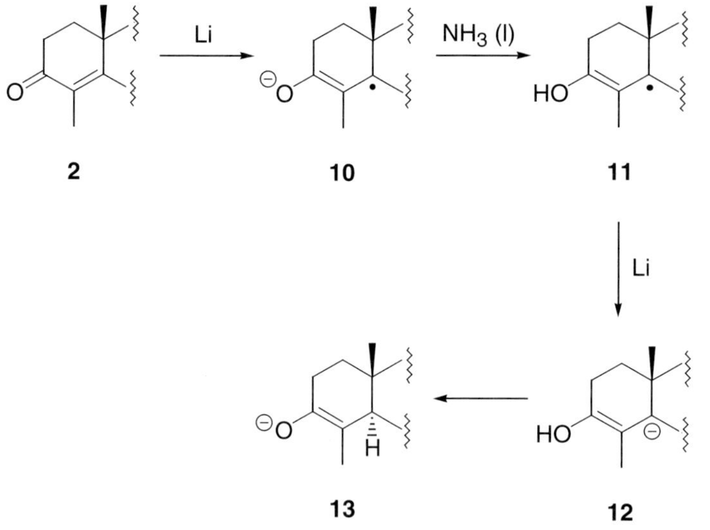

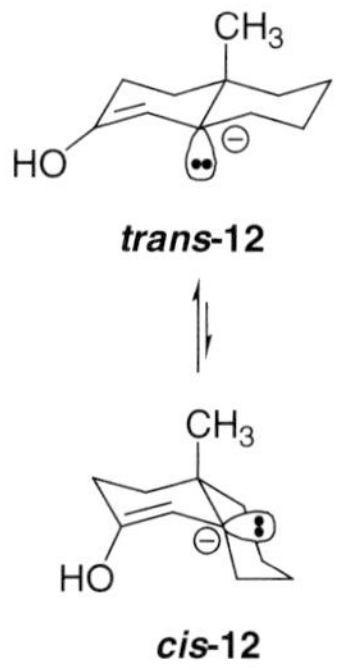

The protonation leads specifically to the *trans*-decalin system, though reduction could apparently give rise to two stereoisomeric products. The guiding principle appears to be that protonation of the intermediate allylic anion **12** takes place axially, orthogonal to the plane of the double bond, and to the most stable conformation of the carbanion which allows the best sp^3-orbital overlap on the β-carbon with the π-orbital system of the double bond.

After reaction of the excess lithium with isoprene the enolate is alkylated with allyl bromide diastereoselectively from the less hindered face, opposite to the axial methyl group at the bridge head, to provide allyl ketone **3** as the single diastereomer.

4 Dysidiolide

Problem

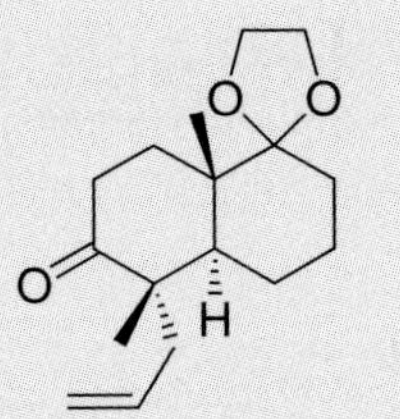

3

1. LDA, PhSSPh, HMPA-THF, r. t., 20 min; MCPBA, CH$_2$Cl$_2$, –78 °C, 2 h; (MeO)$_3$P, C$_6$H$_6$, 80 °C, 13 h, 69 %
2. TMSLi, HMPA-Et$_2$O, –78 °C, 3 h, 64 %
3. KH, DMSO, Ph$_3$PCH$_3$Br, r. t., 2 h, 97 %

4

Hints

- Which is the most acidic proton in **3**?
- **3** is sulfenylated.
- Here MCPBA does not effect an epoxidation.
- An α,β-unsaturated carbonyl compound is formed.
- The last step is a *Wittig* reaction.

Solution

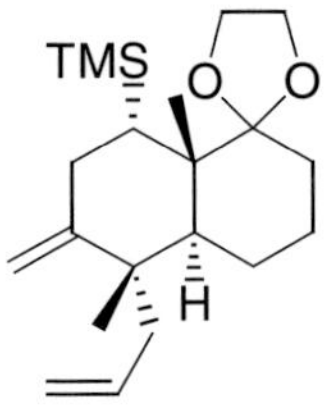

4

Discussion

Conversion of **3** to the corresponding α,β-enone **16** is accomplished by a sulfenylation-dehydrosulfenylation sequence: First, deprotonation adjacent to the carbonyl group of **3** with lithium diisopropylamide (LDA) followed by phenylsulfenylation with diphenyl disulfide furnishes α-phenylthio ketone **14**. The dehydrosulfenylation sequence involves first oxidation followed by thermolysis.

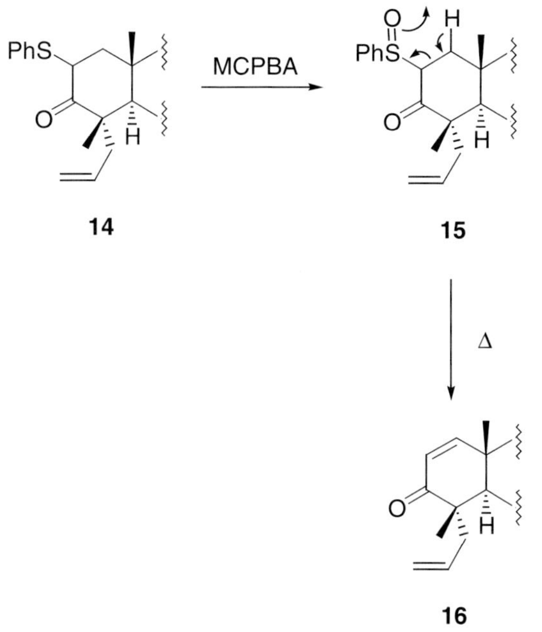

The oxidation of sulfides to sulfoxides is a facile transformation for which many reagents have been employed in the literature. These include hydrogen peroxide, ozone, nitric acid, chromic acid or *tert*-butylhypochlorite. Here, *meta*-chloroperbenzoic acid is used to oxidize α-phenylthio ketone **14**, and successive elimination of the resulting α-phenylsulfinyl ketone **15** by heating at reflux in benzene provides the α,β-unsaturated ketone **16**. Because of their thermal instability sulfoxides easily undergo elimination. The mechanism is explained by *Cram* as stereospecific *cis*-elimination.[7]

Addition of the α,β-unsaturated ketone **16** to excess trimethylsilyllithium[8] rapidly provides the intermediate enolate **17** which upon aqueous work-up yields the corresponding 3-silyl ketone. Interestingly, TMSLi leads like organocopper reagents (e. g. *Gilman*- and *Normant*-cuprates)[9] regioselectively to the 1,4-addition product, whereas organolithium reagents specifically show 1,2-addition to α,β-unsaturated ketones. The stereochemistry results from the less hindered axial attack by the trimethylsilyl anion.

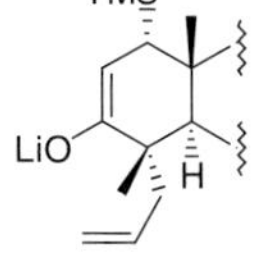

4 Dysidiolide

Final *Wittig*-methylenation with methylenetriphenylphosphorane gives **4** in excellent yield.

Problem

Hints

- *Corey* used the *Sharpless* dihydroxylation protocol. Which reagents are necessary?
- The resulting glycol is cleaved and reduced.

Solution

1. (DHQD)$_2$PYDZ, K$_2$OsO$_4$·2H$_2$O, K$_2$CO$_3$, K$_3$Fe(CN)$_6$, MeSO$_2$NH$_2$, *t*BuOH-H$_2$O, 0 °C, 4 h, 97 %
2. NaIO$_4$, THF-H$_2$O, r. t., 30 min
3. NaBH$_4$, THF-EtOH, 0 °C, 20 min, 94 % (over two steps)

Discussion

18
PYDZ

19
DHQD

In principle, the asymmetric dihydroxylation using the cinchona alkaloid based system evolved by *Sharpless* is especially practical because of the high catalytic turnover, the magnitude of the observed enantioselectivity and the wide range of substrates which may be oxidized.[10] Here, the pyridazine (**18**) linked dihydroquinidine (**19**) ligand (DHQD)$_2$PYDZ affords enhanced selectivity for the dihydroxylation of the mono-substituted double bond relative to the 1,1-disubstituted exocyclic one.[11] Thus, the vinyl group of **4** is regioselectively dihydroxylated to give the corresponding 1,2-diol (see Chapter 8).

Vicinal diols (glycols) are easily cleaved under mild conditions and in good yield with NaIO$_4$ or H$_5$IO$_6$. Mechanistically an intermediate diester of periodic acid is formed first. The successive cleavage of this intermediate **20** leads in an one-step process to the carbonyl **21** and the iodine(V) species **23**.

20 **21** **22** **23**

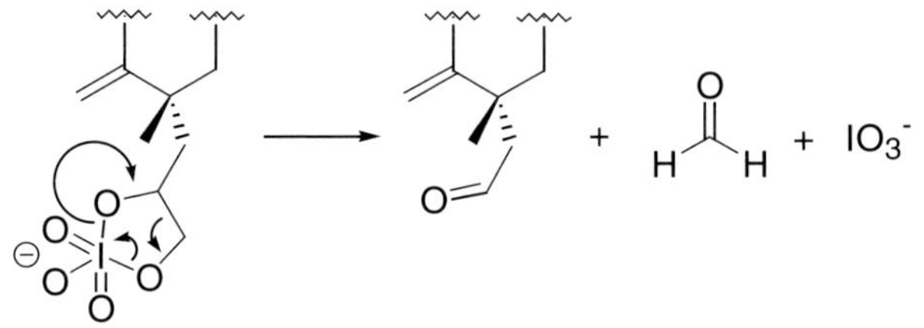

This reaction pathway is suitable for *cis* as well as *trans* diols. In principle, the oxidative cleavage is achieved by lead tetraacetate as well. However, these reagents are complementary, since periodic acid compounds are best used in water and lead tetraacetate in organic solvents. The last step is the reduction of the aldehyde with three equivalents of sodium borohydride to provide alcohol **5**.

Problem

5

1. PPTS, acetone-H_2O, 65 °C, 2 h, 100 %
2. TPSCl, DMAP, 2,6-lutidine, CH_2Cl_2, r. t., 1.5 h, 96 %
3. $(Ph_3P)_3RhCl$, H_2, C_6H_6, 65 °C, 23 h, 76 %

6

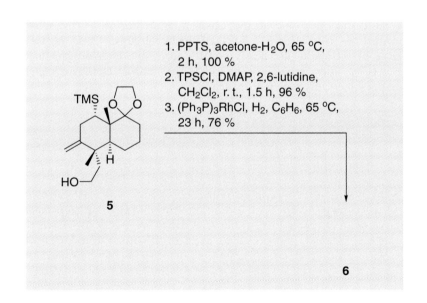

Hints

- Acetals are labile to acidic conditions.
- Which functionality is generally hydrogenated by H_2?

Solution

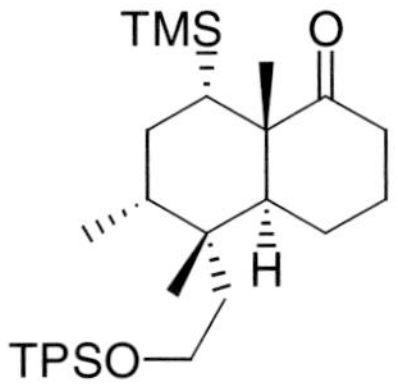

6

Discussion

Acetals have been proven the most serviceable protecting groups for aldehydes and ketones. They are stable to metal hydride reduction, organolithium reagents, strong bases and sodium or lithium in liquid ammonia. The most common method for deprotection is the acid-catalyzed hydrolysis. Thus, pyridinium *para*-toluenesulfonate (PPTS) deketalizes **5** to generate the corresponding carbonyl group. Subsequent protection of the hydroxyl moiety as *tert*-butyl-diphenylsilyl ether and hydrogenation under high pressure (69 bar) using *Wilkinson*'s catalyst gives a nearly 4:1 mixture of diastereomers of **6**. Efforts to improve the diastereoselectivity were unsuccessful. It is assumed that the cyclohexyl ring containing the exocyclic double bond is distorted out of the usual chair conformation by steric interactions between the 1,3-diaxial methyl groups. Thus, the two faces of the olefin are nearly equally accessible to hydrogenation. However, after separation by silica gel radial chromatography diastereomerically pure **6** is obtained.

Problem

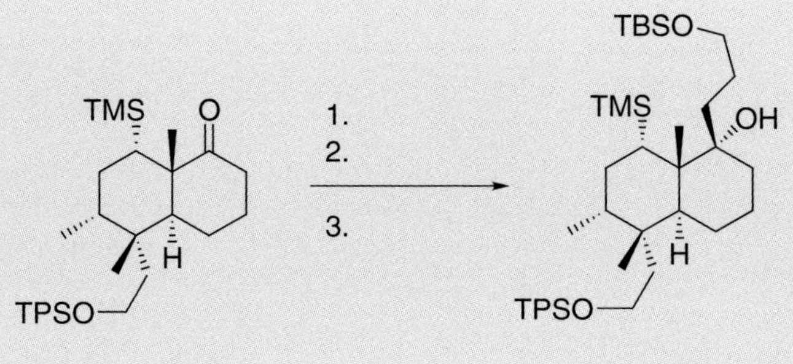

Hints

- Which classes of reagents easily add to carbonyl groups?
- Step 2 introduces an alcohol functionality.
- Final protection by a standard procedure furnishes **7**.

Solution

1. AllylMgBr, Et$_2$O, –78 °C → r. t., 10 min, 99 %
2. BH$_3$·DMS, Et$_2$O, r. t., 2 h; EtOH, NaOH, H$_2$O$_2$, r. t., 3.5 h, 95 %
3. TBSCl, DMAP, 2,6-lutidine, CH$_2$Cl$_2$, r. t., 12 h, 97 %

Discussion

Nucleophilic addition of allylmagnesium bromide affords stereospecifically the corresponding tertiary alcohol. Regioselective hydroboration of the allyl group with borane-dimethyl sulfide complex (BH$_3$·DMS) and oxidative work-up with EtOH, NaOH and H$_2$O$_2$ yields the primary–tertiary diol **24**. Generally, addition of borane to an unsymmetrical alkene gives predominantly the isomer in which boron is bound to the less highly substituted carbon atom. There is evidence suggesting that hydroboration is a concerted process and takes place through a four-membered cyclic transition state formed by addition of a polarized boron-hydrogen bond (in which the boron atom is more positive) to the double bond. Moreover, hydroboration is highly stereoselective and takes place by *syn*-addition of boron and hydrogen. Both the oxidation of the boron-carbon bond to form an alcohol and subsequent protonolysis occur with retention of configuration, thus establishing the *syn*-stereochemistry. Final selective protection of the primary alcohol functionality as *tert*-butyldimethylsilyl ether provides **7**.

Problem

Hints

- BF$_3$ (g) initiates a 1,2-rearrangement.
- An alkene is formed.
- Which silyl protecting group is more labile to acidic conditions?

Solution

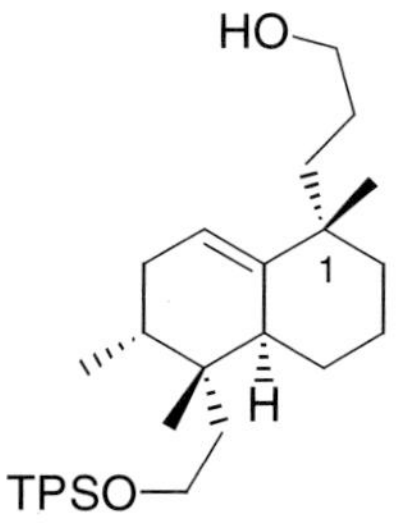

8

Discussion

Treatment of **7** with BF$_3$ (g) and subsequent cleavage of the TBS protecting group furnishes alcohol **8** with the fully substituted bicyclic scaffold of dysidiolide. Interestingly, the unusual quaternary center at C-1 and the endocyclic double bond within the bicyclic core are created by means of a biomimetic carbocation rearrangement. Boron trifluoride initiates heterolysis of the carbon-oxygen bond generating a cationic carbon at C-1 which subsequently induces the stereo-selective methyl 1,2-migration to form **26**. The TMS group facilitates the migration by β-hyperconjugation of TMS in TMSC–C$^+$. In addition, the very facile elimination of the TMS group confirms the location of the double bond.

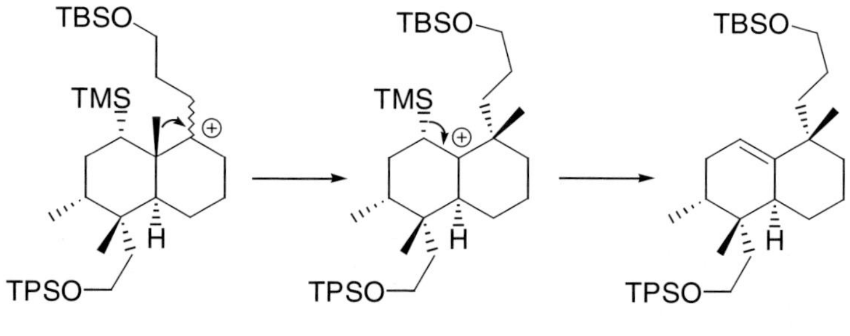

Deprotection of silyl ethers can be accomplished by using a variety of reagents such as TBAF, BF$_3$·OEt$_2$, alkali metal tetrafluoroborate and HF. Here, PPTS is used to distinguish the TBS from the TPS group. Thus, the rearranged **27** cleaves selectively the least bulky silyl group to form mono-protected **8**.[12]

Problem

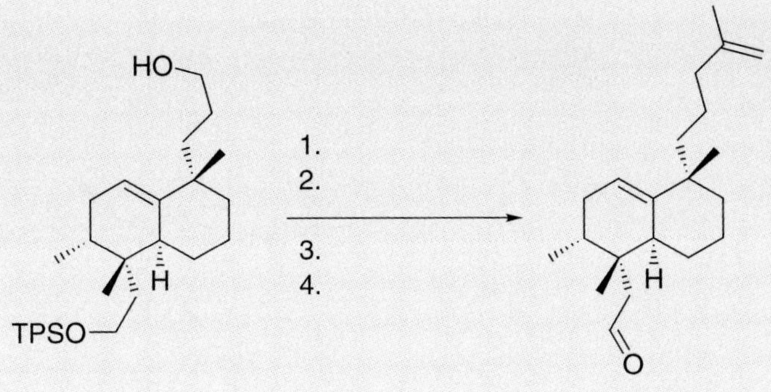

8 → **9**

1.
2.
3.
4.

Hints

- Step 1 is a halogenation.
- The second reaction is a nucleophilic substitution.
- Finally, the protected alcohol is converted to an aldehyde.

Solution

1. I_2, Ph_3P, imidazole, CH_2Cl_2, r. t., 10 min, 97 %
2. 2-Bromopropene, *t*BuLi, CuI, Et_2O, –30 → 0 °C, 30 min, 97 %
3. TBAF, THF, r. t., 8.5 h, 96 %
4. *Dess-Martin* periodinane, pyridine, CH_2Cl_2, r. t., 1 h, 94 %

Discussion

First an iodination of the alcohol is accomplished by treatment with iodine in presence of Ph_3P and imidazole.[13] Successive iodide displacement with the nucleophilic vinyl cuprate derived from 2-lithiopropene (2-bromopropene, *t*BuLi) finished the emplacement of the C-1 side chain. Cleavage of the TPS ether by tetrabutylammonium fluoride and subsequent oxidation with the *Dess-Martin* periodinane provides aldehyde **9**.

4 Dysidiolide

Problem

Hints
- Step 1 yields a separable mixture of diastereomeric carbinols.
- Only the 2'-(R)-diastereomer is further used.
- A photochemical oxidation finishes the synthesis.

Solution
1. 3-Bromofuran, nBuLi, THF, –78 °C, 30 min, 98 %
2. O$_2$, hν, Rose Bengal, iPr$_2$EtN, CH$_2$Cl$_2$, –78 °C, 2 h, 98 %

Discussion

Treatment of **9** with 3-lithiofuran (3-bromofuran, nBuLi) affords a 1:1 mixture of diastereomeric carbinols, which are readily separated by silica gel chromatography. However, the undesired C-2'-epimer of **28** is efficiently converted to the other diastereomer by oxidation of the alcohol with *Dess-Martin* periodinane and subsequent stereoselective oxazaborolidine-catalyzed reduction to give the desired 2'-(R)-carbinol exclusively. This recycling of the undesired 2'-(S)-epimer allows for the transformation of **9** to **28** in 95 % yield overall.

Photochemical oxidation of **28** with singlet oxygen furnishes dysidiolide (**1**) as a white solid. The most common method for generating ^{1}O$_2$ in solution is the dye-sensitized photochemical excitation of triplet oxygen. The mechanism involves the excitation of an appropriate dye with visible light to generate the corresponding excited singlet state. Rapid intersystem crossing (ISC) provides the excited state of the sensitizer, which undergoes energy transfer with triplet oxygen to form singlet oxygen, regenerating the ground state of the sensitizer. Common sensitizers are organic dyes such as Rose Bengal (**29**), methylene blue and certain porphyrin derivatives (see Chapter 1). In addition, to the soluble dye, polymer-bound dyes such as polystyrene-bound Rose Bengal may also be used for the generation of ^{1}O$_2$.

In principle, singlet oxygen undergoes three classes of reaction with alkenes: an ene type reaction forming allylic hydroperoxides, [4+2]-cycloadditions with cisoid 1,3-dienes and 1,2-cycloadditions with electron-rich or strained alkenes. Here, the 3-alkylfuran **30** is regioselectively oxidized by $^1O_2$ in presence of a hindered base to form the corresponding γ-hydroxybutenolide **32**.[14] Actually the mechanism is a *Diels-Alder* reaction with oxygen as dienophile to form endoperoxide **31**.

However, the formation of dysidiolide (**1**) requires the regiospecific removal of the hydrogen at C-1 on the endoperoxide **31**. This is achieved by treatment with a hindered base such as diisopropylethylamine (*Hünig*'s base) at low temperature in order to favor base-catalyzed decomposition rather than thermal decomposition.

4.4 Conclusion

The above-described problem demonstrates the first enantioselective synthesis of dysidiolide, a C_{25} isoprenoid antimitotic agent. The central transformations are the sulfenylation-dehydrosulfenylation sequence to prepare an α,β-enone, the biomimetic cationic 1,2-rearrangement to form stereoselectively the bicyclic scaffold, vinyl cuprate displacement of an iodide furnishing the C-1 side chain and the photochemical oxidation of furan to generate the γ-hydroxybutenolide functionality.

This synthesis is completed in 22 steps with an overall yield of approximately 12 %.

4.5 References

1 B. Baratte, L. Meijer, K. Galaktionov, D. Beach, *Anticancer Res.* **1992**, *12*, 873-880.
2 J. B. A. Millar, P. Russell, *Cell* **1992**, *68*, 407-410.
3 S. R. Magnuson, L. Sepp-Lorenzino, N. Rosen, S. J. Danishefski, *J. Am. Chem. Soc.* **1998**, *120*, 1615-1616.

4   G. P. Gunasekera, P. J. McCarthy, M. Kelly-Borges, E. Lobkorsky, J. Clardy, *J. Am. Chem. Soc.* **1996**, *118*, 8759-8760.
5   a) J. Boukouvalas, Y.-X. Cheng, J. Robichaud, *J. Org. Chem.* **1998**, *63*, 228-229; b) H. Miyaoka, Y. Kajiwara, Y. Yamada, *Tetrahedron Lett.* **2000**, *41*, 911-914; c) D. Demeke, C. J. Forsyth, *Org. Letters* **2000**, *2*, 3177-3179.
6   E. J. Corey, B. E. Roberts, *J. Am. Chem. Soc.* **1997**, *119*, 12425-12431.
7   C. A. Kingsbury, D. J. Cram, *J. Am. Chem. Soc.* **1960**, *82*, 1810-1819.
8   P. F. Hudrlik, M. A. Waugh, M. A. Hudrlik, *J. Organomet. Chem.* **1984**, *271*, 69-76.
9   a) H. Gilman, R. G. Jones, L. A. Woods, *J. Org. Chem.* **1952**, *17*, 1630-1631; b) A. Alexakis, I. Marek, P. Mangeney, J. F. Normant, *Tetrahedron Lett.* **1989**, *30*, 2387-2390.
10 H. C. Kolb, M. S. VanNieuwenhze, K. B. Sharpless, *Chem. Rev.* **1994**, *94*, 2483-2547.
11 E. J. Corey, M. C. Noe, *J. Am. Chem. Soc.* **1996**, *118*, 11038-11053.
12 C. Prakash, S. Saleh, I. A. Blair, *Tetrahedron Lett.* **1989**, *30*, 19-22.
13 P. J. Garegg, B. Samuelsson, *J. Chem. Soc., Perkin Trans. 1* **1980**, 2866-2869.
14 M. R. Kernan, D. J. Faulkner, *J. Org. Chem.* **1988**, *53*, 2773-2776.

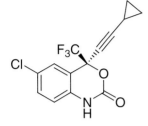

5

Efavirenz (Merck, DuPont 1999)

5.1 Introduction

Efavirenz (DMP 266) (**1**) is an effective non-nucleoside inhibitor of reverse transcriptase of the human immunodeficiency virus (HIV) recently registered by the US Food & Drug Administration (FDA) for treatment of the acquired immunodeficiency syndrome (AIDS).[1,2,3]

Inhibition of HIV reverse transcriptase by nucleosides like azidothymidine (AZT) (**2**) is a proven therapy for delaying the progression to AIDS. However, the rapid viral mutation to resistant strains requires the development of new therapeutic agents. The recent development of both protease inhibitors and non-nucleoside reverse transcriptase inhibitors offers hope of effective treatment especially when coadministered.

Structural features of **1** are an oxazinone moiety and a cyclopropyl acetylene unit connected to a chiral quaternary carbon center.

The importance of this compound gave rise to the development of a short and efficient synthesis, allowing large scale processes, which is presented in this problem.

5.2 Overview

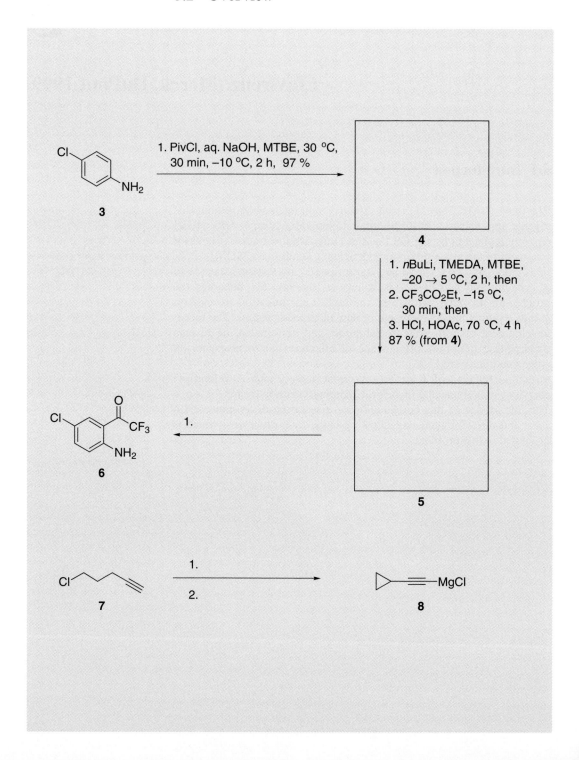

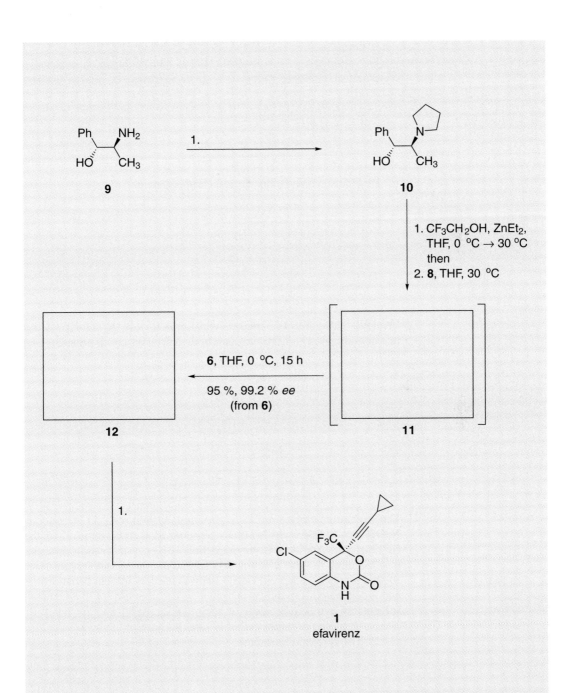

5.3 Synthesis

Problem

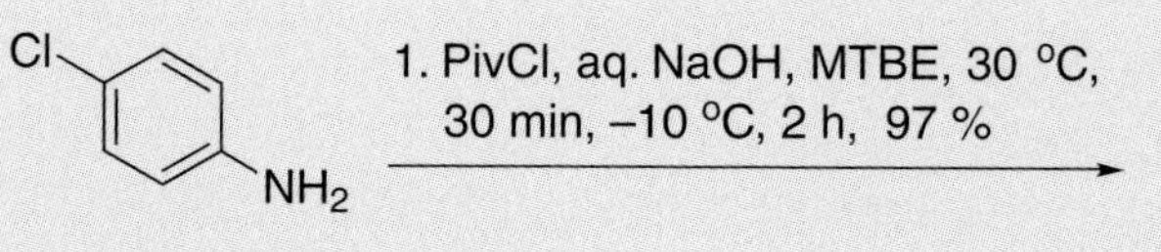

[Reaction scheme: 4-chloroaniline (3) + 1. PivCl, aq. NaOH, MTBE, 30 °C, 30 min, −10 °C, 2 h, 97 % → 4]

Hints

- How does an acid chloride react with an amine?

Solution

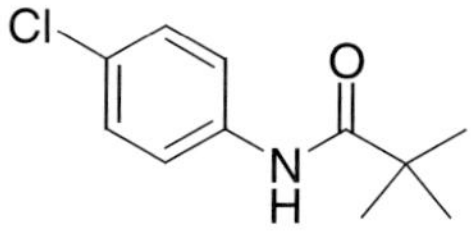

[Structure of 4: N-(4-chlorophenyl)pivalamide]

Discussion

Pivaloyl chloride (13) reacts with aniline (3) to amide 4 by nucleophilic attack of the nitrogen to the carbonyl group under loss of HCl.

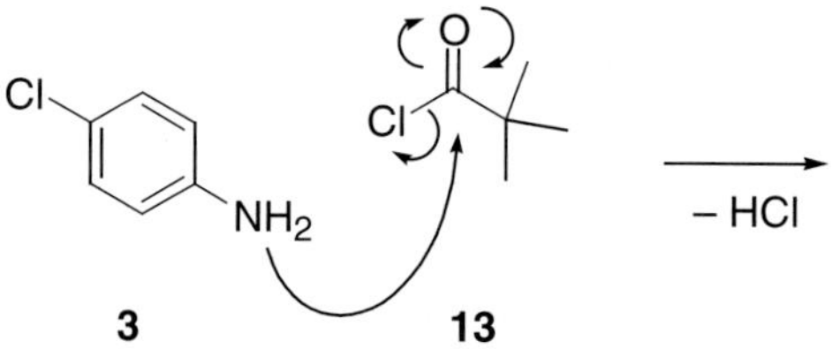

[Mechanism scheme showing 3 + 13 → 4, − HCl]

The reaction is carried out in a two phase system with 30 % aqueous NaOH and MTBE, which allows trapping as well as removal of the HCl evolved. Anilide 4 is obtained directly by crystallization from the reaction mixture. This reaction was carried out on a 1,000 g scale.[2]

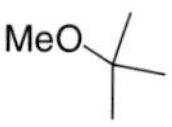

Methyl *tert*-butyl ether (MTBE) (14) was used as solvent because it is not halogenated (at industrial plant scale, the use of halogenated solvents is avoided because of their noxiousness) and its handling compared to that of diethyl ether is safer because of its lower volatility and lack of peroxide formation.

[Structure of 14: MTBE]

Problem

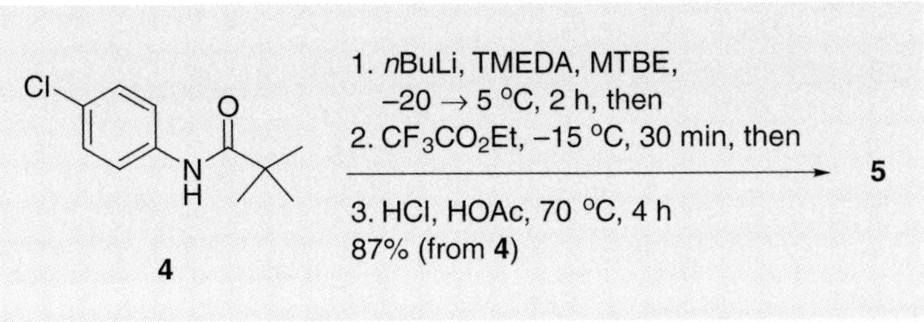

[Scheme: Compound **4** (4-chloro-N-pivaloyl aniline) → **5**
Conditions:
1. nBuLi, TMEDA, MTBE, −20 → 5 °C, 2 h, then
2. CF₃CO₂Et, −15 °C, 30 min, then
3. HCl, HOAc, 70 °C, 4 h
87% (from **4**)]

Hints

- *n*BuLi is a metalating agent.
- The pivaloyl amide group is metal directing. Which regiochemistry of metalation do you expect?
- The resulting organolithium species acts as the nucleophile in the second step. Which bond formation can occur?
- Pivaloyl amides are acid sensitive.
- How does an electrophilic ketone react with water?

Solution

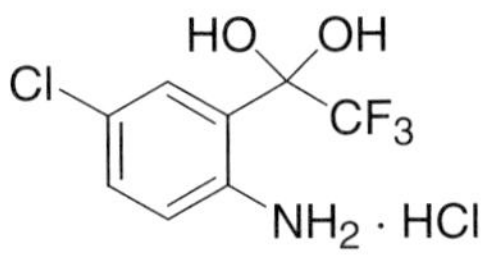

[Structure **5**: 4-chloro-2-(1,1-dihydroxy-2,2,2-trifluoroethyl)aniline hydrochloride, with HO, OH, CF₃, NH₂·HCl substituents]

Discussion

Anilide **4** is lithiated selectively in *ortho*-position to the pivaloyl amide group.[4,5] The organolithium species is generated by reaction of **4** with two equivalents of *n*-butyllithium below 5 °C in MTBE, since the amide proton is also acidic and is deprotonated to yield resonance-stabilized anion **15** before the *ortho*-lithiation of the aromatic system with the second equivalent of *n*-butyllithium takes place. The resulting organolithium species **16** then undergoes nucleophilic attack of ester **17**[6] to give dianion **18**.

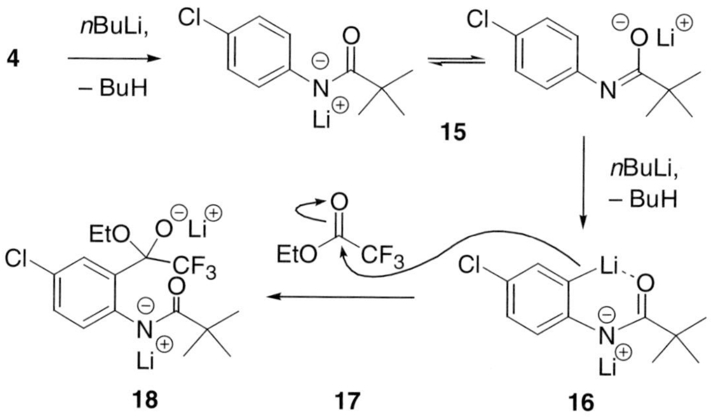

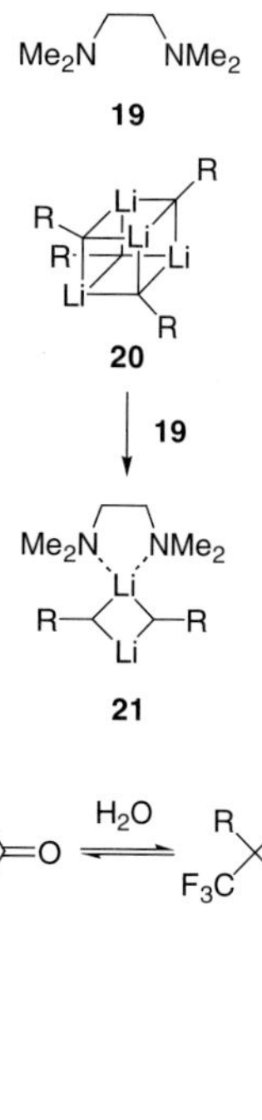

This lithiation reaction is carried out in the presence of the chelating diamine TMEDA **19** because, in hydrocarbon solvents, alkyllithiums are thought to react as aggregates such as **20** or mixtures of aggregates and dissociated species.[7] Donor solvents like THF or even better diamines like **19** effectively break down alkyllithium aggregates, forming monomers and dimers **21** in solution, and thereby significantly increase their basicity and reactivity towards deprotonation.[7] The use of MTBE as the solvent avoids the competitive attack of *n*-butyllithium on THF at the required reaction temperature of < 5 °C.[7c]

Acidic work-up of **18** furnishes ketone **22**. Further addition of HCl and acetic acid at 70 °C results in cleavage of the pivaloyl anilide to yield the unprotected *ortho*-ketoaniline as hydrochloride hydrate **5**, since ketones with neighboring electron-withdrawing groups undergo nucleophilic addition of water. **5** directly crystallized from the reaction mixture and was obtained in 87 % yield with a purity of > 98 %. These three steps were carried out on a 3,700 g scale.

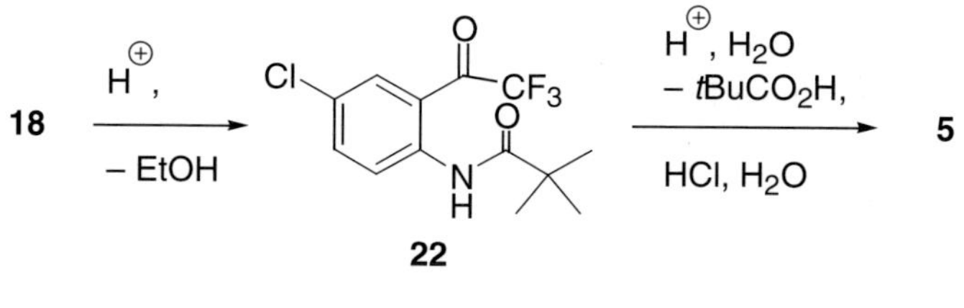

Problem

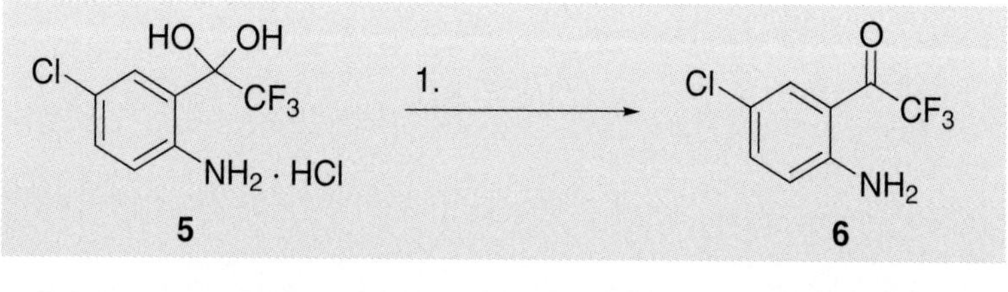

- How can an amine be liberated from its hydrochloride?
- The conditions applied also result in dehydratization of the hydrate.

Hints

1. aq. NaOAc, MTBE, r. t., 30 min, 99 %

Solution

Since hydrochloride hydrate **5** crystallized from the reaction mixture, an additional step was necessary to obtain the desired *ortho*-ketoaniline **6**. The free base was obtained by treatment of **5** with NaOAc in a pH range of 4.0–6.0, which had to be carefully adjusted during the reaction. Hydrate formation of ketones is a reversible process in solution, and the equilibrium fraction of *o*-aminoketone **6** was continuously extracted from the aqueous layer until removal of hydrating water was complete. The extraction was carried out on a 3,000 g scale.

Discussion

Problem

- What are the acidic positions of alkyne **7**?
- By twofold deprotonation, **7** is converted into a nucleophile.
- The chloro substituent is a leaving group.
- How can a terminal alkyne be converted into a *Grignard* reagent?

Hints

1. 2 eq. *n*BuLi, cyclohexane, 0 → 80 °C, 65 %
2. *n*BuMgCl, THF, 0 °C, 1 h

Solution

First, **7** is converted into its dianion **23** by two equivalents of *n*-butyllithium. Not only the terminal hydrogen of alkyne **7** but also the propargylic one is acidic. **23** then undergoes intramolecular nucleophilic attack of the terminal chlorine. NH$_4$Cl work-up yields the desired cyclopropyl acetylene **25**.[8]

Second, **25** is converted into the alkynyl magnesium species **8** by treatment with *n*BuMgCl, because the alkyl *Grignard* reagent is more basic than the alkynyl, so deprotonation of the alkyne **25** under loss of *n*butane takes place.

Discussion

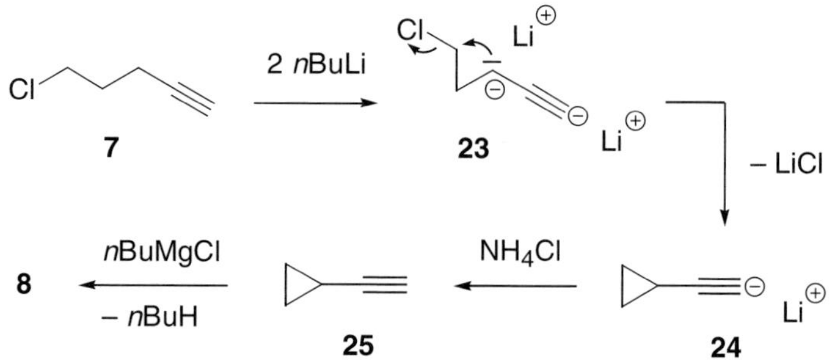

Alkynyl *Grignard* reagent **8** is directly applied in a THF solution for conversion of **10** into **11**.

Problem

Hints
- A nucleophilic substitution is carried out.
- The nitrogen is alkylated in the presence of a mild base.
- The alcohol is not affected by the conditions applied.

Solution

1. 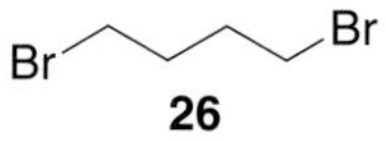 Br~~~Br **26**, NaHCO$_3$, toluene, reflux, 20 h, 96 %

Discussion

Norephedrine **9** is alkylated with dibromide **26** to give pyrrolidine derivative **10**.[2] The hydroxy group of **9** is not alkylated under the conditions applied because of its lower basicity and nucleophilicity.

First, the amino group of **9** is monoalkylated and the resulting ammonium salt **27** is deprotonated with NaHCO$_3$ as base. This yields secondary amine **28**, which then undergoes intramolecular nucleophilic attack to furnish the desired *N*-pyrrolidinyl norephedrine **10** after deprotonation. Intermolecular nucleophilic substitution was not observed under these conditions.

This reaction was carried out on a 1,500 g scale.

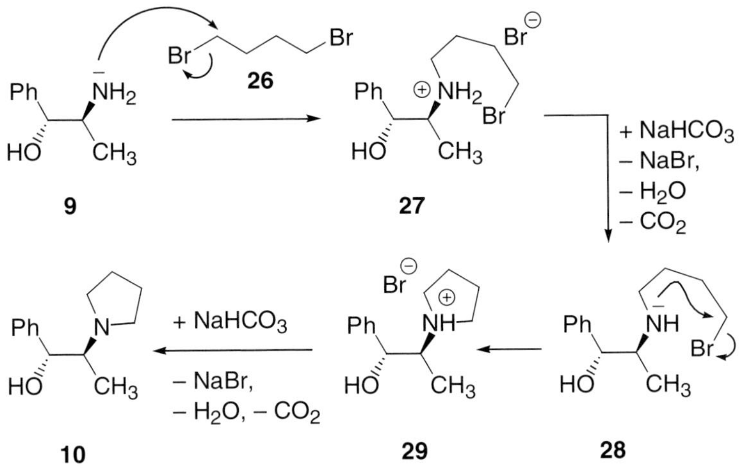

- ZnEt₂ is a basic reagent. How does it react with alcohols?
- A zinc alkoxide is formed in the first step. How will this complex react with alkynyl *Grignard* reagent **8**?
- A transmetalation is carried out.

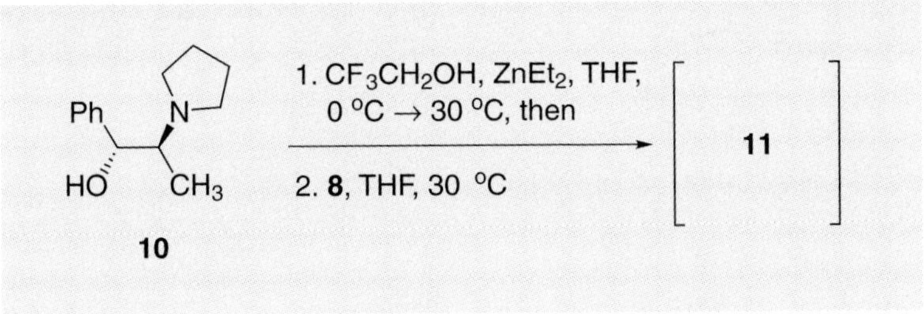

Hints

Solution

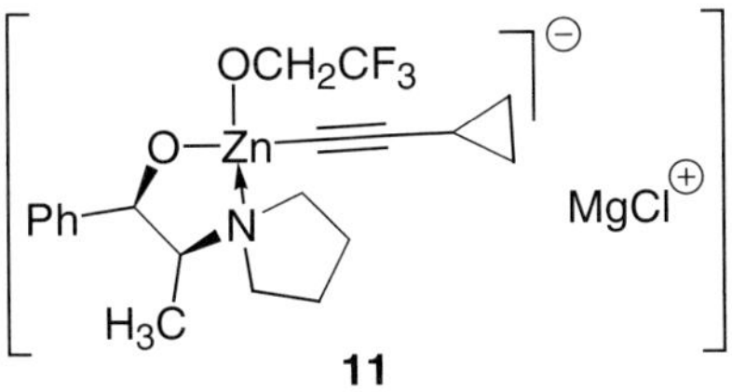

The reaction of basic ZnEt$_2$ with pyrrolidine norephedrine **10** and one equivalent of trifluoroethanol as auxiliary ligand leads to the chiral zinc alkoxide **30**[9] under loss of ethane after deprotonation of the hydroxy groups of **10** and trifluoro ethanol as additive. **30** is not isolated but subjected to addition of cyclopropyl acetylene magnesium

Discussion

chloride (**8**) in the next step. This gives a zincate species, whose proposed structure is **11**. The THF solution of **11** is directly used in the next step. This reaction is also performed on a kg scale and, interestingly, the order of addition of the alcohols and ZnEt$_2$ to the reaction mixture turned out to have no influence on the outcome of selectivity and conversion of this and the following reaction. However, inferior results were obtained with other halides and auxiliary ligands.[1]

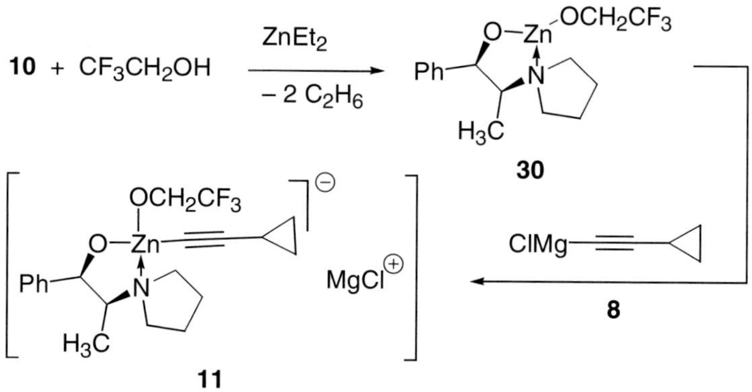

Problem

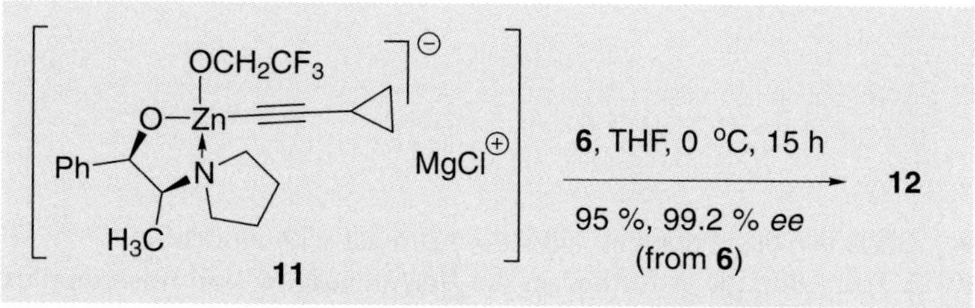

6, THF, 0 °C, 15 h → **12**

95 %, 99.2 % *ee* (from **6**)

Hints

- How does a ketone react with an organozinc compound?
- A tertiary alcohol bearing a stereogenic center is created.

Solution

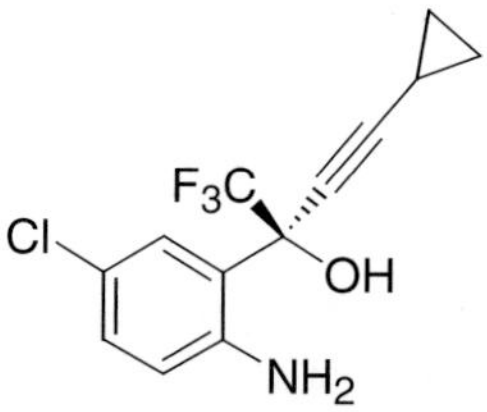

12

In this step, a new type of enantioselective alkynylation of a prochiral ketone promoted by a zinc alkoxide is carried out.[1] By this, the alkyne moiety of zincate **11** undergoes *re*-attack (see Chapter 3) of the ketone **6** furnishing alcohol **12** with the *R*-configuration in 99.2 % *ee* and pyrrolidinyl norephedrine **10** after work-up.

Discussion

Pyrrolidinyl norephedrine **10** as the chiral ligand was recovered and reused nine times.

Even though the enantioselective alkylation of carbonyl compounds is a well investigated reaction giving good to very good optical purities,[10] less is known about the addition of alkynes. There are some examples of asymmetric alkynylations of aldehydes using chiral auxiliaries giving optical purities in the range of 50–97 % *ee*.[11]

An earlier synthesis of efavirenz (**1**) by this research group[2] was based on the addition of 2.2 equivalents of lithium cyclopropyl acetylene (**32**) to the PMB-*N*-protected ketoaniline **31** in the presence of 2.2 equivalents of lithium norephedrine alkoxide **33**. This gave alcohol **34** with an enantiomeric excess of 99 % but the presence of PMB as the *N*-protecting group turned out to be necessary. Cleavage of PMB-groups[12] requires stoichiometric amounts of oxidizing agents such as cerium salts or DDQ and is therefore problematic for environmental aspects. Furthermore, this reaction had to be carried out at low temperatures of –60 °C to achieve high enantioselectivities.

Use of zincate **11** protection of the aniline was not necessary.

Problem

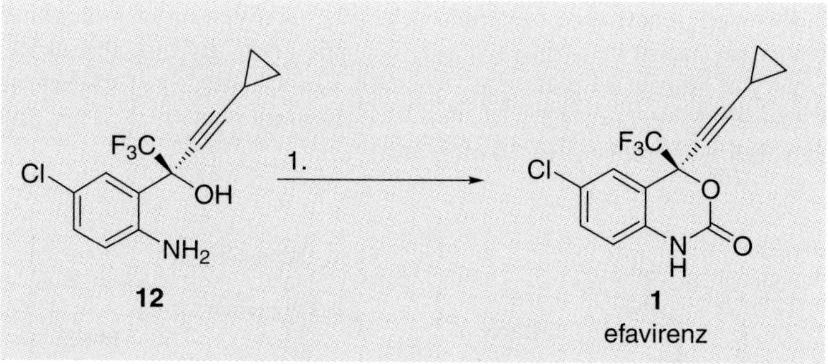

Hints
- What is the simplest reagent to form carbamates?
- Though hazardous, the reagent employed is an industrial bulk chemical.

Solution
1. $COCl_2$, THF, hexane, 0 °C → r. t., 1 h, 95 %

Discussion
Conversion of the amino alcohol **12** into **1** was most conveniently and economically accomplished using phosgene in the absence of base. The reaction presumably proceeds *via* intermediate **35** followed by ring closure. After aqueous work-up ($NaHCO_3$), **1** was crystallized from THF–heptane in excellent yield (95 %) and purity (> 99.5 %, > 99.5 % *ee*).

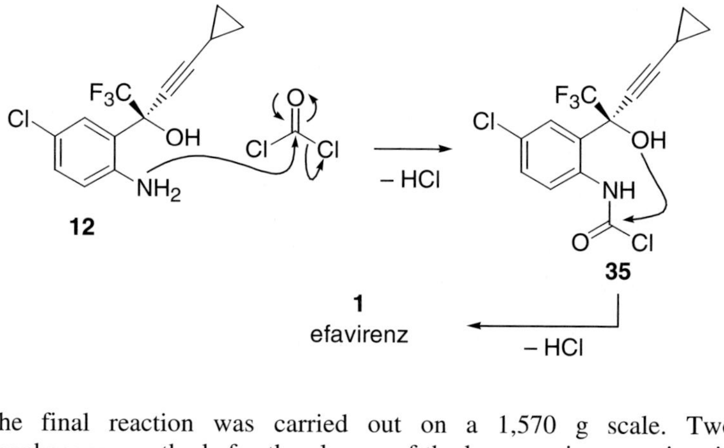

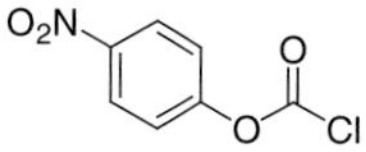

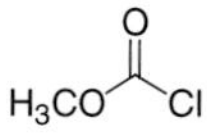

The final reaction was carried out on a 1,570 g scale. Two nonphosgene methods for the closure of the benzoxazinone moiety in **1** with chloroformates **36** and **37** have also been developed but needed an additional step because intermediate formation and ring closure required different reaction conditions.[2]

5.4 Conclusion

A highly efficient, enantioselective industrial synthesis of the HIV reverse transcriptase inhibitor efavirenz is made available for the manufacture of this important compound. A novel, chiral Zn-alkoxide-mediated, enantioselective acetylide addition reaction is used to establish the chiral center in the target with a remarkable level of stereocontrol. The synthesis provides analytically pure efavirenz in an overall yield of 75 % in five steps from 4-chloroaniline.

The advantages of this industry scale synthesis are the use of non-halogenated solvents, formation of inert inorganic salts as waste products, recycling of valuable side products, ambient temperatures, relinquishment of protecting groups and purification by crystallization or filtration.

5.5 References

1   L. Tan, C. Chen, R. D. Tillyer, E. J. J. Grabowsky, P. J. Reider, *Angew. Chem.* **1999**, *111*, 724-727; *Angew. Chem. Int. Ed. Engl.* **1999**, *38*, 711-713.

2   M. E. Pierce, R. L. Parsons, Jr., L. A. Radesca, Y. S. Lo, S. Silverman, J. R. Moore, Q. Islam, A. Choudhury, J. M. D. Fortunak, D. Nguyen, C. Lo, S. J. Morgan, W. P. Davis, P. N. Confalone, C. Chen, R. D. Tillyer, L. Frey, L. Tan, F. Xu, D. Zhao, A. S. Thompson, E. G. Corley, E. J. J. Grabowsky, R. Reamer, P. J. Reider, *J. Org. Chem.* **1998**, *63*, 8536-8543.

3   S. D. Young, S. F. Britcher, L. O. Tran, L. S. Payne, W. C. Lumma, T. A. Lyle, J. R. Huff, P. S. Anderson, D. B. Olsen, S. S. Carrol, D. J. Pettibone, J. A. O'Brien, R. G. Ball, S. K. Balani, J. H. Lin, I.-W. Chen, W. A. Schleif, V. V. Sardana, W. J. Long, V. W. Byrnes, E. A. Emini, *Antimicrob. Agents Chemother.* **1985**, *39*, 2602.

4   V. Snieckus, *Chem. Rev.* **1990**, *90*, 879-933.

5   W. Fuhrer, H. W. Gschwend, *J. Org. Chem.* **1979**, *44*, 1133-1136.

6   J.-P. Begue, D. Bonnet-Delpon, *Tetrahedron* **1991**, *47*, 3207-3258.

7   a) D. R. Hay, Z. Song, S. G. Smith, P. Beak, *J. Am. Chem. Soc.* **1988**, *110*, 8145-8153; b) B. J. Wakefield, *The Chemistry of Organolithium Compounds*, Pergamon, Oxford **1972**; c) R. B Bates, L. M. Kroposki, D. E. Potter, *J. Org. Chem.* **1972**, *42*, 560-566.

8   A. S. Thompson, E. G. Corley, M. F. Huntington, E. J. J. Grabowski, *Tetrahedron Lett.* **1995**, *36*, 8937-8940.

9   For comparison see: D. Enders, J. Zhu, G. Raabe, *Angew. Chem.* **1996**, *108*, 1827-1829; *Angew. Chem. Int. Ed. Engl.* **1996**, *35*, 1725-1728.

10 a) R. Noyori, M. Kitamura, *Angew. Chem.* **1991**, *103*, 34-55; *Angew. Chem. Int. Ed. Engl.* **1991**, *30*, 49-69; b) K. Soai, S. Niwa, *Chem. Rev.* **1992**, *92*, 833-856; c) R. O. Duthaler, A. Hafner, *Chem. Rev.* **1992**, *92*, 807-832; d) D. A. Evans, *Science*, **1989**, *240*, 420; e) T. Shibata, H. Morioka, T. Hayase, K. Choji, K. Soai, *J. Am. Chem. Soc.* **1996**, *118*, 471-472 and references therein.

11 a) S. Niwa, K. Soai, *J. Chem. Soc., Perkin Trans. I* **1990**, 937-943; b) E. J. Corey, K. A. Cimprich, *J. Am. Chem. Soc.* **1994**, *116*, 3151-3152.

12 T. W. Greene, P. G. M. Wuts, *Protective Groups in Organic Synthesis*, (3rd edition), John Wiley & Sons, New York **1999**.

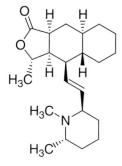

6

(+)-Himbacine (Chackalamannil 1999)

6.1 Introduction

Himbacine (**1**) and himbeline (**2**) are complex piperidine alkaloids isolated from the bark of *Galbulimima baccata*, a species that belongs to the magnolia family.[1] Himbacine has attracted considerable attention as a promising lead in *Alzheimer's* disease research because of its potent muscarinic receptor antagonist property.[2] The senile dementia associated with *Alzheimer's* disease is directly correlated with diminished levels of synaptic acetylcholine in the cortical and hippocampal areas of the brain, and the current form of therapy addresses this issue by inhibiting cholinesterase, which breaks down acetylcholine. Alternatively, biosynthesic enhancement of synaptic acetylcholine levels could be achieved by selectively inhibiting presynaptic muscarinic receptors (M_2), agonist-induced stimulation of which shuts off acetylcholine release.[3] Himbacine is a potent inhibitor of the muscarinic receptor of the M_2 subtype with 20-fold selectivity towards the M_1 receptor. However, to identify a therapeutically useful target, both the selectivity and potency of himbacine need to be optimized through a rigorous structure-activity relationship study in this series. The complex structural features of himbacine make this task difficult. As part of efforts to identify potent and selective M_2 receptor antagonists related to himbacine, a practical synthesis of himbacine and closely related analogs is needed. The following problem is based on a total synthesis realized by *Chackalamannil* in 1999.[4]

1: R = CH_3 himbacine
2: R = H   himbeline

6.2 Overview

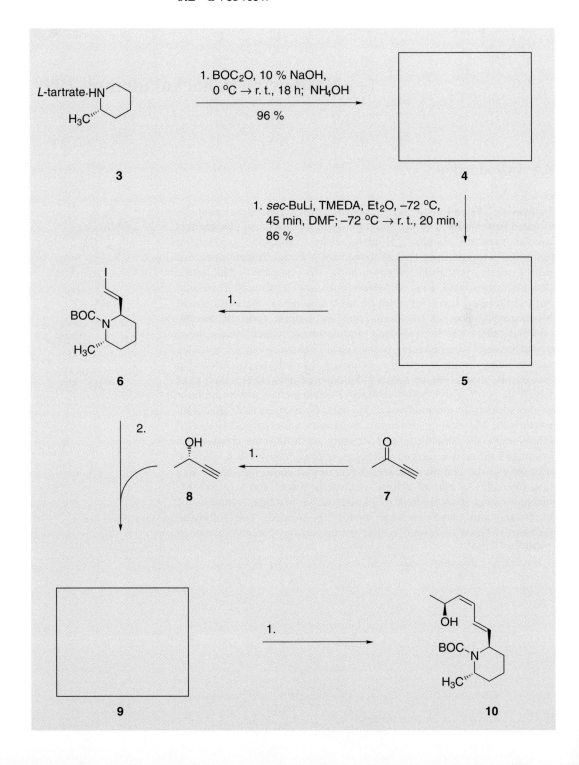

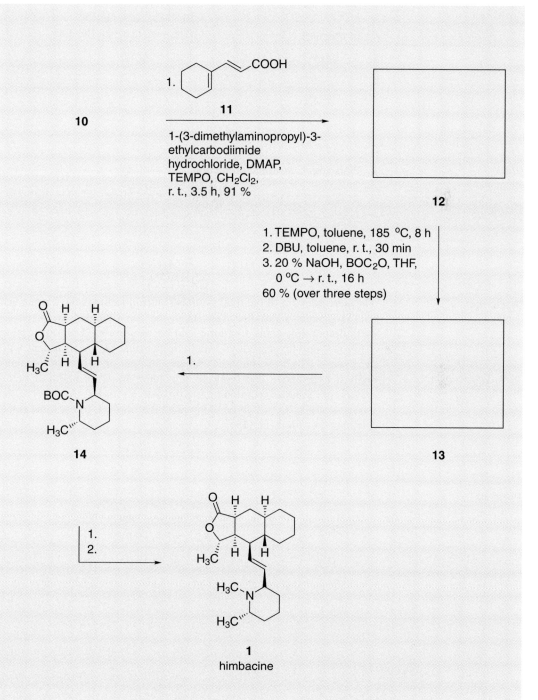

6.3 Synthesis

Problem

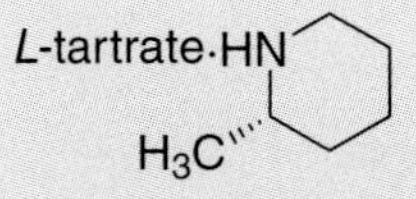

L-tartrate·HN—[piperidine with H₃C at 2-position] (**3**) → **4**

1. BOC$_2$O, 10 % NaOH, 0 °C → r.t., 18 h; NH$_4$OH

96 %

Hints

- This step is a simple protection.

Solution

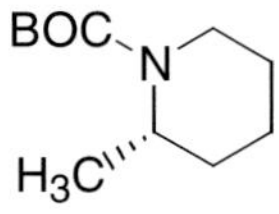

BOC-N—[piperidine with H$_3$C at 2-position] (**4**)

Discussion

The starting point of the total synthesis of himbacine is the tartrate salt of (S)-2-methylpiperidine (**3**). This can be obtained from (±)-2-methylpiperidine using L-tartaric acid for resolution.[5] **3** is converted to (S)-N-tert-butyloxycarbonyl-2-methylpiperidine (**4**) by treatment with an excess of di-tert-butyldicarbonate in the presence of aqueous sodium hydroxide. The work-up procedure involves addition of an excess of aqueous ammonium hydroxide in order to convert the unreacted BOC$_2$O, which co-elutes with the product, to the more polar tert-butyloxy urethane (**15**).

tBuO–C(=O)–NH$_2$

15

Problem

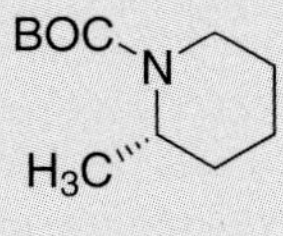

BOC-N—[piperidine with H$_3$C at 2-position] (**4**) → **5**

1. sec-BuLi, TMEDA, Et$_2$O, −78 °C, 45 min, DMF, −78 °C → r.t., 20 min

86 %

6 (+)-Himbacine

- *sec*-BuLi lithiates the α-carbon because of activation by *N*-BOC.
- Dimethylformamide acts as an electrophile.

Hints

Solution

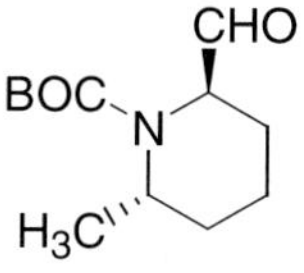

5: BOC-N piperidine with CHO at C-2 and H₃C at C-6

During this reaction an aldehyde functionality is introduced selectively at C-6 of the piperidine. The advantages of the BOC group as activator and director of lithiation are its applicability to carbanion formation at otherwise inactivated α-positions and its convenience for addition and removal.[6] Equatorially substituted 2-piperidines are well recognized to be less stable than their axially substituted isomers because of allylic 1,3-strain.[7] Therefore **4** would be expected to undergo conformational equilibration to **4'**. Directed lithiation *via* **16** and subsequent reaction with dimethylformamide finally yields **5** as an 8:1 mixture of *trans* and *cis* isomers. They can be separated by careful silica gel chromatography.

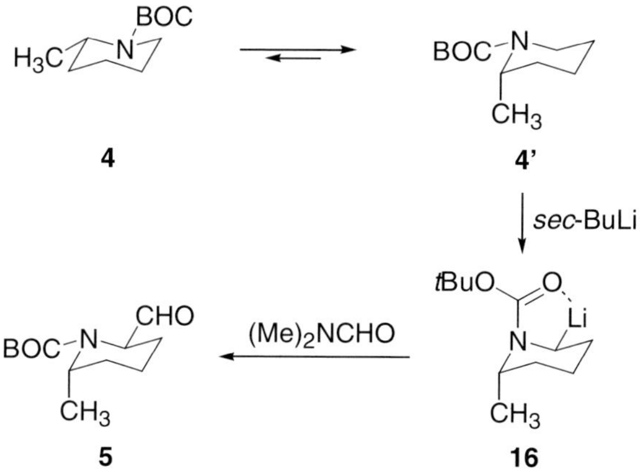

Discussion

4 ⇌ **4'**

↓ *sec*-BuLi

16 → (Me)₂NCHO → **5**

Problem

5 → 1. → **6**

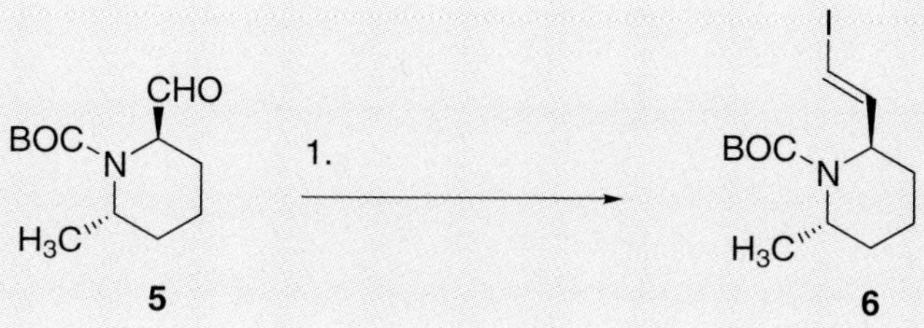

Hints	- The conversion of aldehyde **5** into the vinyliodide **6** is done by a *Wittig* type reaction.
- This reaction involves a chromium reagent.
Solution	1. CHI_3, $CrCl_2$, THF, 0 °C → r. t., 16 h, 50 %
Discussion	Methods for the conversion of an aldehyde into the homologated alkyl halide are quite limited. Treatment of an aldehyde with the *Wittig* reagent $Ph_3P=CHX$ gives a mixture of *Z* and *E* isomers and preparation of the ylide is rather complicated. Conversion of aldehydes with iodoform and chromium dichloride yields the corresponding *E*-alkenyl halides with high stereoselectivity. This reaction is known as the *Takai* reaction.[8] Although ketones are also converted into the corresponding alkenyl halides, they are less reactive than aldehydes. The *E/Z* ratios of the alkenyl halides increase in the order I < Br < Cl and the rates of reaction of the haloform are in the sequence I > Br > Cl. Therefore this method is generally used for the preparation of *E*-vinyl iodides from the corresponding aldehydes. There are two possible reactive species generated from iodoform and $CrCl_2$. One is a chromium dihalocarbenoid (**18**) and the other is a carbodianion species (**19**). The stereoselectivity of this reaction is due to the size of the halogen employed; the use of large halogens increases the formed amount of the thermodynamically more stable *E*-isomer.

$$CHI_3 \xrightarrow{CrCl_2} [Cr^{III}CHI_2] \xrightarrow{CrCl_2} \left[\begin{matrix}Cr^{III}\\ \diagdown CHI\\ Cr^{III}\diagup\end{matrix}\right]$$

17 → **18** → **19**

18 → $\left[\begin{matrix}R\diagdown\diagup CHI_2\\ \vert\\ OCr^{III}\end{matrix}\right]$ **21**

19 → $\left[\begin{matrix}R\diagdown\diagup CHI\cdot Cr^{III}\\ \vert\\ OCr^{III}\end{matrix}\right]$ **20**

21, **20** → R-CH=CH-I **6**

Problem

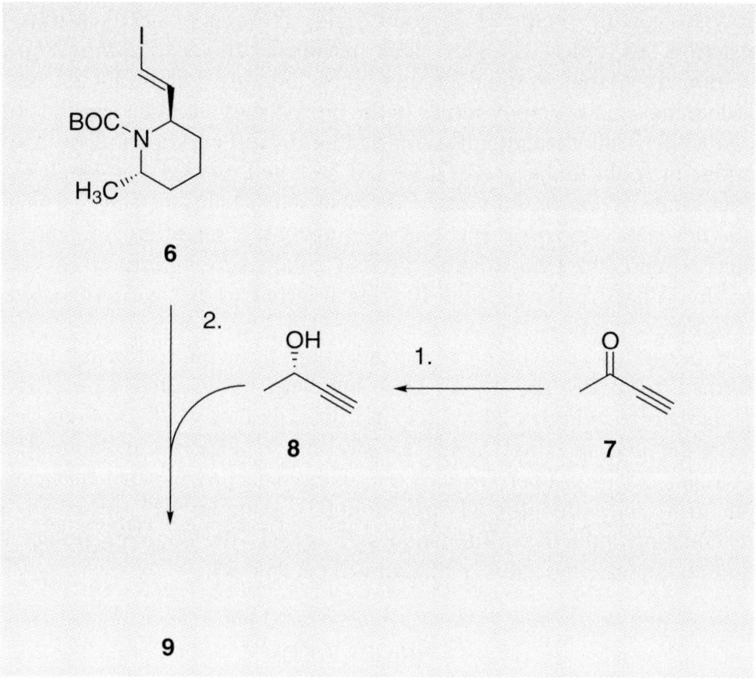

Hints
- Ketone reduction is achieved by using a boron reagent.
- Which metals can catalyze cross coupling reactions?
- What is a *Sonogashira* coupling?

Solution

1. Eapine-Borane, r. t., 16 h, 82 % ee
2. 15 mol% PdCl$_2$(PhCN)$_2$, CuI, THF, r. t., 18 h, 81 %

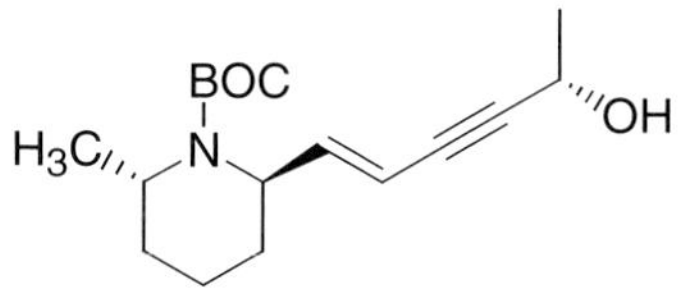

Discussion

The enantioselective reduction of 3-butyn-2-one (**7**) is achieved in 82 % *ee* by use of the Eapine-Borane **22**.[9] Application of the commercial available Alpine Borane® leads to a lower enantioselectivity (77 % *ee*). The mechanism of this reduction is explained in Chapter 13.

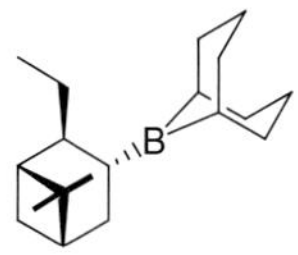

Enynes can be produced in good yields through a metal catalyzed reaction of vinyl iodides and terminal alkynes. *Sonogashira* demonstrated that terminal alkynes react smoothly with bromoalkenes, iodoarenes and bromopyridines in the presence of catalytic amounts of bis(triphenylphosphine)palladium dichloride and cuprous iodide in an amine at room temperature. It should be noted that Pd^0/Cu^I-catalyzed couplings of configurationally defined vinyl halides with terminal alkynes are stereospecific, proceeding with retention of alkene stereochemistry. This valuable Pd^0/Cu^I-catalyzed coupling of sp- and sp^2-hydridized carbon atoms is often referred to as the *Sonogashira* coupling reaction.

The presumed catalytic cycle for this coupling is the following: Once formed from **23**, the highly coordinatively unsaturated 14-electron palladium(0) complex **24** participates in an oxidative addition reaction with the aryl or vinyl halide to give the 16-electron palladium(II) complex **25**. A copper(I)-catalyzed alkynylation of **25** then furnishes an aryl- or vinylalkynyl palladium(II) complex **27**. Finally, a terminating reductive elimination step reveals the coupling prduct **9** and regenerates the active palladium(0) catalyst **24**.

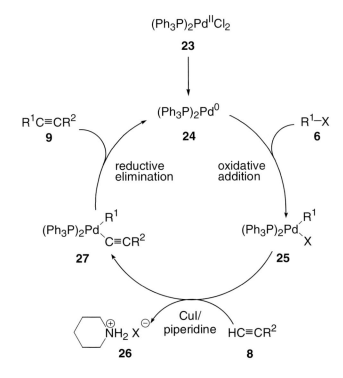

6 (+)-Himbacine

Problem

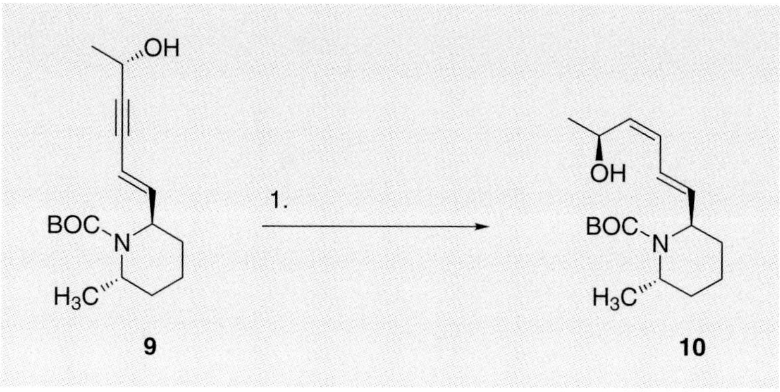

9 → 10 (step 1)

Hints

- Catalytic addition of hydrogen takes place.
- A special catalyst is needed to prevent complete hydrogenation of the triple bond.

Solution

1. H_2, *Lindlar* catalyst, quinoline, $MeOH/CH_2Cl_2$ = 1:2, r. t., 40 min, 95 %.

Discussion

Alkynes are normally completely hydrogenated by using heterogeneous catalysts like Pd, Pt or *Raney* Ni. To break the hydrogenation after addition of one equivalent of hydrogen the catalyst has to be deactivated. In general the *Lindlar* catalyst, which is a combination of $Pd/BaSO_4$ or $Pd/PbO/CaCO_3$/quinoline, is used to achieve this transformation. However, the speed of the hydrogenation has still to be monitored and the reaction has to be stopped after consumption of one equivalent of hydrogen. Otherwise a slow overreduction takes place.

6 (+)-Himbacine

Problem

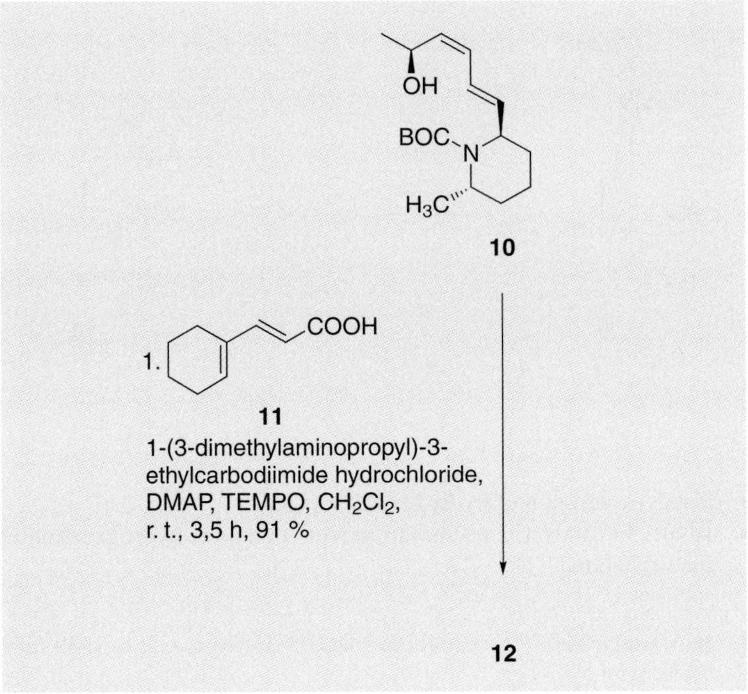

10 — BOC-protected piperidine with OH-bearing side chain

11 — 3-cyclohexenyl cinnamic acid (COOH)

1-(3-dimethylaminopropyl)-3-ethylcarbodiimide hydrochloride, DMAP, TEMPO, CH_2Cl_2, r.t., 3,5 h, 91 %

→ **12**

Hints

- The carbodiimide acts as an acid activating reagent (peptide synthesis).

Solution

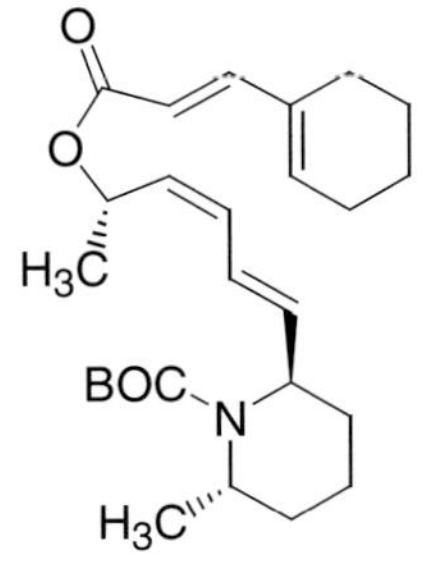

12 — ester product

Discussion

The combination of carboxyl activation by a carbodiimide and catalysis by DMAP provides a useful method for *in situ* activation of carboxylic acids for reaction with alcohols.[10] The reaction proceeds at room temperature. Carbodiimides are widely applied in the synthesis of polypeptides from amino acids. The proposed mechanism for this esterification reaction involves activation of the acid *via* isourea **28** followed by reaction with another acid molecule to form anhydride

31. Compound **31** forms an acylpyridinium species (**32**) by reaction with DMAP. Finally nucleophilic attack by the alcoholate on the acyl group of **32** generates ester **12** and regenerates catalyst DMAP. Since conjugated alkenes are sensitive toward radical reactions, TEMPO is added to the reaction mixture to inhibit radical side reactions.

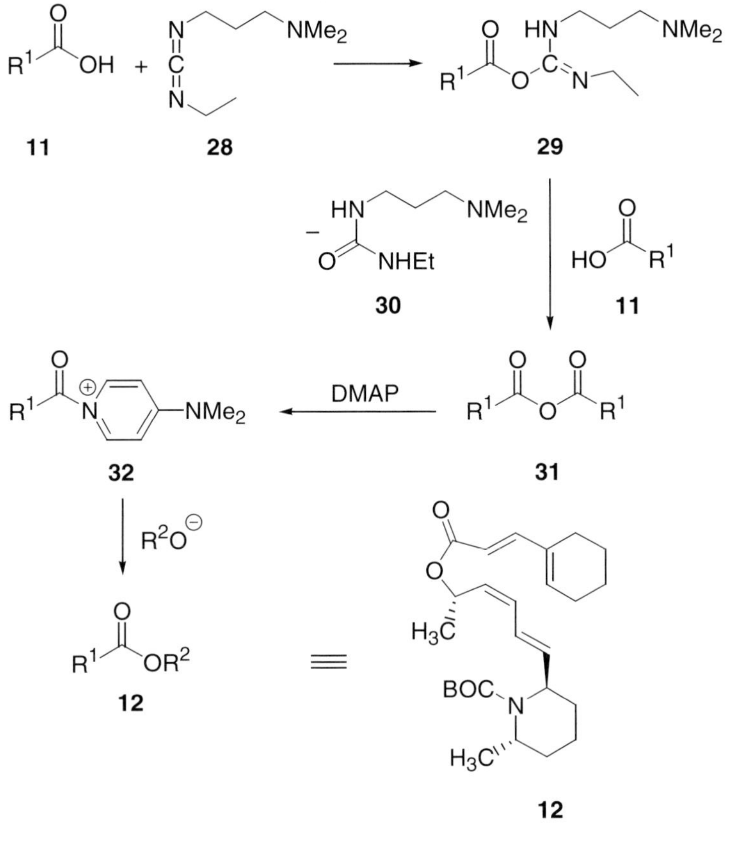

Problem

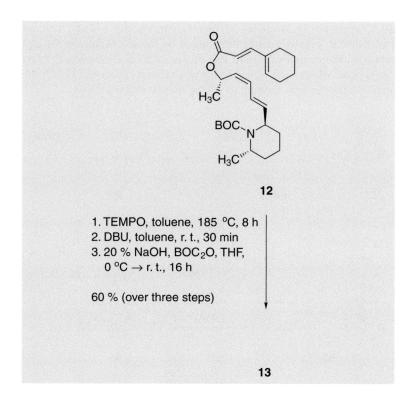

Hints

- An intramolecular [4+2]-cycloaddition takes place.
- DBU leads to an isomerization.
- The last step is necessary for reprotection of the partially deprotected nitrogen.

Solution

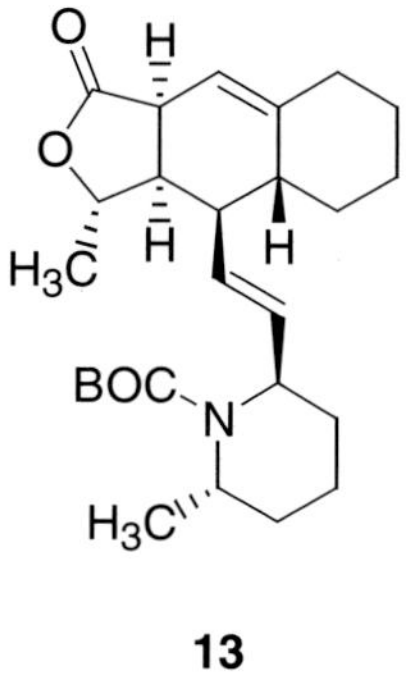

Discussion The first step is an intramolecular *Diels-Alder* reaction. The regioselectivity of this transformation may be explained as follows: The reaction can yield annelated products of type **34** by cyclization in a quasi „*ortho*" sense or bridged products of type **36** by cylization in a

„meta" sense. In our case the short chain length of three atoms excludes the formation of bridged „meta" products.[11] The vinylcyclohexenyl region will act as the diene moiety in preference to the piperidine substituent diene, since it is more likely to adopt the required *cisoid* conformation.

The stereochemistry of the cycloadducts in intramolecular *Diels-Alder* reactions depends upon the different geometry of the possible transition states **37–40** whose nomenclature can be explained as follows: The orientation with the chain connecting the diene and dienophile lying under or above the diene is called e*ndo*. The opposite means *exo*. *E* and *Z* mark the geometry of the diene double bond which is connected with the chain. *Syn* and *anti* describe the arrangement of the hydrogen atoms (or substituents) at the prestereogenic centers which are involved in the C-C bond formation.[12]

In this case the reaction can proceed *via* the *exo-E-anti* transition structure **37** or the *endo-E-syn* transition structure **39** because of the structure of the educt molecule. In a noncatalyzed, high temperature cyclization sterically demanding precursors such as **12** are known to react *via* the *exo-E-anti* transition state **41** to furnish the corresponding *trans*-fused product **42**.[13] In the transition structure the diene is placed under the dienophile because of the α-methyl group at C-5 which leads to formation of compound **42**. Subsequent treatment of the reaction mixture with an excess of DBU effects complete isomerization of **42** to the thermodynamically more stable *cis*-lactone **43**. TEMPO is again used as radical scavenger during this reaction. The prolonged reaction time results in the formation of substantial amounts of *N*-deprotected amine, which is converted to **13** by treatment of the crude reaction mixture with BOC anhydride in the presence of 20 % aqueous sodium hydroxide.

6 (+)-Himbacine

Problem

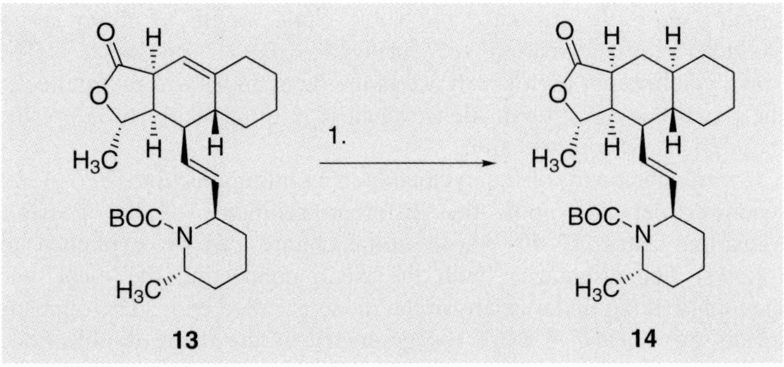

13 → **14**

Hints
- This step is a catalytic reduction.

Solution
1. Raney Ni, H_2, MeOH, r. t., 2.5 h, 72 %

Discussion

Reduction using a large excess of *Raney* nickel in methanol is extremely regioselective; only the internal double bond is affected. This is possibly because of the highly hindered nature of the external disubstituted double bond, which is flanked by the tricyclic ring system and the *N*-BOC substituent. The transformation occurred stereoselectively from the less hindered α-face of the molecule producing the required *R* configuration.

Problem

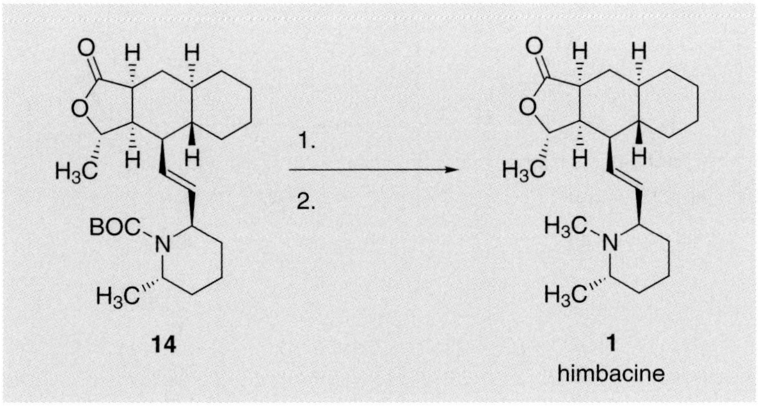

14 → **1** himbacine

Hints
- The amine is deprotected.
- Which methods for *N*-alkylation do you know?

1. TFA, CH$_2$Cl$_2$, r. t., 1 h *Solution*
2. HCHO (37 % in H$_2$O), NaBH$_3$CN, r. t., 1 h
78 % (over two steps)

Direct conversion of **14** to (+)-himbacine is achieved in a one-pot procedure by removing the BOC group with trifluoroacetic acid and reaction of the resulting free amine with aqueous formaldehyde and sodium cyanoborohydride. This reductive elimination furnishes the imine which is *in situ* reduced to the tertiary amine. Another common method for *N*-methylation is the reaction with a base like sodium hydride and methyl iodide. But this method is not suitable for molecules with C-H acidic protons. *Discussion*

6.4 Conclusion

The intramolecular *Diels-Alder* reaction of an appropriately substituted tetraene that bears the entire latent carbon framework and functional group substitution of himbacine has to be mentioned as the key step of this synthesis. It allows an efficient construction of the tricyclic ring system. In conclusion, this total synthesis provides a highly convergent synthesis of (+)-himbacine in about 10 % overall yield and also establishes a practical route to analogs.

6.5 References

1 J. T. Pinhey, E. Ritchie, W. C. Taylor, *Aust. J. Chem.* **1961**, *14*, 106-134.
2 M. J. Malaska, A. H. Fauq, A. P. Kozikowski, P. J. Aagaard, M. McKinney, *Bioorg. Med. Chem. Lett.* **1995**, *5*, 61-66.
3 J. H. Miller, P. J. Aagaard, V. A. Gibson, M. McKinney, *J. Pharmacol. Exp. Ther.* **1992**, *263*, 663-667.
4 S. Chackalamannil, R. J. Davies, Y. Wang, T. Asberom, D. Doller, J. Wong, D. Leone, A. T. McPhail, *J. Org. Chem.* **1999**, *64*, 1932-1940.
5 D. Doller, R. Davies, S. Chackalamannil, *Tetrahedron: Asymmetry* **1997**, *8*, 1275-1278.
6 P. Beak, W. K. Lee, *J. Org. Chem.* **1993**, *58*, 1109-1117.
7 R. Hoffmann, *Chem. Rev.* **1989**, *89*, 1841-1860.
8 K. Takai, K. Nitta, K. Utimoto, *J. Am. Chem. Soc.* **1986**, *108*, 7408-7410.
9 H. C. Brown, P. V. Ramachandran, S. A. Weissman, S. Swaminathan, *J. Org. Chem.* **1990**, *55*, 6328-6333.
10 A. Hassner, V. Alexanian, *Tetrahedron Lett.* **1978**, *46*, 4475-4478.
11 J. Brieger, J. N. Bennett, *Chem. Rev.* **1980**, *80*, 63-97.

12 a) L. F. Tietze, G. Kettschau, *Top. Curr. Chem.* **1997**, *189*, 1-120; b) L. F. Tietze, J. Bachmann, J. Wichmann, Y. Zhou, T. Raschke, *Liebigs Ann./Recueil* **1997**, 881-886.
13 D. Craig, *Chem. Soc. Rev.* **1987**, *16*, 187-238.

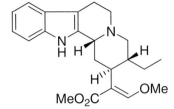

7

Hirsutine (Tietze 1999)

7.1 Introduction

The alkaloids hirsutine (**1**) and its 3α-epimer dihydrocorynantheine (**2a**) belong to the corynanthe group of indole alkaloids. They have been isolated from the plant *Uncaria rhynchophylla* MIQ, which was used for the preparation of the old Chinese folk medicine "kampo". Nowadays, **1** and related compounds attract great attention in medicine because of their growth inhibition of the influenza A subtype H3N2 virus. With an EC_{50} value of 0.40–0.57 µg/mL **1** has shown itself to be 10–20 times more active than the clinically used drug ribavirine (**3**).[1,2] Furthermore, **1** is known for its antihypertensive and antiarrhythmic activity.[3]

The absolute configuration of **1** was established in 1967.[4] Since then, **1**[5] and **2a**[6] have been the subject of several syntheses. A challenge in both cases was the stereochemistry at the three stereogenic centers of the quinolizidine subunit. Tietze and co-workers have previously synthesized indole alkaloids of the corynanthe group,[7] this problem is based on the recent enantioselective total synthesis of **1**.[2,8,12]

1 hirsutine

2a: R = Et
dihydrocorynantheine
2b: R = vinyl
corynantheine

3 ribavirine

7.2 Overview

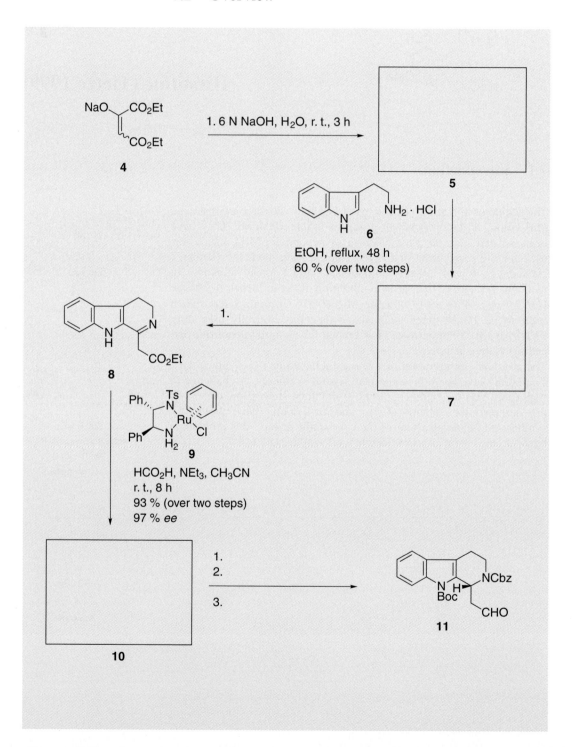

7 Hirsutine

7   Hirsutine

7.3 Synthesis

Problem

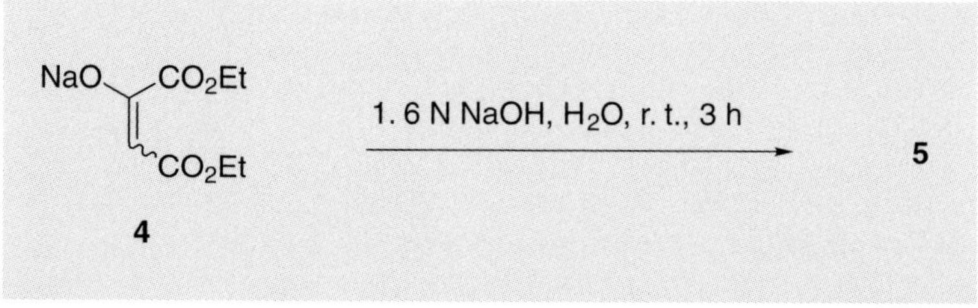

NaO–C(CO₂Et)=CH–CO₂Et (**4**) → **5**   1. 6 N NaOH, H₂O, r. t., 3 h

Hints

- The basic conditions are used for the cleavage of esters.
- Selectively, only one ester group is removed.

Solution

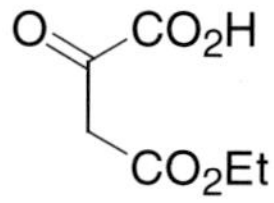

O=C(CO₂H)–CH₂–CO₂Et

5

Discussion

HO–C(CO₂H)=CH–CO₂Et

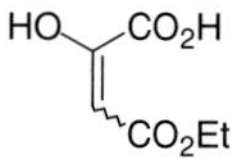

17

Starting from the commercially available sodium salt of diethyl oxalacetate (**4**) partial alkaline saponification results in carbethoxypyruvic acid (**5**).[9] Because of the activation by the α-substituent one ester group is more reactive than the other; thus cleavage of only one ester group is achieved selectively. After acidic work-up an equilibrium mixture of the solid enolic form **17** (ca. 10 %) and the liquid ketonic form **5** is gained.

Problem

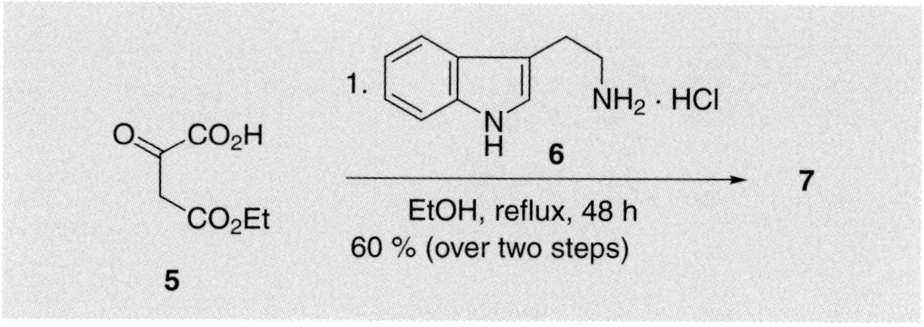

O=C(CO₂H)–CH₂–CO₂Et (**5**) + indole-3-CH₂CH₂NH₂·HCl (**6**) → **7**
1. EtOH, reflux, 48 h
60 % (over two steps)

Hints

- A decarboxylation of **5** takes place.
- The newly formed functional group reacts with the amino group of **6**.

- An iminium ion is generated.
- A cyclization follows.

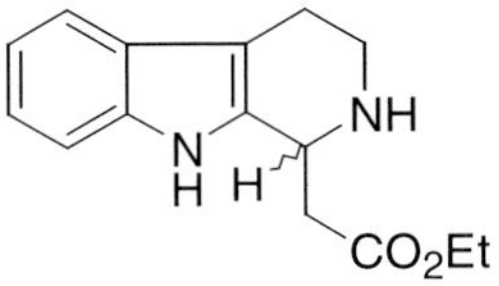

7

Solution

Discussion

Like β-keto carboxylic acids **5** undergoes decarboxylation at "high" temperatures. As a result **5** is transformed into **18** by setting free CO_2. Aldehyde **18** reacts with the hydrochloride salt **6** to iminium ion **19**. The subsequent *Pictet-Spengler* reaction provides tetrahydro-β-carboline **7**.[9]

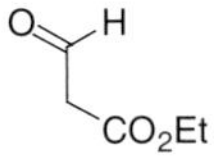

The mechanism of this reaction is still not clear. It has always been thought that the *Pictet-Spengler* reaction proceeds *via* a spiroindolenine intermediate **20** followed by a *Wagner-Meerwein* rearrangement to **21** (way **A**). On the other hand in the case of very reactive electrophiles it was shown that the cyclization can occur by direct attack at the C-2 position of the indole (way **B**). However, after deprotonation of **21** compound **7** is formed.[10]

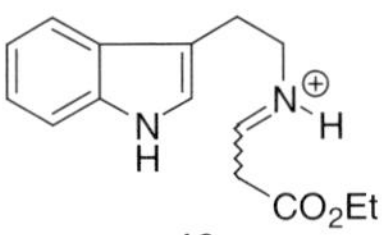

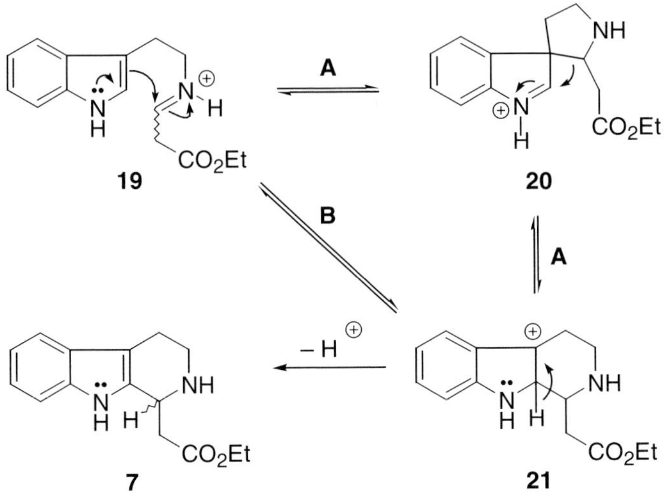

7 Hirsutine

Problem

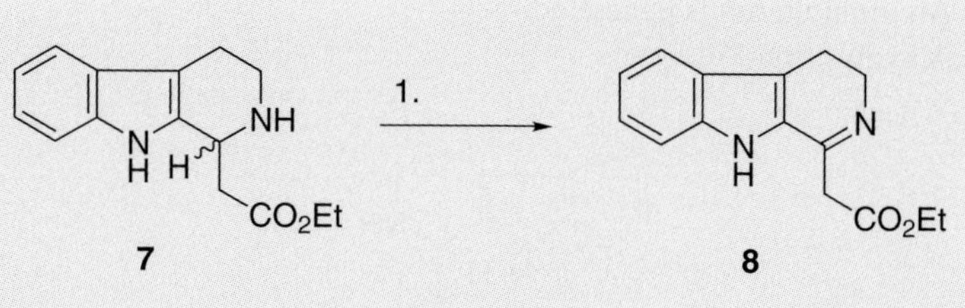

[Scheme: conversion of tetrahydro-β-carboline **7** (with CH₂CO₂Et substituent) to imine **8** via step 1.]

Hints

- The oxidizing agent is a salt.
- The oxidation state of this reagent is VII.

Solution

1. $KMnO_4$, THF, 0 °C, 1 h

Discussion

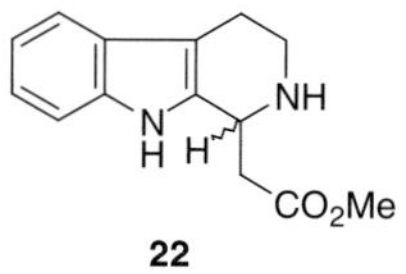

Treatment of tetrahydro-β-carboline **7** with solid potassium permanganate provides imine **8**.[11,12] The ethyl ester group is stable under these conditions in contrast to the methyl ester **22**.

Potassium permanganate is an inexpensive oxidizing agent, which has been widely used in organic syntheses for the transformation of various functional groups. $KMnO_4$ is generally dissolved in aqueous solutions, mixtures of water and miscible organic solvents, or in nonpolar aprotic solvents (such as CH_2Cl_2 or C_6H_6) with the aid of crown ethers or less expensive quaternary ammonium salts as phase transfer agents. The polar aprotic THF can also be used as solvent. In acidic media $KMn(VII)O_4$ is reduced to water soluble manganese(II) or -(III) salts, which allow easy work-up. In any case, in basic media reduction creates mainly manganese dioxide, which is difficult to remove.

[Structure **22**: tetrahydro-β-carboline with CH₂CO₂Me substituent]

Problem

[Scheme: imine **8** treated with Ru catalyst **9** (Ph, Ph, Ts, NH₂, Ru, Cl ligands), HCO_2H, NEt_3, CH_3CN, r.t., 8 h → **10**, 93 % (over two steps), 97 % ee]

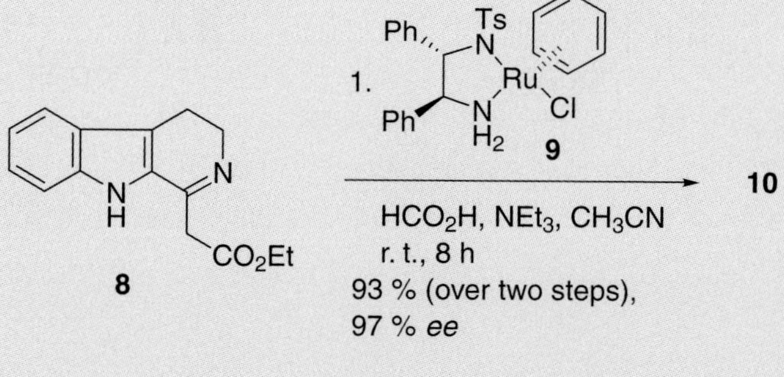

- Ru-complex **9** is used catalytically.
- Formic acid serves as hydrogen source.
- The imine is reduced.

Hints

Solution

10

Asymmetric hydrogenation of cyclic imine **8** using two mol % of chiral Ru-complex **9** in a formic acid-triethylamine mixture, as developed by *Noyori* and co-workers, results in the desired stereoisomer **10** with an excellent optical purity of 97 % ee.[12,13]

Formic acid as a stable organic hydrogen donor is an alternative to the flammable molecular hydrogen. Although Ru(II)-complexes are known to catalyze the reversible process **C**, the reduction of the imine is a result of transfer hydrogenation by formic acid. Interference of molecular hydrogen does not take place.

A chiral Ru hydride **23** is formed and it is assumed that the hydrogen transfer occurs *via* metal-ligand bifunctional catalysis. The N-H linkage may stabilize a transition state **24** by formation of a hydrogen bond to the nitrogen atom. Stereochemistry is determined by formal discrimination of the enantiofaces at the sp^2 nitrogen atom of the cyclic imine.

Another fact is the quite impressive functional group selectivity of this method. Because of their greater reactivity imines can be reduced in the presence of ketones, although chiral Ru-complex **9** catalyzes the transfer hydrogenation of ketones. Besides this catalytic enantioselective reduction of imines others are known.[8]

Discussion

$HCO_2H \rightleftharpoons H_2 + CO_2$
C

23

24

25

Problem

10 → **11**

Hints

- In the first two steps protection of both nitrogen atoms proceeds. Which has to be protected first?

7 Hirsutine

- In the last step the ester group is reduced to an aldehyde without overreduction to the corresponding alcohol.

Solution

1. CBZCl, NEt$_3$, CH$_2$Cl$_2$, 0 °C → r. t., 24 h
2. (BOC)$_2$O, DMAP, CH$_3$CN, r. t., 5 h
3. DIBAH, CH$_2$Cl$_2$, –78 °C, 2 h

64 % (over three steps)

Discussion

The first procedure is the selective protection of an amine in the presence of an indole-NH group or an alcohol.[14] Indole-NH (pK$_a$ ≈ 17) shows a similar acidity to that of alcohols (pK$_a$ ≈ 18). In the next step the benzyloxycarbonyl (CBZ) protected compound **26** is treated with di-*tert*-butyldicarbonate (BOC)$_2$O and DMAP to form **27**. *Tert*-butylchloroformate is unstable and therefore cannot be used for the preparation of BOC derivatives. Conversion of **10** with (BOC)$_2$O in the presence of DMAP would lead to the double BOC protected compound **28**.

DMAP **29** is a powerful catalytic activating agent for the reaction of nucleophiles with esters, amides, anhydrides or carboxylic chlorides. It is much more reactive than pyridine (pK$_a$ = 5.23) because of the enhanced basicity (pK$_a$ = 9.70) and nucleophilicity.[15]

Finally stoichiometric addition of DIBAH in a nonpolar solvent at low temperature yields the desired aldehyde **11**. Overreduction to the corresponding alcohol **33** is prevented under these conditions because of the formation of the relative stable intermediate **34**, which decomposes only in the course of aqueous work-up. At higher temperature and in polar solvents such as THF formation of alcohol **33** occurs. Thus, grade of reduction of esters with DIBAH can be controlled by temperature and solvent.

Problem

Hints

- In the first step a nucleophile is added to the carbonyl group.
- The second step is an elimination.
- Is the elimination acid or base catalyzed?

Solution

1. PMBOH, PTS, Na$_2$SO$_4$, r. t., 12 h
2. Ph$_2$P(O)OH **36**, 170 °C

60 % (over two steps)

Discussion

Enol ether **13** is prepared from butanal **12** by acetalization with alcohol PMBOH **35**. The resulting acetal **40** is subjected to elimination with phosphinic acid **36**. Acetalization proceeds via nucleophilic attack of the alcohol on the protonated aldehyde **37**, dehydratization of the hemiacetal **38** and further nucleophilic attack on the carbenium ion **39**. Since all steps are reversible, the created water has to be removed to achieve quantitative turnover. This is carried out by the use of water binding agents or solvents (dry Na$_2$SO$_4$, CaCl$_2$, orthoesters) or azeotropic distillation.

When acetal **40** is heated with the phosphinic acid **36**, the elimination of PMBOH **36** proceeds via the carbenium ion **39**, which is also an intermediate of the acetalization. This elimination is reversible as well, so the mixture of enol ether **13** and PMBOH **35** has to be removed by distillation from the reaction mixture.

Other proton sources may be used but PTS turned out to be too weak and KHSO$_4$ led to decomposition of **40**. Enolether **13** was obtained as a 1:1 mixture of E and Z isomers but the configuration of the double bond has no influence on the next transformation.

7 Hirsutine

Problem

[Structures: 11 (tetrahydro-β-carboline with NBOC, NCBZ, CHO substituents) + 13 (CH₃CH=CH-CH₂-OPMB) + 14 (Meldrum's acid)]

1. H₂N⌒NH₂ · 2 AcOH **41**, benzene, ultrasound, 60 °C, 4 h, 90 %, asymmetric 1,3 induction: > 20 : 1

↓

15

Hints

- In the first part of this one-pot multi-component reaction a carbonyl group reacts with a CH-acidic compound.
- This condensation forms a heterodiene.
- Then an electrocyclic reaction proceeds. What is the dienophile?
- At the end of this domino reaction, the cycloadduct primarily formed reacts with loss of acetone and CO_2.

Solution

[Structure of 15: tetrahydro-β-carboline fused with dihydropyranone bearing OPMB]

15

Discussion

This three-component reaction as the key step of this synthesis is called a domino-*Knoevenagel*-hetero-*Diels-Alder* reaction. Domino reactions are defined as processes of two or more bond forming reactions in which a subsequent transformation takes place by virtue of the functionalities introduced in a former transformation.[2,16] The tetrahydrocarbolinaldehyde **11** first reacts in a *Knoevenagel* type condensation with *Meldrum*'s acid **14** and ethyleneammonium diacetate **41** as the catalyst to oxabutadiene **44** which then undergoes a *Diels-Alder* reaction with enol ether **13**.

Catalytic amounts of the neutral ammonium salt ethyleneammonium diacetate **41** are suitable to deprotonate C,H-acidic *Meldrum*'s acid **14**. The resulting carbanion **42** undergoes nucleophilic attack to the aldehyde **11** followed by elimination of water which leads to oxabutadiene **44**.

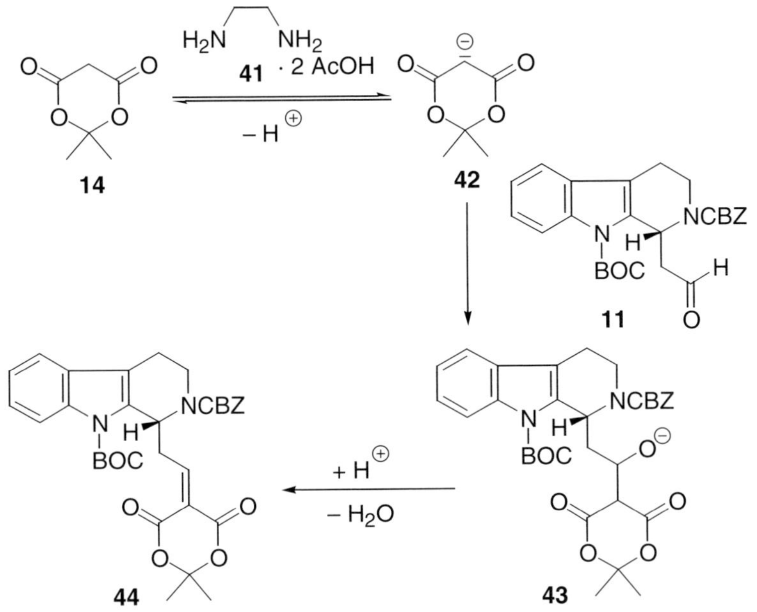

Hetero-*Diels-Alder* reaction of **44** with enol ether **13** as the dienophile gives cycloadduct **45**, which is not isolable but reacts with the water formed in the condensation step with loss of acetone and CO_2 to lactone **15**. A suggested mechanism for the formation of lactone **15** is a retro *Diels-Alder* reaction which leads to the ketene intermediate **46**. Ketene **46** adds to the water formed in the previous condensation step, yielding β-keto-carboxylic acid **47**, which then undergoes decarboxylation to **48**.

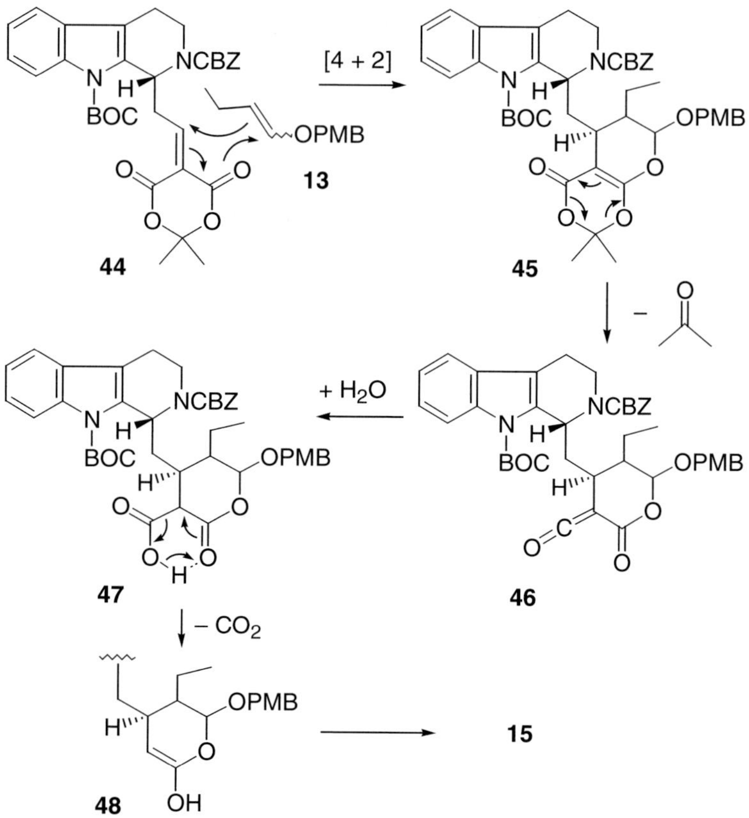

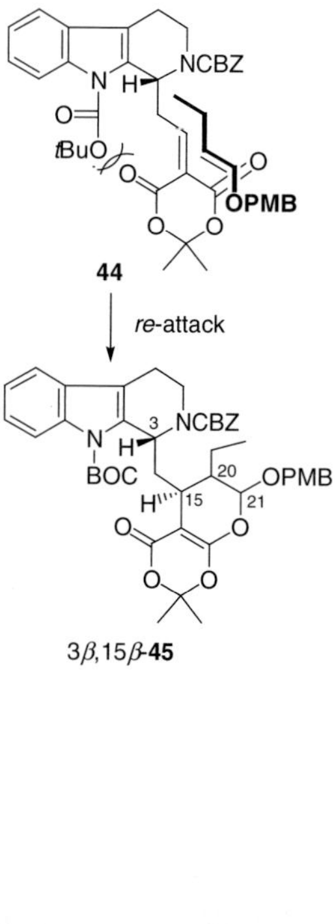

7 Hirsutine

Lactone **15** is formed as the only product because of the high stereoselectivity of this reaction: Since the heterodiene **44** obtained from the *Knoevenagel* condensation is considered to react from the *E*-configuration at the carbon-carbon double bond (*Z*-heterodienes are less reactive)[7a,17], the sterically highly demanding BOC protecting group on the indole nitrogen favors a *re*-attack of the enol ether *syn* to the hydrogen at C-3, giving a new stereogenic center with the α-hydrogen at C-15 (see Chapter 6). The *trans* formation of H and H-15 is furnished with an asymmetric 1,3 induction of >1 : 20. The configuration at C-20 and C-21 is not yet important since these stereogenic centers are transformed stereoconvergently in the next step. Application of ultrasound provides thorough mixing of the reaction components, and is necessary because of the poor solubility of *Knoevenagel* product **44** in benzene.

7 Hirsutine

Problem

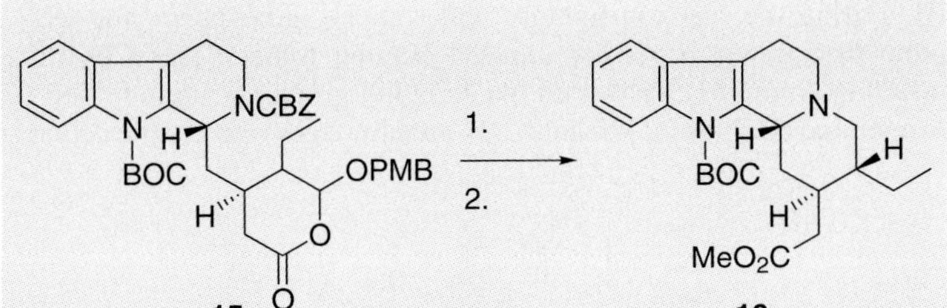

Hints

- First the lactone is converted into an ester by solvolysis.
- The solvolysis also liberates an aldehyde.
- The CBZ protecting group is labile towards hydrogenation.
- How do an amine and an aldehyde react?
- Finally, a double bond is hydrogenated under stereoelectronic control.

Solution

1. K_2CO_3, MeOH, r. t., 20 min, then
2. H_2, Pd/C, MeOH, r. t., 4 h
67 % (from **15**)

Discussion

The first step of this domino reaction opens the lactone **15** by reaction with potassium carbonate in methanol leading to hemiacetal **49**. This is not stable under the reaction conditions and decomposes yielding aldehyde **50** and PMBOH **35** (see formation of acetal **40** described above).

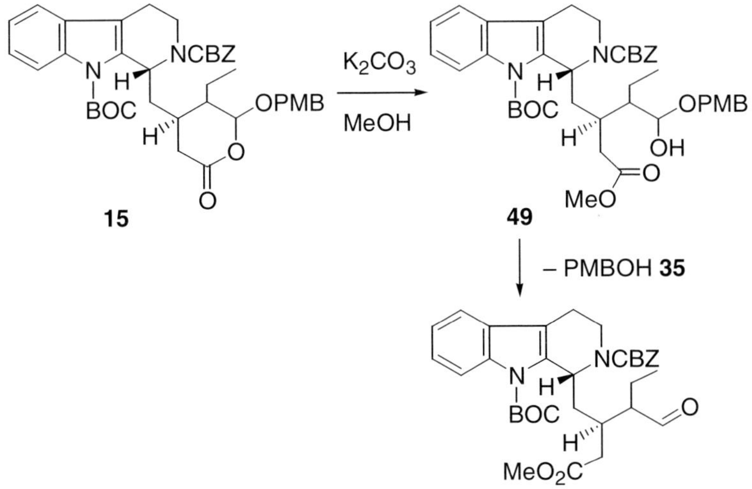

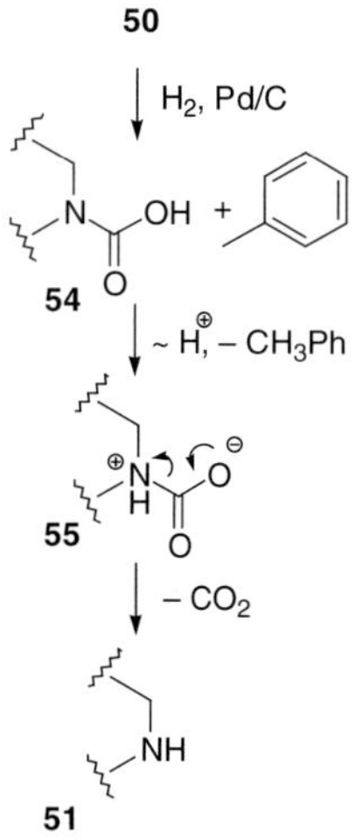

By stirring the reaction mixture under an H_2 atmosphere, the CBZ-protecting group is readily cleaved leaving toluene and CO_2. The resulting amine **51** reacts with the carbonyl group *via* iminium ion **52** to enamine **53**. Now, C-20 and C-21 are no longer stereogenic centers.

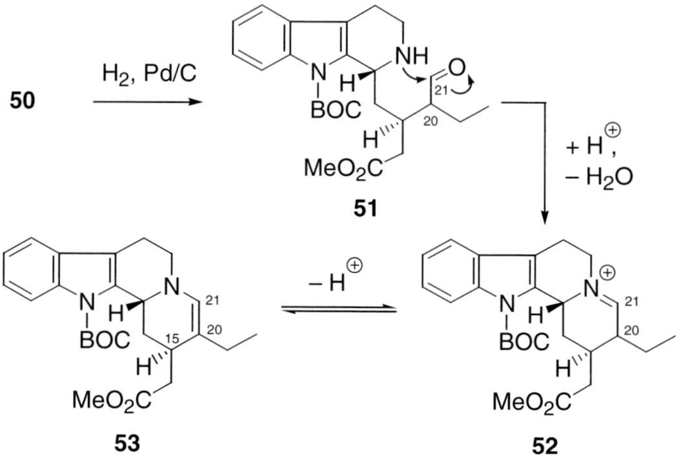

After the condensation, the double bond of enamine **53** is hydrogenated stereoselectively yielding enantiopure quinolizidine **16** with *trans* configuration of the substituents at C-15 and C-20 as the only product.

The high *trans* selectivity of this transformation may be explained by means of stereoelectronic effects. Therefore the attack of the hydrogen atom on enamine **53** has to proceed *anti* toward 15-H under formation of the 15,20-*trans*-compound **16**.[18] Conformation **56** with pseudoequatorial orientation of the ethyl group at C-20 is assumed to be energetically most favored. An attack on C-20 *anti* toward 15-H proceeds *via* a more favorable chair-like transition state, whereas an attack *syn* toward 15-H results in a less favorable boat-like transition state.

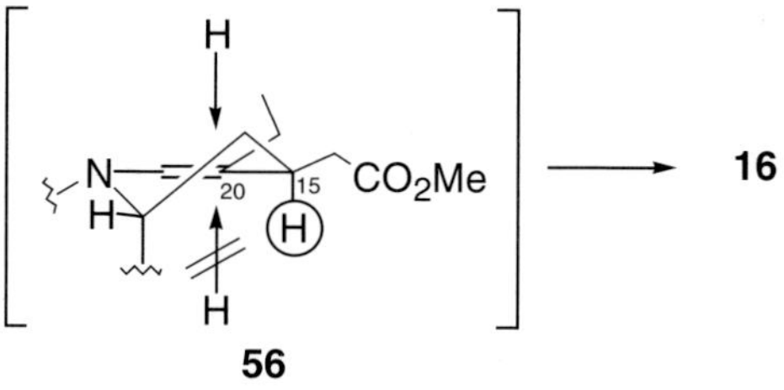

7 Hirsutine

Problem

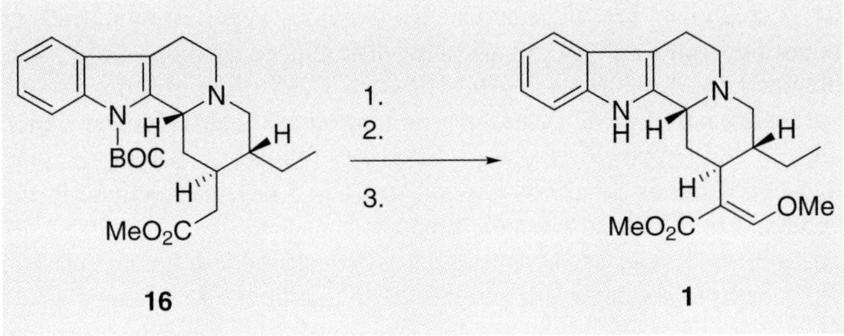

Hints

- The first step removes a protecting group. What conditions are necessary?
- Then a condensation is carried out.
- The final step employs conditions which are suitable to convert a carboxylic acid into its methyl ester.

Solution

1. TFA, CH_2Cl_2
2. Ph_3CNa, HCO_2Me, THF, Et_2O
3. 1 eq. MeOH, HCl, CH_2Cl_2

Discussion

The *tert*-butyloxycarbonyl (BOC) group is cleaved using TFA in an aprotic solvent like CH_2Cl_2. The cleavage proceeds *via* protonation of the carbonyl group of carbamate **16**, subsequent elimination of carbenium ion **58** and liberation of unstable free carbaminic acid **60** which breaks down yielding the unprotected indole **61** with loss of CO_2. Cation **58** is deprotonated giving 2-methyl propene **59**.

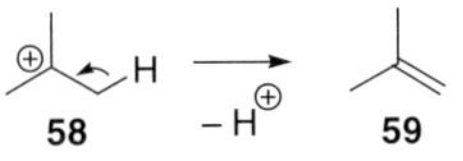

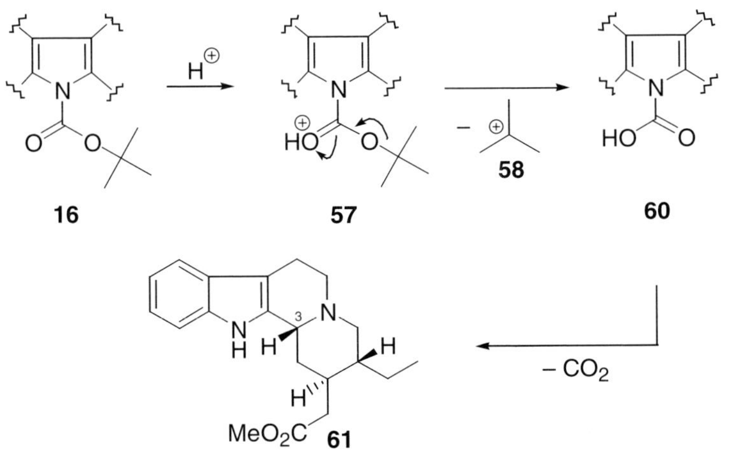

61 is a known key intermediate in previous reported syntheses of hirsutine.[5] Interestingly, 3α-**61** with different configuration at C-3 is obtained by domino-*Knoevenagel*-hetero-*Diels-Alder* reaction of the tetrahydrocarboline **62** containing an unprotected indole moiety which allows the opposite facial selectivity following solvolysis and hydrogenation as described above. 3α-**61** is a key intermediate in the synthesis of dihydrocorynantheine **2a**.[6]

Second, methyl ester **61** is subjected to crossed ester condensation with methyl formate and triphenyl methyl sodium **64** as base.[5,6] Ester **61** is deprotonated and the resulting anion **65** adds to methyl formate. The anion formed is stabilized by loss of methanolate yielding the desired product as its conjugated base. As a *Claisen* type ester condensation, all steps but the last are reversible since the resulting aldehydo ester **67** is more acidic than methanol; therefore the last deprotonation step occurs irreversibly. The use of LDA[19] as a base is also possible for this transformation.

In the final step, hydroxymethylene ester **69** is methylated by means of one equivalent of MeOH in CH_2Cl_2 saturated with HCl gas resulting in formation of the enol ether moiety of hirsutine **1**.[5c] As a vinylogous carboxylic acid, the hydroxy group of **69** can be methylated applying esterification conditions. Thus, activation of **69** occurs by protonation, followed by addition of MeOH as a nucleophile to the double bond. Resulting intermediate **72** then undergoes rotation of the bond formed and H^+-shift. Loss of water and H^+ gives hirsutine (**1**) as the desired product.

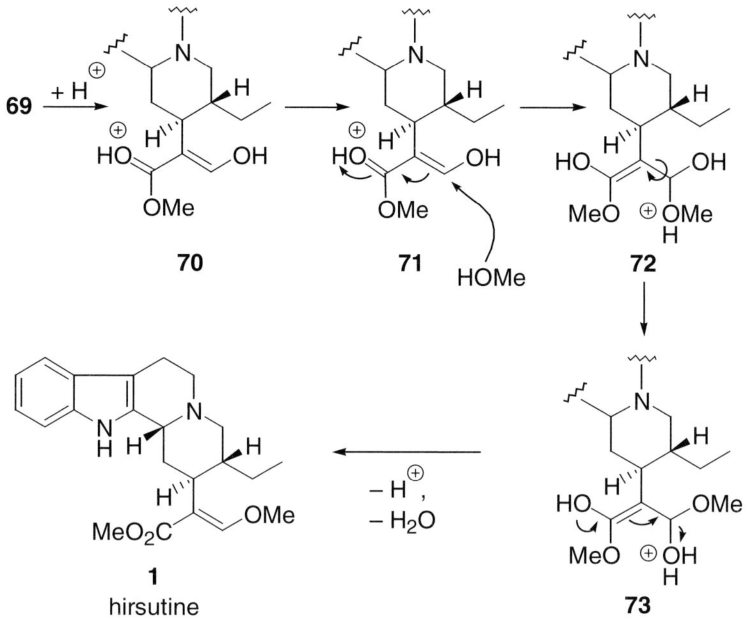

A moderate yield of this reaction may be due to side reactions like dimethyl acetal formation at the hydroxymethylene group.[19] Alkylating reagents like dimethyl sulfate or diazomethane[6d] have been employed as well to furnish the enol ether moiety in **1** but proved to give lower yields.

7.4 Conclusion

The biologically interesting alkaloid hirsutine (**1**) was synthesized in a highly efficient and selective manner from the three structurally simple precursors tetrahydrocarboline **11**, *Meldrum's* acid **14** and enol ether **13** in a sequence of a domino-*Knoevenagel*-hetero-*Diels-Alder* reaction, solvolysis of the lactones obtained, and subsequent hydrogenation. One of the three stereogenic centers of hirsutine (**1**) was introduced by enantioselective hydrogenation of an imine using Ru catalysis; this first stereocenter induced the second in the *Diels-Alder* reaction, while the third was built up utilizing a stereoelectronic effect in the hydrogenation of a cyclic enamine.

7.5 References

1   H. Takayama, Y. Limura, M. Kitajima, N. Aimi, K. Konno, *Bioorg. Med. Chem. Lett.* **1997**, *7*, 3145-3148.
2   L. F. Tietze, A. Modi, *Med. Res. Rev.* **2000**, *20*, 304-322.

3   a) H. Masumiya, T. Saitoh, Y. Tanaka, S. Horie, N. Aimi, H. Takayama, H. Tanaka, K. Shigenobu, *Life Sci.* **1999**, *65*, 2333-2341; b) S. Horie, S. Yano, N. Aimi, S. Sakai, K. Watanabe, *Life Sci.* **1992**, *50*, 491-498.

4   W. F. Trager, C. M. Lee, *Tetrahedron* **1967**, *23*, 1043-1047.

5   a) M. Lounasmaa, J. Miettinen, P. Hanhinen, R. Jokela, *Tetrahedron Lett.* **1997**, *38*, 1455-1458; b) R. T. Brown, M. F. Jones; M. Wingfield, *J. Chem. Soc., Chem. Commun.* **1984**, 847-848; c) E. Wenkert, Y. D. Vankar, J. S. Yadav, *J. Am. Chem. Soc.* **1980**, *102*, 7971-7972; d) N. Aimi, E. Yamanka, J. Endo, S. Sakai, J. Haginiwa, *Tetrahedron* **1973**, *29*, 2015-2021; e) N. Aimi, E. Yamanka, J. Endo, S. Sakai, J. Haginiwa, *Tetrahedron Lett.* **1972**, *11*, 1081-1084.

6   a) E. E. van Tamelen, J. B. Hester, Jr., *J. Am. Chem. Soc.* **1969**, *91*, 7342-7348; b) R. L. Autrey, P. W. Scullard, *J. Am. Chem. Soc.* **1968**, *90*, 4917-4923; c) J. A. Weisbach, J. L. Kirkpatrick, K. R. Williams, E. L. Anderson, N. C. Yim, B. Douglas, *Tetrahedron Lett.* **1965**, *39*, 3457-3463; d) E. E. van Tamelen; J. B. Hester, Jr., *J. Am. Chem. Soc.* **1959**, *81*, 3085; e) D. Staerk, P.-O. Norrby, J. W. Jaroszewski, *J. Org. Chem.* **2001**, *66*, 2217-2221.

7   a) L. F. Tietze, J. Bachmann, J. Wichmann, Y. Zhou, T. Raschke, *Liebigs Ann.,/Receuil* **1997**, 881-886; b) L. F. Tietze, J. Bachmann, J. Wichmann, O. Burkhardt, *Synthesis* **1994**, 1185-1194; c) L. F. Tietze, J. Wichmann, *Angew. Chem.* **1992**, *104*, 1091-1092; *Angew. Chem. Int. Ed. Engl.* **1992**, *31*, 1079-1080; d) L. F. Tietze, J. Bachmann, W. Schul, *Angew. Chem.* **1988**, *100*, 983-985; *Angew. Chem. Int. Ed. Engl.* **1988**, *27*, 971-973.

8   L. F. Tietze, Y. Zhou, *Angew. Chem.* **1999**, *111*, 2076-2078; *Angew. Chem. Int. Ed. Engl.* **1999**, *38*, 2045-2047.

9   B. G. Kline, *J. Am. Chem. Soc.* **1959**, *81*, 2251-2555.

10 a) G. Casnati, A. Dossena, A. Pochini, *Tetrahedron Lett.* **1972**, *11*, 5277-5280; b) E. D. Cox, J. M. Cook, *Chem. Rev.* **1995**, *95*, 1797-1842; c) P. Ducrot, C. Rabhi, C. Thal, *Tetrahedron* **2000**, *56*, 2683-2692.

11 S. P. Burke, R. L. Danheiser (eds.), *Handbook of Reagents for Organic Synthesis - Oxidizing and Reducing Agents*, John Wiley & Sons, Chichester **1999**.

12 L. F. Tietze, Y. Zhou, E. Töpken, *Eur. J. Org. Chem.* **2000**, 2247-2252.

13 a) N. Uematsu, A. Fujii, S. Hashiguchi, T. Ikariya, R. Noyori, *J. Am. Chem. Soc.* **1996**, *118*, 4916-4917; b) R. Noyori, S. Hashiguchi, *Acc. Chem. Res.* **1997**, *30*, 97-102; c) K. J. Haack, S. Hashiguchi, A. Fujii, T. Ikariya, R. Noyori, *Angew. Chem.* **1997**, *1091*, 297-300; *Angew. Chem. Int. Ed. Engl.* **1997**, *36*, 285-288.

14 a) P. J. Kocienski, *Protecting Groups*, Georg Thieme Verlag, New York **1994**; b) T. W. Greene, P. G. M. Wuts, *Protective Groups in Organic Synthesis*, (3rd edition), John Wiley & Sons, New York **1999**.

15 A. J. Pearson, W. J. Roush (eds.), *Handbook of Reagents for Organic Synthesis - Activating Agents and Protecting Groups*, John Wiley & Sons, Chichester **1999**.

16 a) L. F. Tietze, U. Beifuss, *Angew. Chem.* **1993**, *105*, 137-170; *Angew. Chem., Int. Ed. Engl.* **1993**, *32*, 131-164; b) L.F. Tietze, *Chem. Rev.* **1996**, *96*, 115-136; c) M. J. Parsons, C. S. Penkett, A. J. Shell, *Chem. Rev.* **1996**, *96*, 195-206; d) A. Bunce, *Tetrahedron* **1995**, *51*, 13103-13159; e) N. Hall, *Science* **1994**, *266*, 32-34.

17 L. F. Tietze, G. Kettschau, *Top. Curr. Chem.* **1997**, *189*, 1-120.

18 P. Deslongchamps, *Stereoelectronic Effects in Organic Synthesis*, Pergamon Press, Oxford **1983**.

19 I. Ninomiya, T. Naito, O. Miyata, T. Shimeda, E. Winterfeldt, R. Freund, T. Ishida, *Heterocycles* **1990**, *30*, 1031-1077.

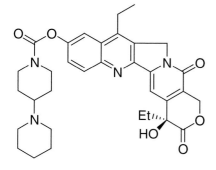

8

(+)-Irinotecan® (Curran 1998)

8.1 Introduction

(+)-Irinotecan (**1**) is a derivative of the pentacyclic quinoline alkaloid camptothecin (**2**); the latter was first isolated from the heartwood of the tree species *Camptotheca acuminata* (*Nyssacea*) by *Wall* et al. in 1966.[1] Two years later *A. T. McPhail* and *G. A. Sim* determined the structure of **2** by X-ray analysis.[2]
During the last ten years (20S)-camptothecin (**2**) and its derivatives have represented some of the most promising agents for the treatment of solid tumors by chemotherapy.[3] Their biochemical activity is caused by interfering with the process of unwinding the supercoiled DNA helix through inhibition of the cellular enzyme topoisomerase I. In replicating cancer cells, the formation of a ternary complex of topoisomerase I, DNA, and compound **1** leads to apoptosis and programmed cell death. In contrast to **2**, **1** shows high water solubility and no problems with administration in clinical trials. Today it is being used in several countries for cancer therapy.
Curran et al. have developed an efficient access to **2** and its analogs,[3] which provides the first direct synthesis of the prodrug irinotecan (**1**) without passing through the toxic molecule **3** as in prior studies.[4] The latter is the active antitumor agent, which is formed *in vivo* by metabolism from **1**.

8.2 Overview

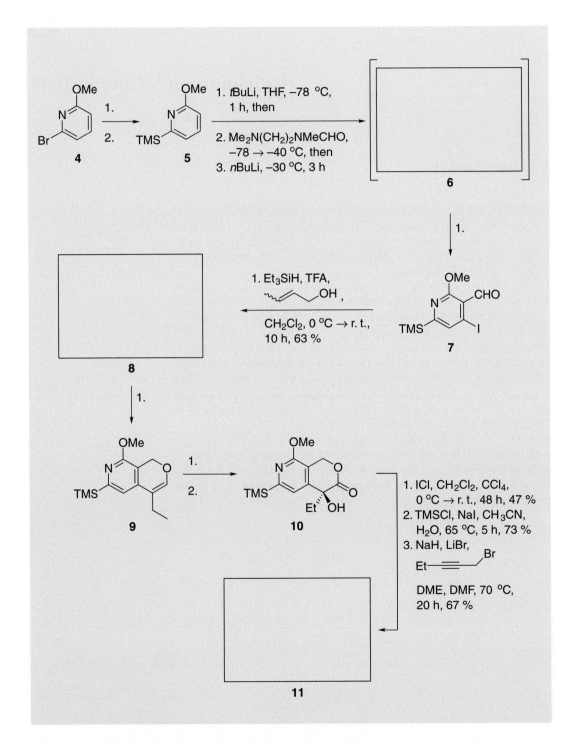

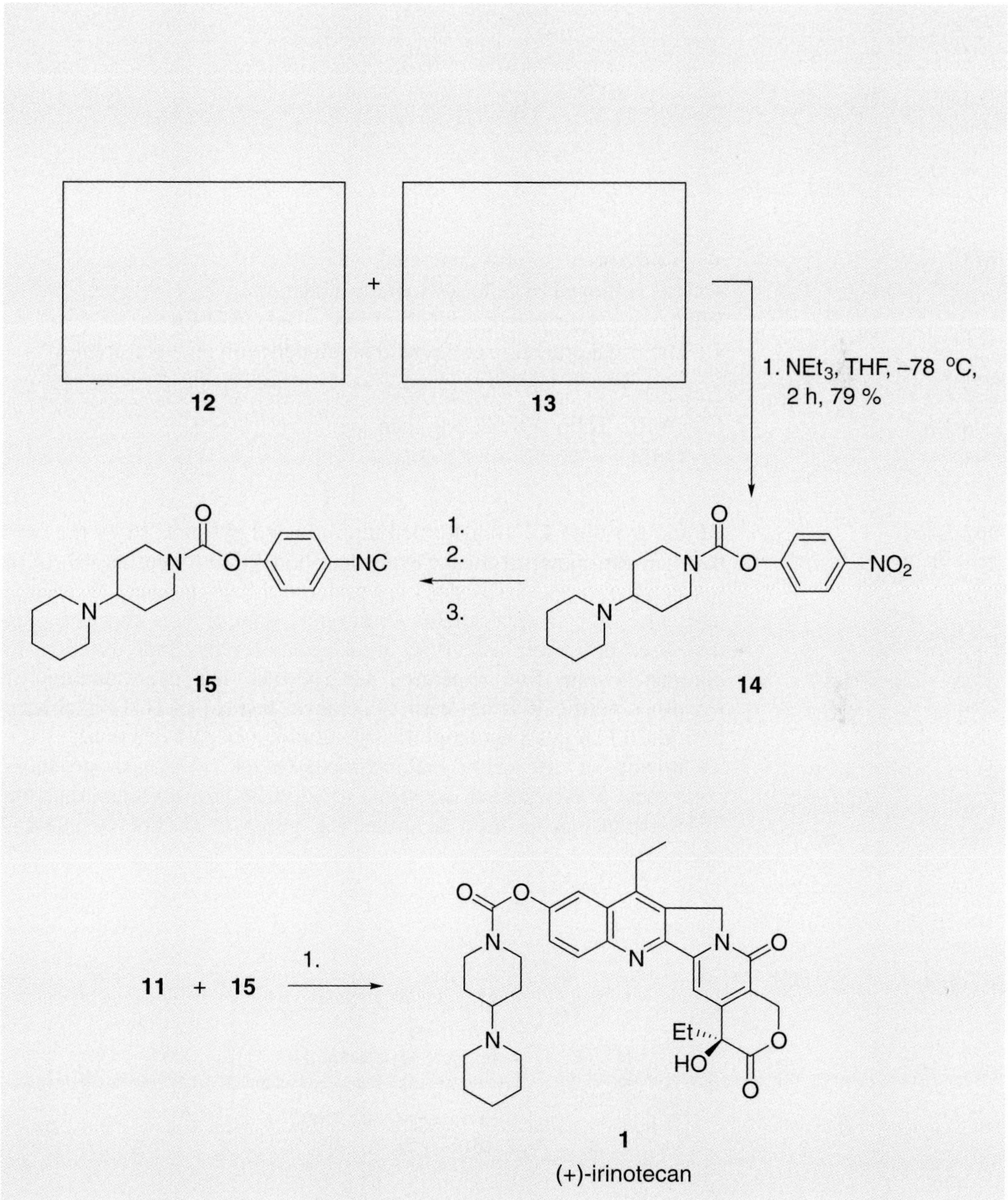

8 Irinotecan

12 + **13** 1. NEt₃, THF, −78 °C, 2 h, 79 %

15 1. 2. 3. ← **14**

11 + **15** 1. → **1**
(+)-irinotecan

8.3 Synthesis

Problem

[Scheme: 6-bromo-2-methoxypyridine **4** → (1., 2.) → 6-TMS-2-methoxypyridine **5**]

Hints

- A lithium organyl is generated.
- It is achieved by halogen-metal exchange.
What reagent do you use: nBuLi, LDA, nBuLi/KOtBu?
- The organometallic compound is trapped with an electrophile.

Solution

1. nBuLi, THF, –78 °C, 1 h, then
2. TMSCl, –78 °C → r. t., 92 %

Discussion

[Structures: **16** (6-Li-2-methoxypyridine); **17** (pyridine) → LiR → **18** (2-R-pyridine)]

Bromo pyridine **4** is transformed into lithiated pyridine **16** by the very fast halogen-metal exchange with the alkyl lithium species nBuLi at low temperatures.[5] LDA does not undergo a halogen-metal exchange and use of nBuLi/KOtBu (*Schlosser* conditions) furnishes an undesired potassium salt. Both are reagents for the deprotonation. In contrast to the five-membered heterocycles the deprotonation of pyridines with LiR is not easy to achieve. Instead of H-/Li-exchange pyridine **17** shows a nucleophilic substitution (*Ziegler* reaction).

Treatment of **16** with chlorotrimethylsilane (TMSCl) produces compound **5**. Arylsilanes are stable to alkyl lithium reagents; thus the TMS group can be used as protecting group in further metalation-chemistry.

Problem

[Scheme: **5** → 1. tBuLi, THF, –78 °C, 1 h, then 2. Me$_2$N(CH$_2$)$_2$NMeCHO, –78 → –40 °C, then 3. nBuLi, –30 °C, 3 h → **6**]

Hints

- First a directed deprotonation takes place.
Which regiochemistry do you expect?

- Me$_2$N(CH$_2$)$_2$NMeCHO is an electrophile.
- The lithium of nBuLi coordinates to the two nitrogen atoms of the ethylenediamine moiety.
- A second deprotonation of the pyridine yields intermediate **6**.

Solution

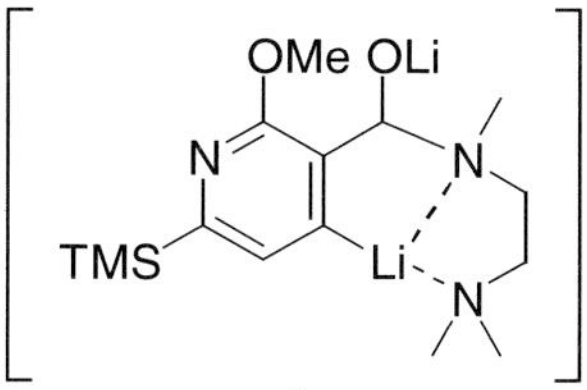

6

Discussion

In the metalation reaction of pyridine **5** with tBuLi the methoxy substituent exhibits a directing effect. Solvated (tBuLi)$_4$ aggregate **19** coordinates to the oxygen of the methoxy group to give **20**. Successive deprotonation yields the *ortho*-lithiated species **21**, which is stabilized by O-Li coordination.

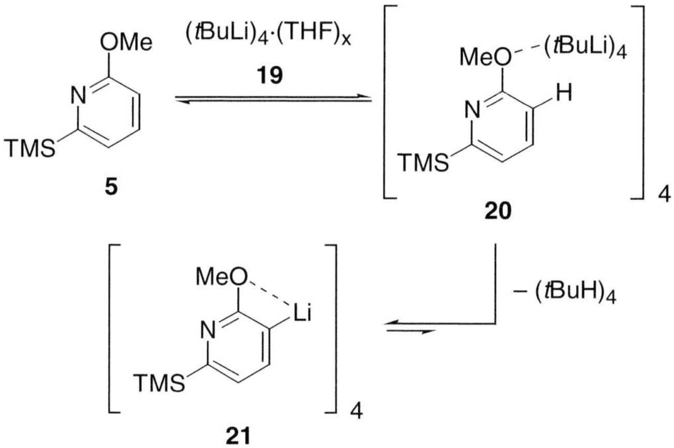

This reaction type, known as directed *ortho* metalation reaction, requires strong bases, normally an alkyllithium reagent (most lithium dialkylamides are of insufficient kinetic basicity). Alkyllithium bases show high solubility in organic solvents due to association into aggregates of defined structure, typically as hexamers in hydrocarbon solvents e. g. hexane or tetramers-dimers in polar solvents e. g. THF (see Chapter 5).

Addition of *N'*-formyl-*N,N,N'*-trimethylethylenediamine to **21** yields α-amino alkoxide **22**. A further directed *ortho* metalation takes place by coordination of nBuLi to the ethylenediamine moiety and subsequent deprotonation. As result intermediate **6** is formed.

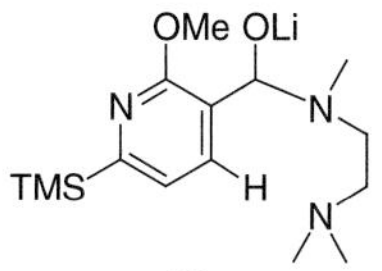

22

126   8   Irinotecan

Problem

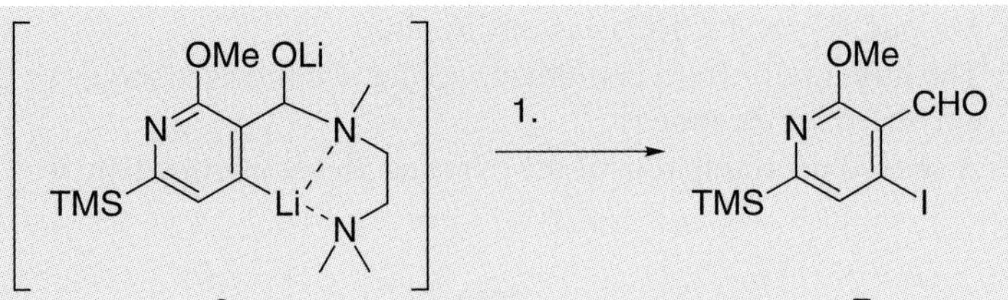

Hints
- Species **6** reacts with an electrophile.
- What iodonium sources do you know?
- The used reagent is very cheap.

Solution

1. I$_2$, −78 → 0 °C, 1 h, 49 % (from **5**)

Discussion

Lithiated pyridine **6** is trapped with iodine. Other iodonium sources are ICN, ICl, and *N*-iodosuccinimide (NIS see Chapter 16). Aqueous work-up destroys the deprotonated *N,O*-acetal and provides aldehyde **7**.

Application of silanes and metalation chemistry offers an access to substituted aromatic compounds, which are difficult to prepare by classical electrophilic substitution because of harsh reaction conditions and formation of regioisomers.

Problem

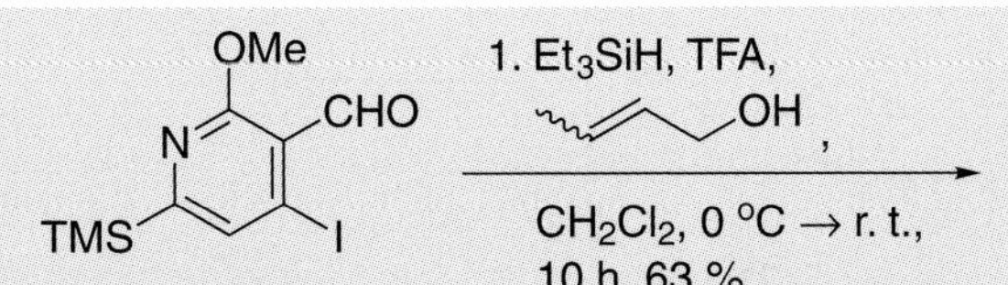

Hints
- In the course of the reaction the crotyl alcohol reacts first with the aldehyde. Note that the media is acidic.
- Water is set free.
- Et$_3$SiH is a reducing agent.
- An ether bond is formed.

Solution

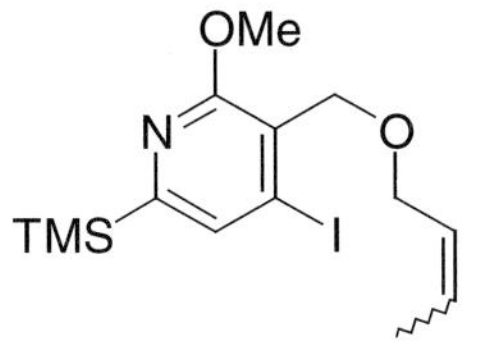

8

Discussion

Et$_3$SiH reduces aldehydes or ketones in alcoholic acidic media to ethers.[6] The mechanism of the reductive etherification involves the following steps. After activation of the aldehyde **7** attack of the nucleophilic crotyl alcohol takes place. Elimination of water provides cation **26**, which is reduced by Et$_3$SiH to ether **8**.

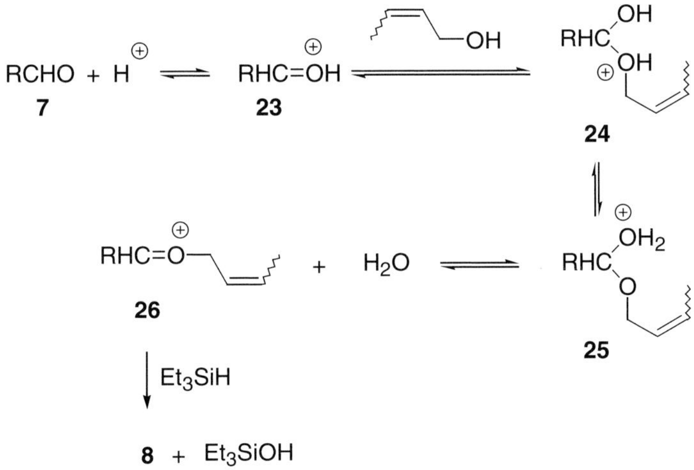

RCHO + H$^+$ ⇌ RHC=OH$^+$ ⇌ **24**
7   **23**

RHC=O$^+$—crotyl + H$_2$O ⇌ **25**
26

↓ Et$_3$SiH

8 + Et$_3$SiOH

Other silane byproducts than triethylsilanol or ether formation from two molecules of crotyl alcohol are not observed. Side reaction which decreases the yield of **8** may be the reduction of aldehyde **7** to the corresponding alcohol **27**, if water instead of the crotyl alcohol attacks the activated aldehyde **23**.

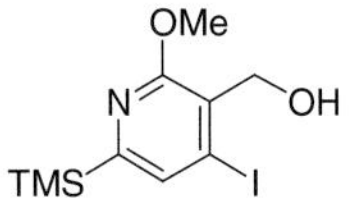

27

Problem

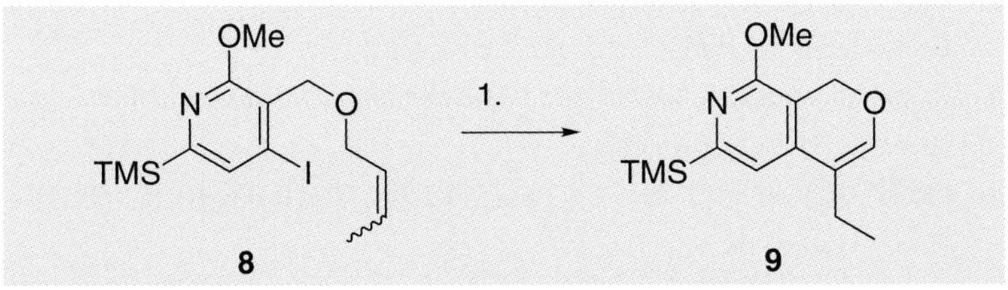

8 → 1. → **9**

8 Irinotecan

Hints
- The intramolecular ring-closure is catalyzed by a metal.
- The used metal is a subgroup element of the periodic table. It is not Ni!
- Finally a double bond isomerizes.

Solution
1. Pd(OAc)$_2$, K$_2$CO$_3$, Bu$_4$NBr, DMF, 85 °C, 18 h, 69 %

Discussion

In order to prepare the bicyclic compound **9**, an intramolecular *Heck* reaction under *Grigg's* conditions (Pd(OAc)$_2$, K$_2$CO$_3$, Bu$_4$NBr) is carried out (see Chapter 13).[7] Starting from aryl iodide **8** a six-membered ring formation occurs providing intermediate **28**, which reacts to alkene **29**. Finally double bond isomerization to compound **9** is forced by the formation of a thermodynamically more stable enol ether.

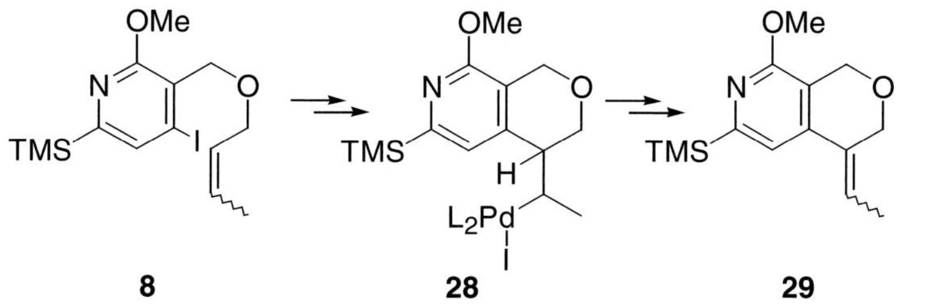

Problem

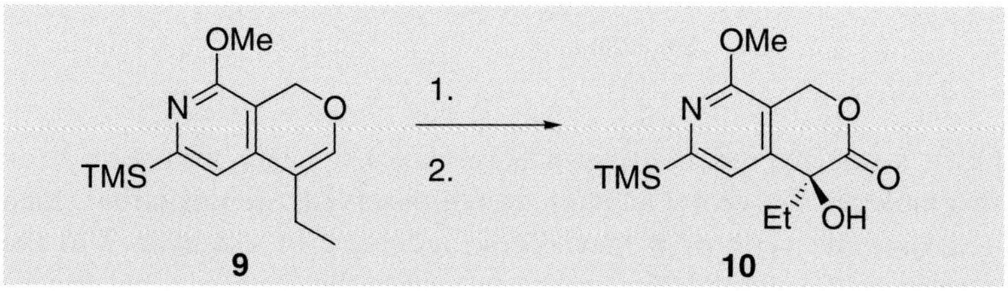

Hints
- In the first step the alkene group is dihydroxylated.
- The used method was developed by the same research group as the asymmetric epoxidation of allylic alcohols.
- In the second step a lactole is oxidized under mild conditions.

Solution
1. OsO$_4$, K$_3$Fe(CN)$_6$, K$_2$CO$_3$, MeSO$_2$NH$_2$, (DHQD)$_2$PYR, *t*BuOH, H$_2$O, 0 °C, 12 h
2. I$_2$, CaCO$_3$, MeOH, H$_2$O, r. t., 32 h
85 %, 94 % *ee* (over two steps)

Transformation of alkene **9** into diol **30** is a *Sharpless* asymmetric dihydroxylation.[8] Its catalytic cycle with $K_3Fe(CN)_6$ as co-oxidant is shown below.

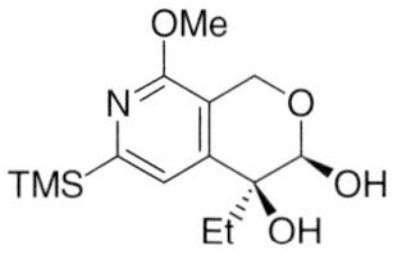

Discussion

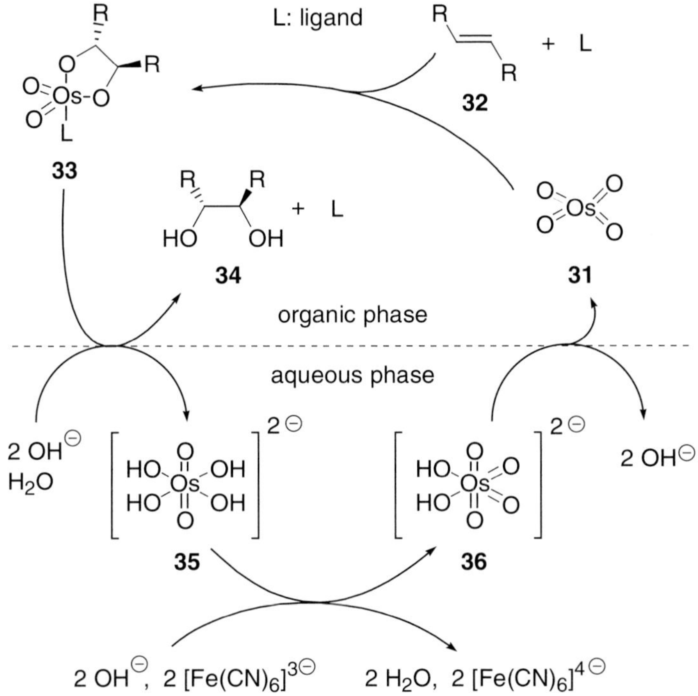

Employment of the heterogeneous solvent system (tBuOH/H_2O) has the advantage that the olefin osmylation and the osmium reoxidation occur in different phases. Osmium tetroxide **31** is the only oxidizing agent which enters the organic phase. Dihydroxylation of the alkene **32** in the presence of a chiral ligand forms osmium(VI) glycolate **33**. Hydrolysis of **33**, which is the turnover limiting step in this cycle, provides diol **34** and a water soluble inorganic osmium(VI) species **35**. Addition of methane sulfonamide ($MeSO_2NH_2$) accelerates this process.

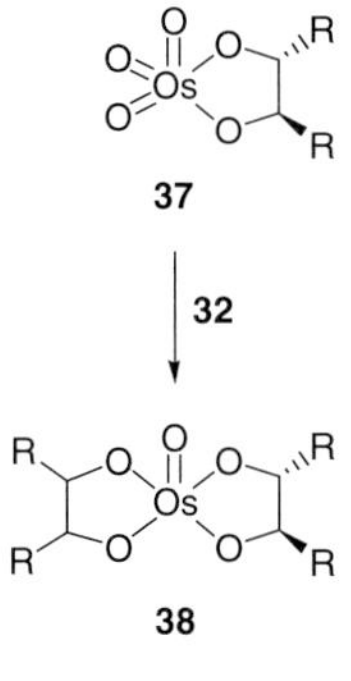

Use of organic solvent soluble co-oxidants such as *N*-methylmorpholine *N*-oxide (NMO) leads to the oxidation of osmium(VI) glycolate **33** to osmium(VIII) species **37**, which oxidizes alkene **32** without the presence of a chiral ligand. Thus the enantiomeric excess of diol **34** is lowered.

The chiral ligand consists of a diphenylpyrimidine (PYR) **39**, which is connected to two dihydroquinidine (DHQD) molecules **40**. Dihydroquinidine (DHQD) **40** and dihydroquinine (DHQ) **41** are diastereomers. However, in the asymmetric dihydroxylation, they behave like pseudo-enantiomers, giving diols of opposite configuration.

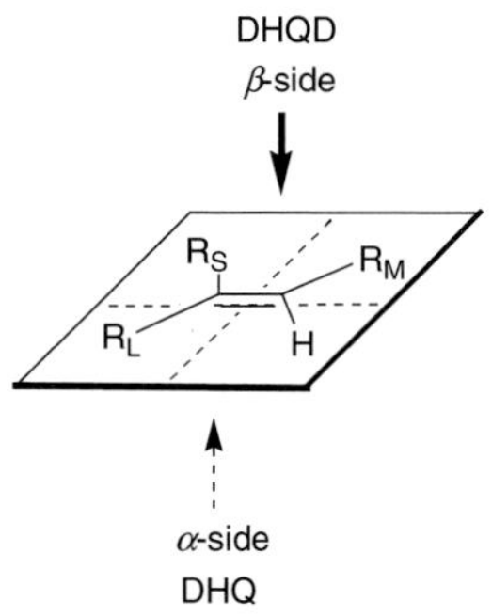

8 Irinotecan

PYR 39

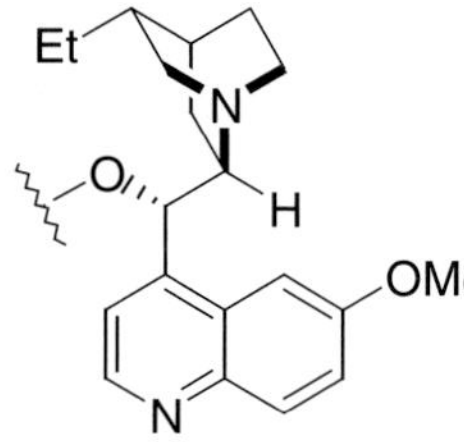

DHQD 40

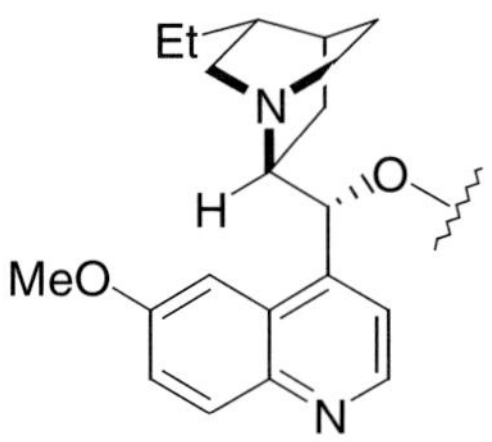

DHQ 41

'mnemonic' device **42**

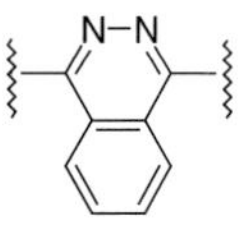

PHAL 43

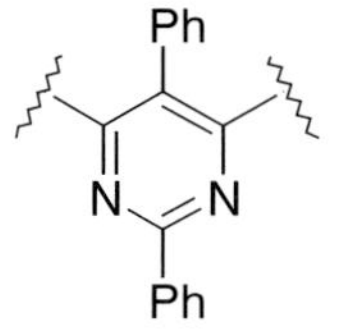

The facial selectivity of the dihydroxylation can reliably be predicted using the "mnemonic" device **42**. The smallest substituent on the olefin (generally the hydrogen) is always placed in the south-east quadrant (H), which is the most hindered space in the asymmetric environment. The south-west quadrant (R_L) is especially attractive for large aliphatic groups in the case of PYR **39** and for aromatic groups in the case of PHAL **43**. Use of DHQD **40** causes dihydroxylation from the β-side. Treatment of enol ether **9** with the common (DHQD)$_2$PHAL-system provides only 32 % ee.

Usually, racemerisation of semi-acetal such as **30** is possible. This circumstance is not important in this case since lactole **30** is immediately oxidized to lactone **10**. Iodine is the oxidizing source and CaCO$_3$ traps resulting HI.

Problem

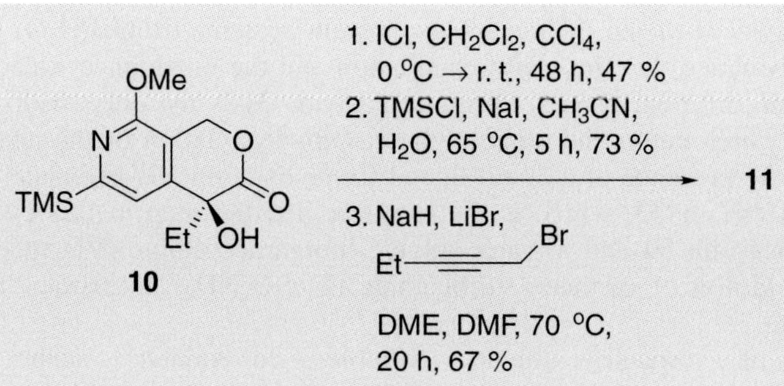

1. ICl, CH$_2$Cl$_2$, CCl$_4$, 0 °C → r. t., 48 h, 47 %
2. TMSCl, NaI, CH$_3$CN, H$_2$O, 65 °C, 5 h, 73 %
3. NaH, LiBr, Et—≡—Br
DME, DMF, 70 °C, 20 h, 67 %

10 → **11**

Hints

- In the first step an *ipso* substitution takes place.
- In the second step a strong *Lewis* acid is formed *in situ*.
- The *Lewis* acid is TMSI.
- TMSI cleaves a C-O bond.
- Finally an alkylation occurs.

8   Irinotecan

11

Solution

Treatment of aryl silane **10** with the iodonium source ICl in an iododesilylation reaction yields compound **44**, which proceeds by an *ipso* substitution mechanism.[9] Activation towards electrophilic attack arises from β stabilization of resonance form **45** by Si. Silicon is arranged β to the positive charge and the carbon-silicon bond can overlap with the empty π orbital (hyperconjugation).

Discussion

44

10   **45**   **46**   **44**

Conversion of the iododesilylation was not complete. Besides 45 % of desired compound **44** also 47 % of starting material **10** was recovered. Model studies for the silane-iodine exchange have shown that other iodonium sources such as I_2, I_2/AgBF$_4$, or NIS give inferior results. A reason for the incomplete conversion may be the formation of an iodopyridinium salt **47** resulting in ring deactivation.[5]

In the second step the strong *Lewis* acid TMSI is generated *in situ* from TMSCl and NaI, which cleaves the OMe ether bond. The resulting hydroxypyridine **48** isomerizes immediately to pyridone **49**. Ether cleavage has also been achieved by use of HI giving similar yield.

After deprotonation of amide **49** with NaH a nucleophilic substitution occurs leading to the *N*-propargylated molecule **11**. The last three steps were carried out in the dark, because aryl iodides can decompose by sunlight under formation of iodine.

47

48

49

Problem

12 + 13 → 1. NEt$_3$, THF, −78 °C, 2 h, 79 % → **14**

8 Irinotecan

Hints
- A carbamate is generated.
- Compound **12** is a commercially available diamine.

Solution

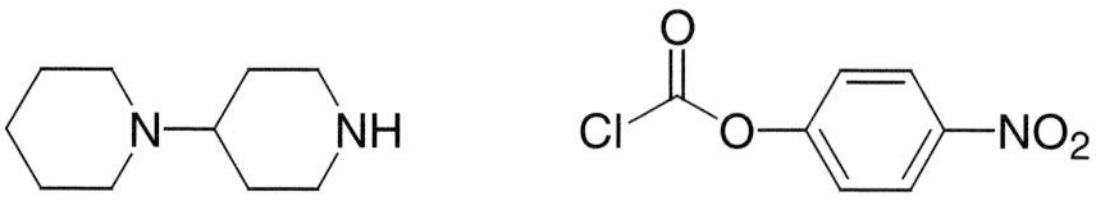

[Structure of compound **12**: piperidine–N–piperidine–NH diamine]

[Structure of compound **13**: 4-nitrophenyl chloroformate]

Discussion

Coupling of amine **12** and chloroformate **13** is similar to the preparation of amides from amines and carboxylic acid chlorides (see Chapter 14). Secondary amine **12** attacks the activated carbonyl group. Elimination of HCl, which is trapped by NEt$_3$, provides carbamate **14**.

Problem

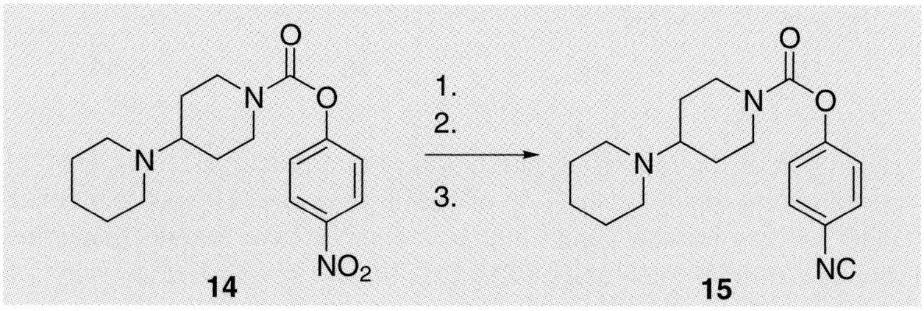

[Reaction scheme: compound **14** (carbamate with NO$_2$ group) → via steps 1, 2, 3 → compound **15** (carbamate with NC group)]

Hints
- Reduction of the nitro group takes place in the first step. What is the reagent of choice: LiAlH$_4$ or H$_2$?
- An amino group arises.
- Step two is the formation of an amide bond.
- Finally the isocyano group is formed by elimination of water.

Solution

1. H$_2$, 10 % Pd/C, AcOEt, r. t., 12 h, 91 %
2. Formic acid, DCC, py, CH$_2$Cl$_2$, 0 °C, 3 h, 83 %
3. Triphosgene (**52**), NEt$_3$, CH$_2$Cl$_2$, 0 °C → r. t., 2 h, 79 %

Discussion

Use of H$_2$ and Pd/C provides the transformation of a nitro into an amino group. Compound **50** cannot be synthesized by treatment with LiAlH$_4$, because the carbamate group would be attacked, too.
In the next step coupling of amine **50** with formic acid in the presence of DCC leads to amide **51** (see Chapter 3).

Finally treatment with triphosgene (**52**) [bis(trichloromethyl) carbonate] causes a dehydratization of the amide group leading to isonitrile **15**.[10] Triphosgene (**52**) is a solid alternative to the gaseous and very toxic phosgene (intermediate **54**) in many reactions for reasons of safe handling. As shown below, NEt$_3$ serves as co-nucleophile and traps HCl, which is set free during the reaction.

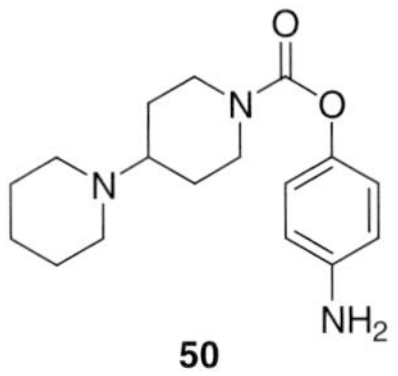

(+)-irinotecan **1**

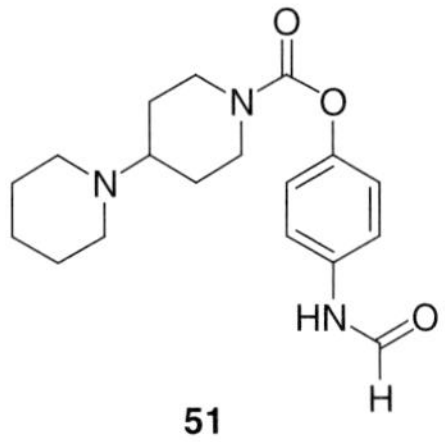

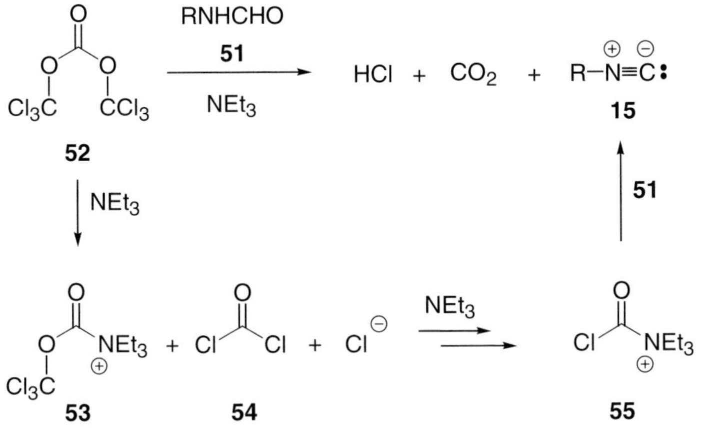

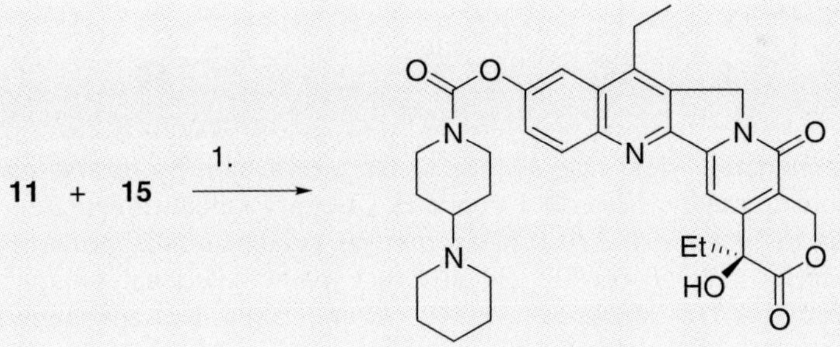

Problem

11 + **15** →

Hints

- The reaction is not carried out in the dark.
- Irradiation with a 275 W GE sunlamp is performed.
- It is a radical domino annulation.

Solution

1. Me$_6$Sn$_2$, C$_6$H$_6$, $h\nu$ (275 W GE sunlamp), 80 °C, 9 h, 31 %

8 Irinotecan

Discussion

In this protocol the pyridine- and the pyrrolidine-ring of **1** are built up in a one-pot radical domino reaction.[11] Photolysis of iodopyridone **11** in the presence of hexamethylditin provides radical **56**, which attacks the reactive isonitrile **15**. The resulting radical **57** reacts with the alkyne group in a 5-*exo-dig* cyclization (see Chapter 11). Next, the newly formed vinyl radical **58** cyclizes onto the aryl ring generating species **59**. Final oxidation *via* a so far unknown mechanism yields **1** with 31 % yield. For the generation of radical **56** by photolysis two ways (**A** and **B**) are possible.

The moderate yield may be due to the purification by reverse-phase chromatography, because **1** contains a tertiary amine. Preparation of other derivatives of **2** has shown that the radical annulation normally proceeds with 50–60 % yield and that many functional groups are tolerated (free alcohols, amines, esters, chlorides and terminal alkenes). Also $(Me_3Si)_4Si$ may be a useful substitute for hexamethylditin, because tin residues are toxic and difficult to separate from the products.

8.4 Conclusion

Anti-tumor compound (20*S*)-irinotecan (**1**) was prepared in 13 steps with an overall yield of 1.15 % for the longest linear synthesis. The short and selective preparation of aryl iodide **11** features two key steps – *ortho* metalation and *Sharpless* asymmetric dihydroxylation. In only one step **11** is transformed into the target molecule **1** by application of a radical domino annulation with isonitrile **15**. This method gives access to the broad family of camphotecin derivatives because of the quite impressive generality of the substrates that can be employed.

8.5 References

1. M. E. Wall, M. C. Wani, C. E. Cook, K. H. Palmer, A. T. McPhail, G. A. Sim, *J. Am. Chem. Soc.* **1966**, *88*, 3888-3890.
2. A. T. McPhail, G. A. Sim, *J. Chem. Soc.(B)* **1968**, 923-928.
3. H. Josien, S.-B. Ko, D. Bom, D. P. Curran, *Chem. Eur. J.* **1998**, *4*, 67-83.
4. D. P. Curran, S.-B. Ko, H. Josien, *Angew. Chem.* **1995**, *107*, 2948-2950; *Angew. Chem. Int. Ed. Engl.* **1995**, *34*, 2683-2684.
5. a) T. L. Gilchrist, *Heterocyclic Chemistry*, 2nd ed., Longman Group UK, London **1992**; b) V. Snieckus, *Chem. Rev.* **1990**, *90*, 879-933.
6. M. P. Doyle, D. J. DeBruyn, D. A. Kooistra, *J. Am. Chem. Soc.* **1972**, *94*, 3659-3661.
7. R. Grigg, V. Sridharan, P. Stevenson, S. Sukirthalingam, T. Worakun, *Tetrahedron* **1990**, *46*, 4003-4018.
8. a) M. Beller, C. Bolm, *Transition Metals for Organic Synthesis*, Wiley, New York **1998**; b) H. C. Kolb, M. S. VanNieuwenhze, K. B. Sharpless, *Chem. Rev.* **1994**, *94*, 2483-2547; c) D. P. Curran, S.-B. Ko, *J. Org. Chem.* **1994**, *59*, 6139-6141.
9. A. R. Bassindale, A. R. Taylor, *The Chemistry of Organic Silicon Compounds* (Eds: S. Patai, Z. Rappoport), Wiley, New York **1989**.
10. H. Eckert, B. Forster, *Angew. Chem.* **1987**, *99*, 922-923; *Angew. Chem. Int. Ed. Engl.* **1987**, *26*, 894-895.
11. a) D. P. Curran, H. Liu, *J. Am. Chem. Soc.* **1991**, *113*, 2127-2132; b) D. P. Curran, H. Liu, H. Josien, S.-B. Ko, *Tetrahedron* **1996**, *52*, 11385-11404; c) W. R. Bowman, C. F. Bridge, P. Brookes, *J. Chem. Soc., Perkin Trans. 1*, **2000**, 1-14; d) D. P. Curran, *Aldrichimica Acta* **2000**, *33*, 104-110.

(+)-Laurallene (Crimmins 2000)

9.1 Introduction

A number of sesquiterpenoids and nonterpenoid C-15 acetogenins have been isolated from red algae of the genus *Laurencia*.[1] Two basic structural types of halogenated eight-membered ring ethers have been found. The lauthisan type **2** contains a *cis* α,α′-disubstituted oxocene and the *R*-configuration at carbons C-6 and C-7 while the laurenan structural subclass **1** possesses the *S*-configuration at C-6 and C-7 enforcing a *trans*-disubstituted pattern at the ether oxygen. Many synthetic efforts have been made towards laurencin (**2**); however in 2000 the *Crimmins* group presented the first total synthesis of the molecule (+)-laurallene (**1**),[2] which was isolated from *Laurencia nipponica* in 1979 by *Fukuzawa* and *Kurosawa*.[3] This problem based on this synthesis. The biogenetic precursor of several members of the laurenan structural subclass – such as laureatin (**3**) and **1** – is 3Z,6S,7S-laurediol (**5**). Moreover biosynthetic studies have demonstrated that enzyme lactoperoxidase (LPO) directly transforms **5** into (+)-prelaureatin (**4**) and in a second step in additional presence of a bromine distributor **4** is converted into **1**. At a screening program for insecticidal compounds of marine origin it was found that the methanol extracts possess a strong insecticidal activity against mosquito larvae (*Culex pipiens pallens*). The measured LC$_{50}$ (2.86 ppm) and IC$_{50}$ values (0.06 ppm) for **3** display the significant larvacidal activity.[4]

138   9   (+)-Laurallene

9.2 Overview

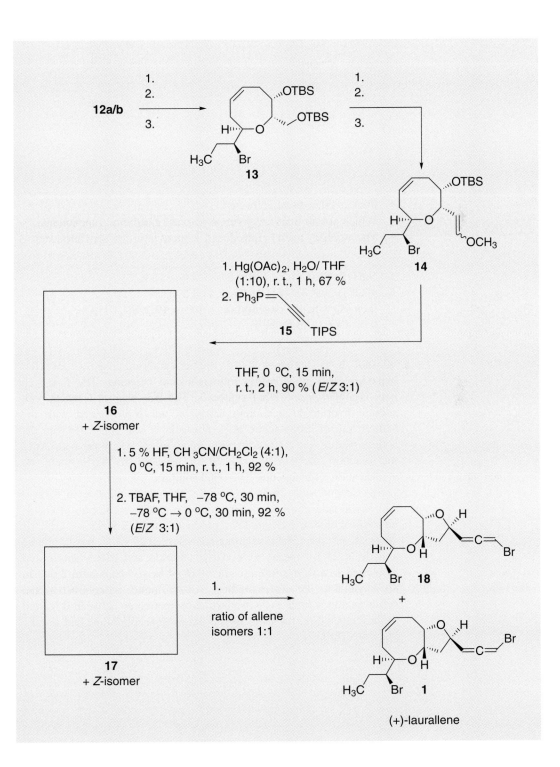

9.3 Synthesis

Problem

[Scheme: compound **6** (CH2=CH-CH(OH)-CH2-OBn, with H on stereocenter) → compound **7** via steps 1, 2; compound **7** is the glycolate ester bearing the (R)-4-benzyl-1,3-oxazolidine-2-thione auxiliary]

Hints

- First an ether-bond is formed.
- Which possibilities are given to activate a carboxyl functionality?
- The auxiliary itself is introduced without a further activation step.

Solution

[Structure **19**: (R)-4-benzyloxazolidine-2-thione]

1. NaH, BrCH$_2$CO$_2$Na, THF, 0 °C → r. t., 15 h, 93 %
1. (COCl)$_2$, cat. DMF, CH$_2$Cl$_2$, r. t., 2 h
   then (R)-4-benzyloxazolidine-2-thione **19**, NEt$_3$, CH$_2$Cl$_2$,
   0 °C → r. t., 13 h, 88 %

Discussion

[Structure **20**: glycolic acid ether intermediate]

[Structure **21**: mixed pivalic anhydride]

Alcohol **6** is prepared by a copper-catalyzed reaction of (R)-benzylglycidyl ether with vinylmagnesium bromide. The first step here is a *Williamson* ether synthesis. The free alcohol **6** reacts with sodium hydride to a sodium alkoxide, which is treated with the sodium salt of bromoacetic acid. The acid is also converted into the sodium salt to avoid the formation of an ester as side product. After the reaction carboxylic acid **20** is released in 93 % yield by acidification with aqueous 10 % HCl solution.

In the second step **20** is activated with oxalyl chloride to form a carboxylic acid chloride and treated *in situ* with a solution of (R)-4-benzyloxazolidine-2-thione **19** to produce **7** in 88 % yield.

Alternatively the *Crimmins* group transformed **20** into a mixed pivalic anhydride **21** by means of pivaloyl chloride and triethylamine and converted it with the lithium salt of (R)-4-benzyloxazolidin-2-one in 89 % yield to the corresponding oxazolidinone. In contrast to the sulfur analog, which is synthesized at temperatures between 0 °C and room temperature, the formation of an acyloxazolidinone had to take place at −78 °C.

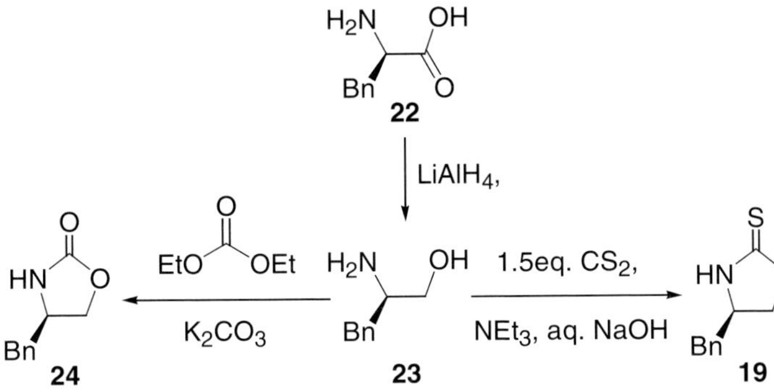

9 (+)-Laurallene

Substrate for this kind of *Evans* auxiliary is the aminoacid (R)-phenylalanine, which is reduced to the alcohol **23** and exposed either to carbon disulfide in combination with triethylamine and sodium hydroxide to synthesize **19** or to diethylcarbonate and potassium carbonate to form oxazolidin-2-one **24**.
Under more drastic conditions e. g. five equivalents carbon disulfide in a basic potassium hydroxide medium **23** is transformed in 16 h to a 1,3-thiazolidine-2-thione **25**.[5]

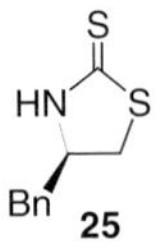

Problem

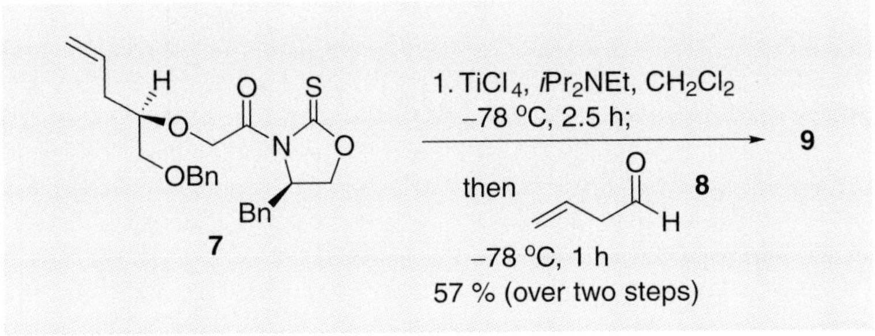

Hints

- Which reaction-type is described here?
  1) *Evans* alkylation
  2) Aldol addition
- Which diastereomer is the favored product?
- What is the coordination number of titanium?

Solution

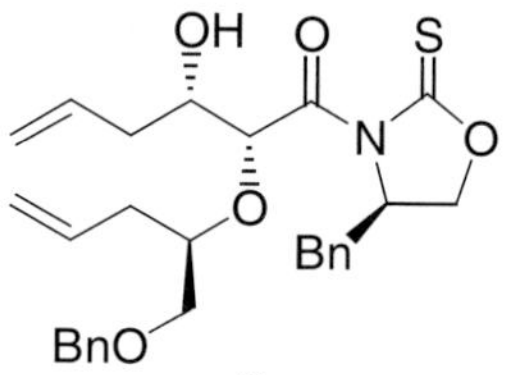

9

Discussion

This reaction is a formal asymmetric aldol addition following a modified *Evans* protocol. The enolate **26** is formed at 0 °C in the presence of one equivalent of titanium tetrachloride as *Lewis* acid and two equivalents diisopropylethylamine (*Hünig*'s base) as proton acceptor. Selectively the Z-enolate is formed. The carbon-carbon bond formation takes place under substrate control of the *Evans*-auxiliary, whose benzyl group shields the *si*-face of the enolate.

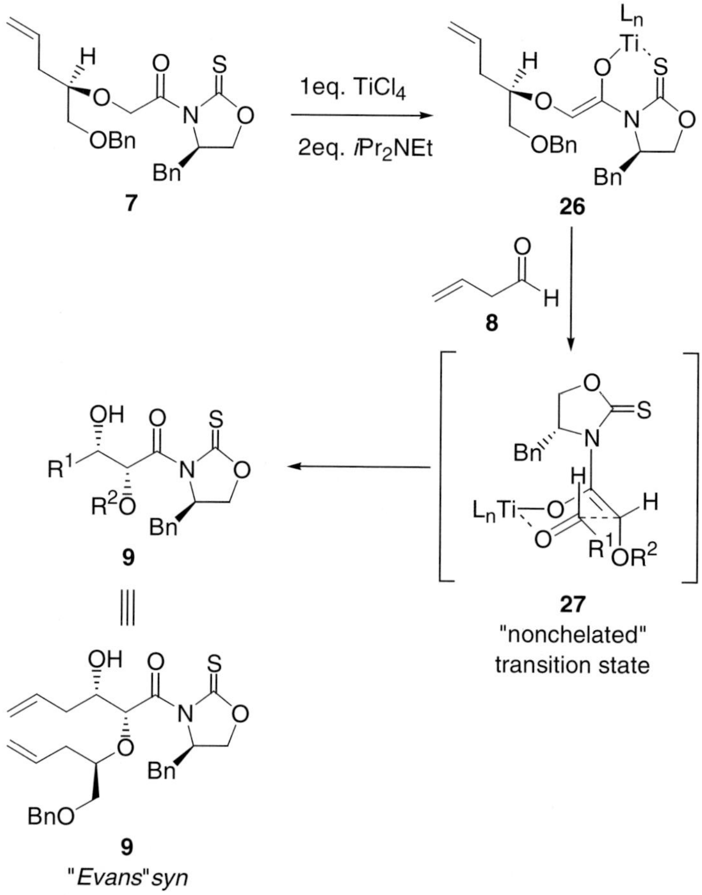

27
"nonchelated" transition state

9
"*Evans*" syn

The reaction proceeds *via* the *re*-side of the enolate and the *re*-side of the aldehyde **8**. The afore mentioned transition state **27** leads to the „*Evans*"-*syn*-aldol product **9** in about 57 % yield and an excellent selectivity of >95:5.[6]

If one chloride ion is lost, the titanium enolate can also proceed through the more rigid transition state **28**. Here both the carbonyl group of the aldehyde **8** and the thionyl group of the auxiliary are coordinated to the metal center. The product of this reaction is the „non *Evans*"-*syn*-aldol-adduct **29**.[7] In this point titanium is advantageous over boron, where only a „nonchelated" transition state like **27** is possible.[8] It is suggested, that, depending on the amount of base that is used, both transition states are accessible. Employing one equivalent of base, **28** is formed, but a second equivalent participates by coordinating to the titanium, disfavoring the chelation of the thiocarbonyl to the metal. Besides *i*Pr$_2$NEt even bases like tetramethylethylenediamine (TMEDA) and (−)-sparteine have shown good results.[9] Reaction of the oxazolidinone derivative takes place in a higher yield of 65 % but a poorer diastereoselectivity. Besides TiCl$_4$ other common *Lewis* acids used in such reactions are Bu$_2$BOTf (Chapter 10) or Sn(OTf)$_2$. Which transition state is formed depends on the reaction conditions.

Problem

Hints

- Which functionality is generated after the reduction with LiBH$_4$?
- At least one TBS-ether is built. Which conditions for this protection do you know?

Solution

Discussion

The first step is a mild reductive removal of the chiral auxiliary with lithium borohydride (LiBH$_4$) to afford the diol **30**. The auxiliary can be recovered in this step.

An advantage of the oxazolidinethione is that even the inexpensive sodium borohydride (NaBH$_4$) can be used to generate an alcohol. LiBH$_4$ and NaBH$_4$ are less reactive substances but therefore show much greater chemoselectivity. They are commonly used as reducing reagents for aldehydes and ketones at 0 °C or room temperature e. g. in the presence of an ester. In contrast lithium aluminumhydride (LiAlH$_4$) will reduce almost any carbonyl group and handling is much more difficult.

Furthermore other functionalities instead of an alcohol can be introduced in this step. In the presence of imidazole and a hydroxylamine salt a transformation into the *Weinreb* amide **32** takes place. Also the oxazolidinethione can be directly reduced to the aldehyde with diisobutylaluminum hydride (DIBAH).

In the second step a *tert*-butyldimethylsilyl-ether (TBS-ether) is built. This protecting group is quite stable to a variety of organic reaction conditions and is cleaved only under strong acidic or strong basic conditions, under *Lewis* acid catalysis and in the presence of a fluorine source. The TBS-triflate/2,6-lutidine system is one of the most powerful methods to introduce the silyl-protecting group even to secondary and tertiary alcohols. Selective silylation of primary over secondary alcohols is achieved with TBS-chloride together with basic activators like triethylamine or dimethylaminopyridine (DMAP).[10]

Problem

Hints

- What kind of cycloaddition takes place?
- The eight-membered ring is closed in a catalytic reaction. What kind of possible catalysts do you know?

1. (Cy$_3$P)$_2$Cl$_2$Ru=CHPh, CH$_2$Cl$_2$, reflux, 2 h, r. t., overnight open to air, 95 %

Solution

Discussion

In this olefin metathesis reaction the diene **10** is treated with 8 mol % of the *Grubbs*-ruthenium-catalyst **33** resulting in smooth conversion to the oxocene structure **11** in 2 h and 95 % yield. The reaction takes place in a highly diluted solution of refluxing dichloromethane (0.005 M) to have optimal conditions for an intramolecular reaction sequence. In contrast to the formerly used highly active *Schrock*-catalyst **34**, catalyst **33** shows remarkable tolerance towards many different functional groups, is air stable and thus easier to generate and to handle. Titanium and tungsten-based cataysts have also been developed, but are less used.[11]

Nowadays new imidazolin-2-ylidene substituted Ru-catalysts like **35** are under investigation, which show increasing activity even at low temperatures. Advantageously they show no sensitivity to air or to moisture. Potential activity is expected to result from higher *Lewis* basicity and steric demand. They give good results even in the synthesis of tetrasubstituted cycloalkenes.[12]

This intramolecular reaction results in the formation of a cyclic system, and therefore it is called ring-closing metathesis (RCM). In this process a diene **36** is treated with a metal alkylidene **37**. Two competing pathways are available *via* the intermediate metal alkylidene **38**: **A**) RCM will occur to afford cyclic adducts **39** and **B**) intermolecular reaction can occur to form polymeric structures **40** (acyclic diene metathesis polymerization (ADMET)). The reaction is also complicated because of the possibility of ring-opening metathesis (ROM), the retro reaction of path **A**, and ring opening metathesis polymerization (ROMP) (path **C**).[13]

Which products are obtained is a result of thermodynamic and kinetic parameters.

The following catalytic cycle is postulated.[14] The (commercially available) initiating catalyst **33** is transformed into the propagating catalyst **41** running the following mechanism one time. **41** then enters the cycle by loss of one phosphine ligand. Henceforth **42** catalyzes the transformation of **10** to **11**.

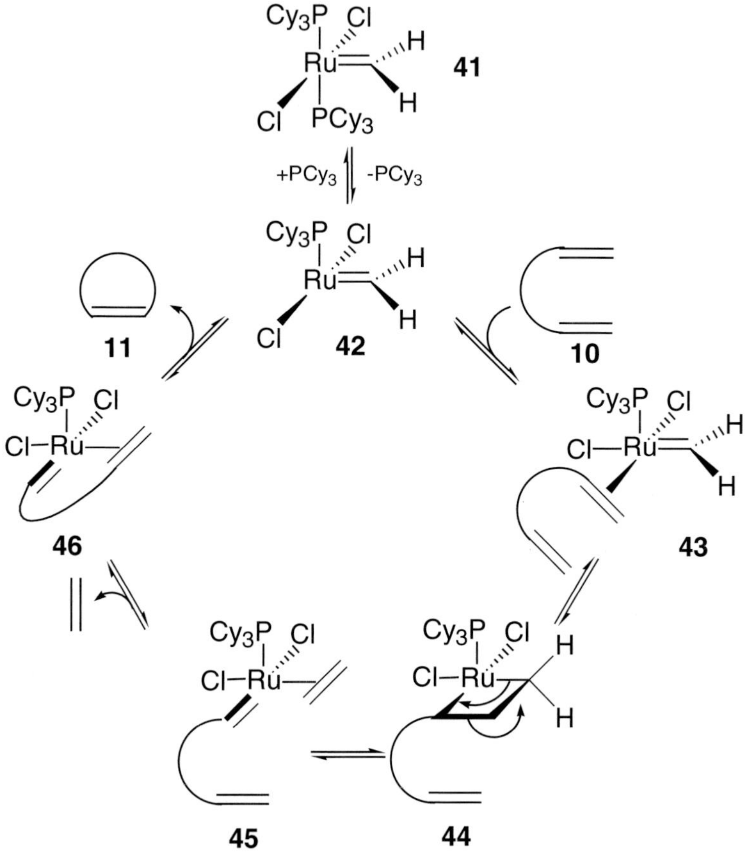

In this dissociative pathway (which is assumed to be the major one today) first a phosphine is displaced from the metal center to form an active 14-electron-intermediate **42**. After alkene coordination *cis* to the alkylidene fragment the 16-electron-olefine-complex **43** undergoes [2 + 2]-cycloaddition to give a metallacylobutane **44**. Compound **44** breaks down in a symmetric fashion to form carbene complex **45**. The ethylene is replaced in the conversion to complex **46**. In the next steps (they are not further discribed above), another intramolecular [2 + 2]-cycloaddition joins up the eight-membered ring **11** regenerating the catalyst **42**. Each step of the reaction is thermodynamically controlled making the whole RCM reversible. With additional excess of phosphine added to the reaction mixture an associative mechanism is achieved, in which both phosphines remain bound.

The *trans* substituted pattern of the substituents flanking the ether oxygen is the result of a *gauche* effect of the TBS-oxygens at C-6 and C-7. Two conformations **A** and **B** are possible.[15]

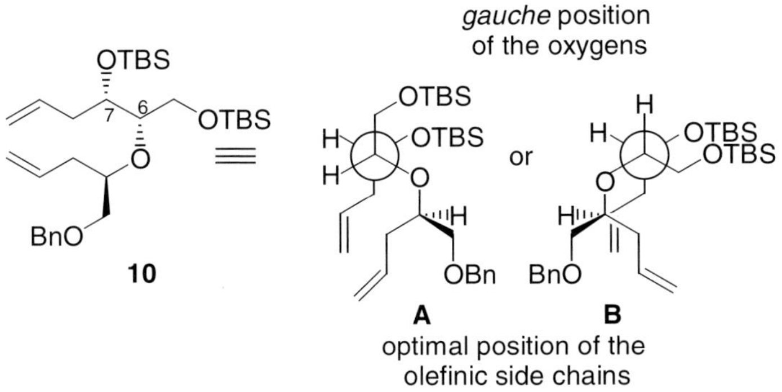

In conclusion, six-, seven- or as here eight-membered rings as well as macrocycles can be synthesized. Stereogenic centers are not touched and the process is compatible with many other functional groups like esters, amines, alcohols and epoxides.

Problem

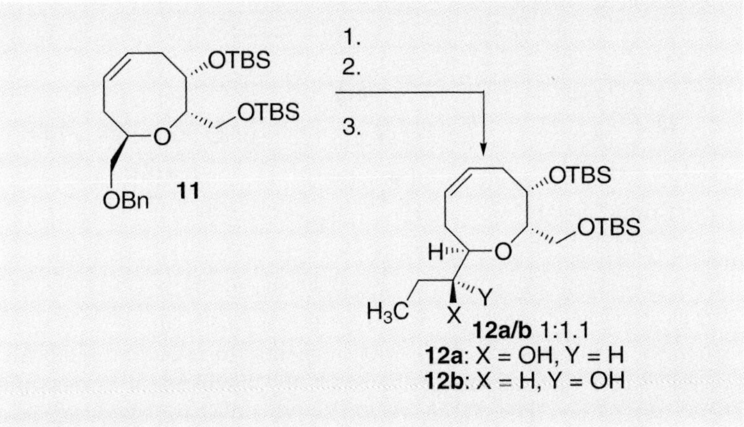

Hints

- First the benzyl ether is cleaved.
- What kind of functionality do you need to introduce an ethyl-side-chain?

- A common oxidation method is used.
- In Step 3 two products are obtained. Which model is useful to describe the attack of the nucleophile?

Solution

1. Na, NH$_3$, THF, −78 °C, 1.5 h, 95 %
2. (COCl)$_2$, DMSO, NEt$_3$, CH$_2$Cl$_2$, −78 °C, 40 min, r. t., 1 h
3. EtMgBr, CH$_2$Cl$_2$, 0 °C, 1 h, 90 % (over two steps)

Discussion

There are different methods to cleave benzyl ether bonds. The most common one is hydrogenolysis with palladium on carbon or platinum as catalysts under H$_2$ atmosphere. The standard solvents are ethanol or ethyl acetate. Pd is the preferred and milder one, because the use of Pt at any rate results in aromatic ring hydrogenation. Also a number of methods have been developed in which hydrogen is generated *in situ*, e. g. from cyclo-hexene, -hexadiene or formic acid (see Chapter 7).

In this case the *Birch*-reduction system (sodium in liquid ammonia) is used. Normally this system is employed to reduce aromatic rings to 1,4-dihydrobenzenes. Here it is advantageous because in contrast to catalytic hydrogenation it does not touch olefinic double bonds and the expected alcohol is generated in 95 % yield (see universal mechanism on the left).

Oxidation of the alcohol under *Swern* conditions (see Chapter 2) followed by addition of commercially available ethylmagnesium-bromide provided the secondary alcohol with nearly no stereocontrol (1:1.1 in favor of isomer **12b**) in 90 % overall yield.

Generally additions of *Grignard* reagents to α–chiral-aldehydes proceed highly diastereoselectively. Of the three possible transition states (A) *Cram*-, (B) *Felkin Anh*-, the (C) *Cram*-chelate-transition state **47** is favored here. The reason is the α-oxygen which should be chelated by the magnesium cation as well as the carbonyl group, leading to **12b** as the main product. If R is small (e. g. Me or TMS) the medium group takes its place perpendicular to the carbonyl group. The attack of the nucleophile proceeds along the least hindered trajectory taking into account the *Bürgi-Dunitz* angle of about 107° measured from the C=O bond.

But this – and other examples[16] – indicate, that chelation control is apparently limited in α-alkoxyaldehydes where the chelating ether oxygen resides in a medium ring, probably because of reduced *Lewis* basicity of the ether oxygen.

Attempted use of diethylzinc in the presence of a chiral catalyst to control the stereochemistry at this carbon atom was not effective.

Problem

Reagents for conversion of **12a/b** (1:1.1; **12a**: X = OH, Y = H; **12b**: X = H, Y = OH) to **13**:
1. (COCl)$_2$, DMSO, NEt$_3$, CH$_2$Cl$_2$, −78 °C, 40 min, r.t., 1 h
2. L-Selectride®, THF, −78 °C, 30 min; 1 M NaOH, 30 % H$_2$O$_2$, r.t., 1 h; 83 % (over two steps)
3. CBr$_4$, P(oct)$_3$, C$_6$H$_6$, r.t., 30 min, 70 °C, 1 h, 88 %

Hints

- *Crimmins* took the mixture of secondary alcohols and first reoxidized the stereogenic center.
- L-Selectride® is a selective reducing reagent.
- The last step is similar to an *Appel*-reaction.

Solution

Structure **13**.

Discussion

In the first step *Crimmins* and co-workers used the *Swern* oxidation protocol to provide a ketone as prochiral sp^2-center.

The formed ketone is selectively reduced to an alcohol with lithium-tris-*sec*-butylborohydride (L-Selectride®) at −78 °C in only 30 min.. Only one diastereomer is built. Because lithium belongs to the non-chelating cations, the reaction proceeds through transition state **48** proposed by *Felkin* and *Anh*.[17] To remind: In the most stable conformation the largest group of the stereogenic center is perpendicular to the carbonyl group.

Transition state **48**.

9 (+)-Laurallene

The reaction of the secondary alcohol with trioctylphosphine and carbon tetrabromide results in the formation of the **13**. The conditions are similar to those established by *Appel* et al. for the conversion of alcohols to chlorides. The transformation proceeds through an S_N2 mechanism resulting in inversion of the configuration.[18]

Problem

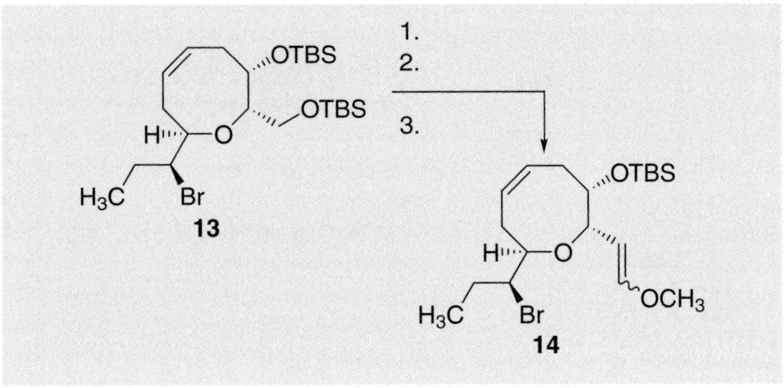

Hints
- Under which conditions is only the primary TBS-ether cleaved?
- Secondly the molecule is prepared for a chain extension.
- Finally a *Wittig* reaction takes place.

Solution
1. HF-pyridine, pyridine, THF, r. t., 4 h, 90 %
2. (COCl)$_2$, DMSO, CH$_2$Cl$_2$, −78 °C, 40 min, r. t., 1 h
3. Ph$_3$P=CHOCH$_3$, KOtBu, THF, 0 °C, 25 min
75% (over two steps)

Discussion

The basic HF-pyridine/pyridine system represents a mild method to convert only the primary TBS-ether into the alcohol. The excess pyridine acts as buffer. Use of tetrabutylammoniumfluoride (TBAF) would result in cleavage of both TBS-ethers.[10]

In the second step the alcohol is transformed to an aldehyde by means of the *Swern* oxidation. Other reagents to oxidize alcohols to aldehydes are e. g. *Dess-Martin*-periodinane and chromium reagents like PCC or PDC.

9 (+)-Laurallene

At least the methoxymethylenation proceeds in good yield to give a mixture of vinylethers **14**. The one-carbon extension follows the *Wittig* protocol. The ylide **51** is formed *in situ* from (methoxymethyl)triphenylphosphonium chloride (**49**) and potassium *tert*-butoxide at 0 °C in THF. **51** belongs to the semistable ylides, because the methoxy group is able to stabilize a carbanion only over an inductive effect. This results in a mixture of *E*- and *Z*-isomers. No further definition of the ratio of diastereomers is made, because the reaction serves the purpose of chain elongation and just in the next step configuration of the double bond is destroyed again. How to reach only the *E*- or *Z*-isomer is described later in this chapter.

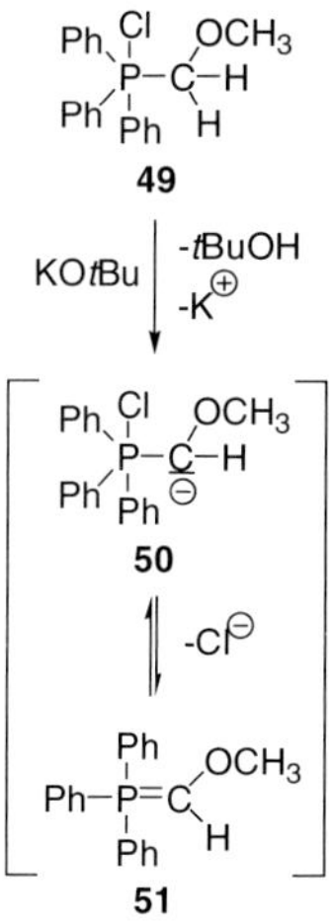

Moreover the reaction runs the normal *Wittig* mechanism over the four-membered oxaphosphetane ring and collapses to the alkene mixture. (For mechanistic details see Chapter 10.)

Problem

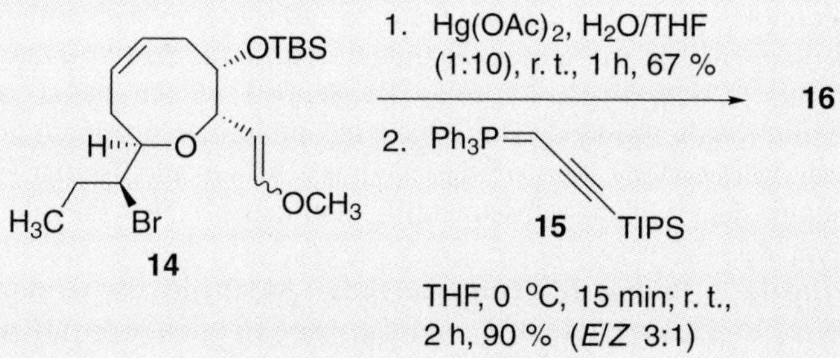

Hints

- The vinylether system is forced to collapse to form an aldehyde.
- Therefore the transition metal adds to the double bond to result in the *Markovnikov* product.
- Even the water participates while a work-up with potassium iodide completes the reaction.
- Which reaction proceeds adding a triphenylphosphine species to an aldehyde?

Solution

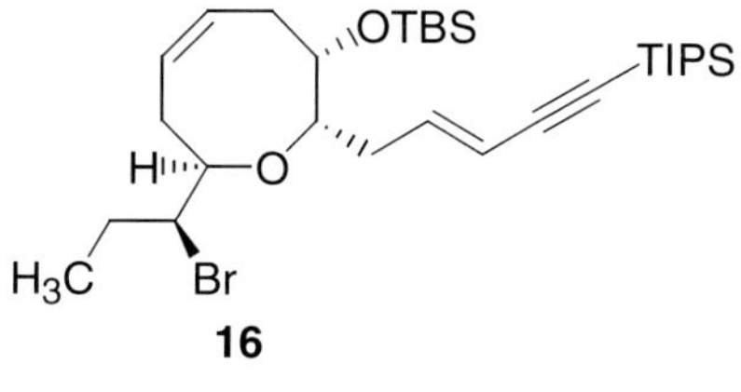

Discussion

In the first reaction step the enol ether is transformed into an aldehyde. This sequence is named solvomercuration/demercuration.[19]

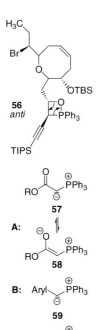

Therefore mercury(II) acetate interacts as an electrophilic transition metal with the nucleophilic alkene to form the three-membered ring **52**. This mercurinium ion is opened by relatively feeble nucleophiles like alcohols – or in this reaction water. Similar to a hydroboration the attack happened at the more substituted end of the mercurinium ion according to *Markovnikov*'s rule. To get rid of the metal, solid potassium iodide is added. This means insoluble mercury(II) iodide is formed, followed by loss of the methoxy group and formation of enol ether **54**, which subsequently tautomerizes to the desired aldehyde **55**.

Next again a *Wittig* reaction succeeds. Aldehyde **55** is directly converted to a 3:1 mixture of *E*- and *Z*-enynes **16** upon exposure to the *in situ* generated *Wittig* salt **15**. The *E*-isomer is built as main product, passing the *anti*-configurated oxaphosphetane **56**.

What drives the reaction to an *E*- or *Z*-alkene?
Generally the stereoselectivity of the [2 + 2]-cycloaddition depends on the ylide. They are divided into three types: **A**) those with conjugating or anion stabilizing substituents adjacent to the negative charge, for example a carbonyl group (so called stabilized ylides **57**), **B**) those with substituents having only slight stabilizing properties (semi-stabilized ylides **59**) and **C**) those without a stabilizing neighbor group (unstabilized ylides **60**). The extra stabilization of the first sort of ylides is represented by an alternative enol-type structure **58**. These ylides can be stored for months and need not to be generated *in situ* like **59** and **60**. To anticipate the solution a general rule is:

- with stabilized ylides the *Wittig* reaction is *E*-selective.
- with unstabilized ylides the *Wittig* reaction is *Z*-selective.
- semistabilized ylides produce mixtures of *E*- and *Z*-olefines.

9 (+)-Laurallene

To explain this phenomenon, there are two seperate processes to consider. The first and most important one for this reason is the formation of the oxaphosphetane. The addition of the ylide to the aldehyde can, in principle, produce two isomers with either a Z or E substituted double bond. The following elimination step is stereospecific, with the oxygen and phosphorus departing in a *syn*-periplanar transition state. With unstabilized ylides the *syn* diastereomer of the oxaphosphetane similar to **61** is formed preferentially. This step is kinetically controlled and therefore irreversible, and predominantly the Z-alkene **62** that results reflects this.

On the other hand the stability of **57** causes the reaction leading to a reversible oxaphosphetane where the isomers **63** and **65** can interconvert *via* the starting material. The stereoselectivity in this step is thermodynamically controlled. The more stable four-membered ring is *anti* **65**, with the bulky groups on opposite sides of the ring. The product of this reaction after elimination of triphenylphosphine oxide is only the *E*-alkene **66**.

This explains the moderate selectivity of only 3:1 obtained here. Because of the triple bond adjacent to the negative charge **15** rather belongs to the semistabilized ylides **C** than to category **A**, resulting in the production of *E/Z*-mixtures. Here neither the kinetically nor the thermodynamically controlled pathway dominates.

Some stabilized ylides are too stable to be very reactive. In this case phosphonates are used instead of phosphonium salts. For the *Horner-Wadsworth-Emmons*-reaction see Chapter 2.

Problem

Hints

- The aim of these steps is the removal of both silyl protecting groups.
- Which conditions belong to which protecting group and for what reason?

Solution

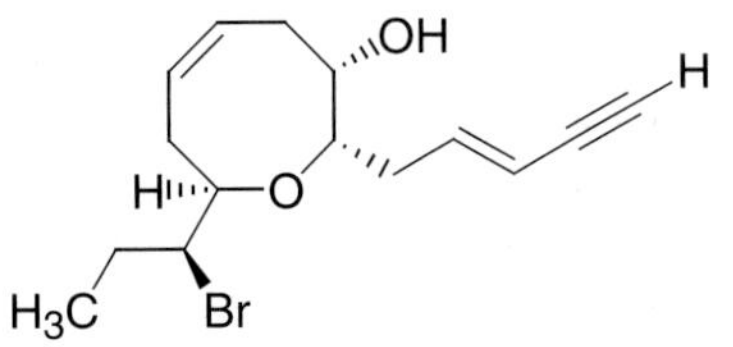

17
E-prelaureatin

Discussion

Removal of the tri-*iso*-propylsilyl (TIPS) and *tert*-butyldimethylsilyl (TBS) protecting groups could be accomplished concomitantly with TBAF in tetrahydrofuran at 0 °C, but here competing elimination of the secondary bromide was observed. Better overall yields and cleaner conversion was observed when TBS ether was cleaved with 5 % aqueous HF in acetonitrile at 0 °C followed by removal of the acetylenic TIPS with TBAF under milder conditions of −78 °C.[10] The diastereomers are not separated before the desilylation process; therefore even a 3:1 mixture of *E*- and *Z*-enyne is obtained.

Prelaureatin **4** and its *E*-isomer **17** are likewise goals in natural product synthesis. *Crimmins* and co-workers developed an own synthetic route to **4**. The reaction sequence is similar up to aldehyde **55**. Afterwards a *Z*-vinyl-iodide is selectively formed and the alkyne is introduced via a *Sonogashira* reaction.

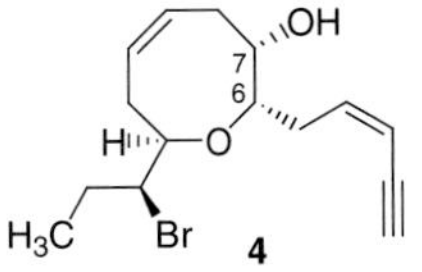

4
Z-prelaureatin

Problem

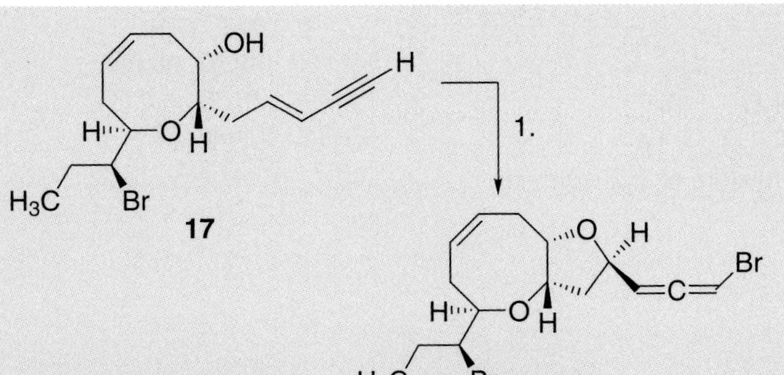

17

1
(+)-laurallene
+ bromallene isomer (ratio 1:1)

- To introduce a bromo functionality a reagent delivering a cationic bromine species is needed.
- Cyclization to furan takes place without any further activation step.

Hints

1. TBCO **67**, CH_2Cl_2, r. t., 24 h, 53 %

Solution

To generate the title bromo allene **1** the *E*-enyne was treated with a bromine cation. According to a procedure described by *Murai* the reagent 2,4,4,6-tetrabromocyclohexadienone (TBCO) **67** is used as the bromine source.[20]

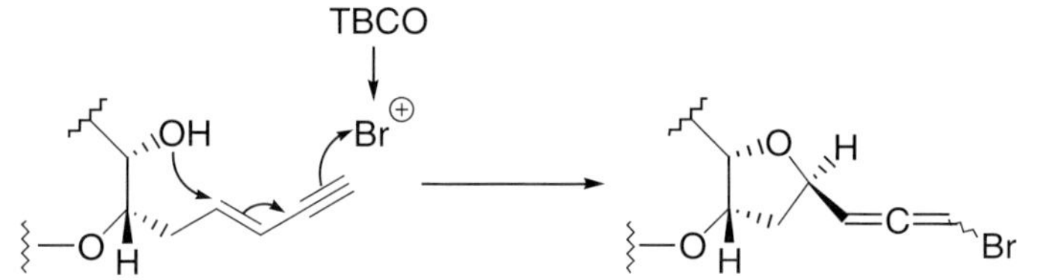

Discussion

The conversion to the allenic structure proceeds *via* nucleophile attack of the triple bond of **17** to the bromocation and produces laurallene **1** and its bromo allene isomer **18** in 53 % overall yield. They were separated by HPLC to give pure (+)-laurallene in 24 %.

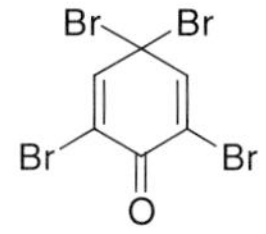

9.4 Conclusion

The first total synthesis of (+)-laurallene illustrates a route to a α,α'-disubstituted cyclic eight-membered ether systems. The strategy is based on the introduction of the stereoselectivity through an asymmmetric aldol-addition with subsequent ring closing olefin-metathesis reaction. Both the lauthisan and the laurenan type structural class are accessible. The synthesis of (+)-laurallene is achieved in 20 steps and approximately 4 % yield. Further application of the aldol-ring closing metathesis strategy for the construction of medium rings is in progress.

9.5 References

1. D. J. Faulkner, *Nat. Prod. Rep.* **1999**, *16*, 155-198.
2. M. T. Crimmins, E. A. Tabet, *J. Am. Chem. Soc.* **2000**, *122*, 5473-5476.
3. A. Fukuzawa, E. Kurosawa, *Tetrahedron Lett.* **1979**, *32*, 2797-2800.
4. K. Watanabe, K. Umeda, M. Miyakado, *Agric. Biol. Chem.* **1989**, *53*, 2513-2515.
5. D. Delaunay, T. Toupet, M. LeCorre, *J. Org. Chem.* **1995**, *60*, 6604-6607.
6. R. E. Gawley, J. Aube, *Principles in Organic Synthesis*, Tetrahedron Organic Chemistry Series 14, Pergamon, **1996**, 178-182.
7. M. T. Crimmins, B.W. King, E. A. Tabet, *J. Am. Chem. Soc.* **1997**, *119*, 7883-7884.
8. D. J. Ager, I. Prakash, D. R. Schaad, *Aldrichimica Acta* **1997**, *30*, 3-12.
9. M. T. Crimmins, K. Chaudhary, *Org. Lett.* **2000**, *2*, 775-777.
10. a) T. W. Greene, P. G. W. Wuts, *Protective Groups in Organic Synthesis*, 3rd Ed., John Wiley & Sons, New York, **1999**; b) P. J. Kocienski, *Protecting Groups*, Georg Thieme Verlag, Stuttgart, New York, **1994**.
11. A. Fürstner, *Angew. Chem.* **2000**, *112*, 3140-3172; *Angew. Chem. Int. Ed. Engl.* **2000**, *39*, 3012-3043.
12. M. Scholl, T. M. Trnka, J. P. Morgan, R. H. Grubbs, *Tetrahedron Lett.* **1999**, *40*, 2247-2250.
13. R. H. Grubbs, S. J. Miller, G. C. Fu, *Acc. Chem. Res.* **1995**, *28*, 446-452.
14. a) E. L. Dias, S. T. Nguyen, R. H. Grubbs, *J. Am. Chem. Soc.* **1997**, *119*, 3887-3897; b) M. S. Sanford, M. Ulman, R. H. Grubbs, *J. Am. Chem. Soc.* **2001**, *123*, 749-750.
15. E. L. Eliel, S. H. Wilen, *Stereochemistry of Organic Compounds*, John Wiley and Sons, New York, **1994**.
16. M. T. Crimmins, K. A. Emmitte, *Org. Lett.* **1999**, *1*, 2029-2032.
17. J. W. Burton, J. S. Clark, S. Derrer, T. C. Storck, J. G. Bendal, A. B. Holmes, *J. Am. Chem. Soc.* **1997**, *119*, 7483-7498; K. Tsushima, A. Murai, *Tetrahedron Lett.* **1992**, *33*, 4345-4348.
18. R. Appel, *Angew. Chem.* **1975**, *87*, 863-874; *Angew. Chem. Int. Ed. Engl.* **1975**, *14*, 801-812.
19. R. C. Larock, *Solvomercuration/Demercuration Reactions in Organic Synthesis*, Springer, Berlin, **1986**.
20. J. Ishikara, Y. Shimada, N. Kanoh, Y. Takasugi, A. Fukuzawa, A. Thurai, *Tetrahedron* **1997**, *53*, 8371-8382.

10

Myxalamide A
(Heathcock 1999)

10.1 Introduction

Myxalamide A (**1**) is one of a growing number of polyene antibiotics, which includes the phenalamides as well as the myxalamides. The four known myxalamides were isolated from the gliding bacteria *Myxococcus xanthus*, and the most abundant of the group, myxalamide B (**3**), was found to be a potent electron-transport inhibitor and exhibits antibiotic and antifungal activity.[1] Myxalamide B **3** was shown to inhibit NADH oxidation at complex I in beef heart submitochondrial particles with an IC$_{50}$ of 170 pm/mg of protein. The phenalamides were isolated more recently from *Myxococcus stipiatus*.[2] Phenalamide A1 (**2**) (stipiamide) exhibits antifungal and antiviral properties as well as the ability to reverse P-glycoprotein-mediated multidrug resistance.[3]

The myxalamides and the phenalamides have attracted some synthetic attention. A partial synthesis of myxalamide D (**4**) was reported by *Cox* and *Whiting* in which an *anti* aldol reaction was used to set the C-12/C-13 stereochemistry.[4] In addition, a total synthesis of **2** was described by *Andrus*.[3] In this approach, an asymmetric alkylation set the distal stereocenter (C-16), while an asymmetric crotylboration created the *anti* stereorelationship at C-12/C-13. In contrast to the other synthetic efforts, the approach to **1** described in this chapter would enable a single asymmetric reaction early in the synthesis to produce the *anti* C-12/C-13 relationship as well as set the distal C-16 stereocenter.[5]

10.2 Overview

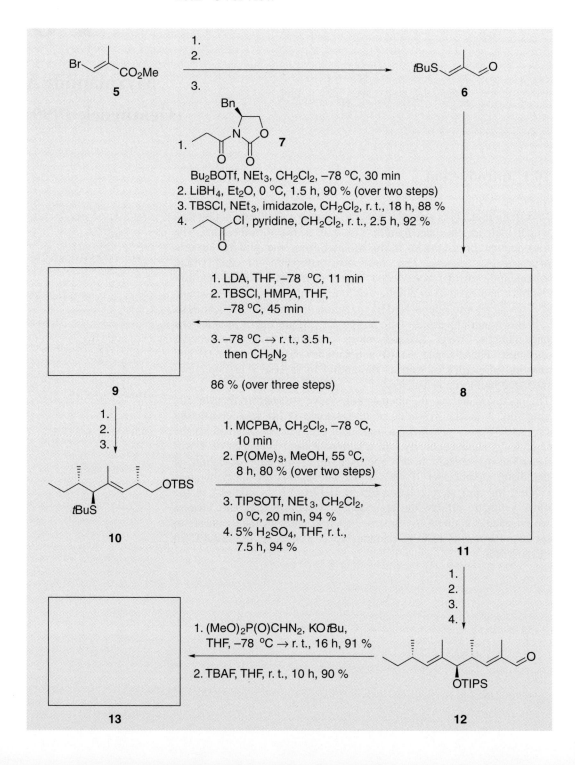

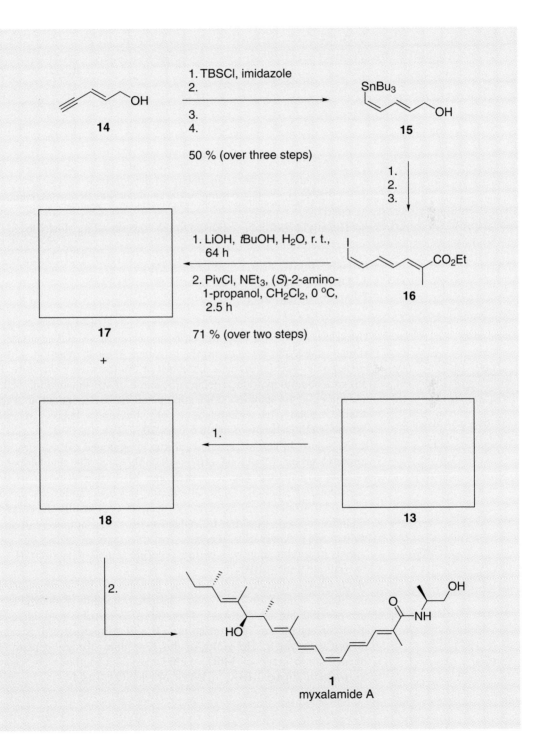

10.3 Synthesis

Problem

5: Br-CH=C(CH₃)-CO₂Me → 6: tBuS-CH=C(CH₃)-CHO

Hints

- α,β-unsaturated esters can undergo 1,4-additions.
- Which agent reduces esters without reduction of the α,β-double bond?
- The allylic alcohol is oxidized to the corresponding aldehyde.

Solution

1. NaH, tBuSH, THF, r. t., 40 min, 87 %
2. DIBAH, CH$_2$Cl$_2$, 0 °C, 70 min, 97 %
3. MnO$_2$, hexane, r. t., 15 h, 86 %

Discussion

The addition of a nucleophilic carbon species to an α,β-unsaturated ketone, aldehyde, ester or nitrile is called *Michael* reaction. Other nucleophiles such as amines, alkoxides or thiolates react similarly. In this case sodium hydride generates the anion of *tert*-butanethiol which undergoes an 1,4-addition at the conjugated ester. Regeneration of the double bond by elimination of bromine leads to the thermodynamically more stable *E*-ester **20**.

Reduction of an α,β-unsaturated carbonyl compound can occur at the carbonyl group, giving an allylic alcohol, or at the double bond, yielding a saturated ketone, which can be further reduced to the corresponding alcohol. These alternative reaction modes are called 1,2- and 1,4-reduction, respectively. DIBAH is generally the reagent of choice to accomplish the 1,2-reduction of α,β-unsaturated carbonyl compounds.[6] Alternative reagents for this type of reductions are NaBH$_4$ in combination with cerium chloride (*Luche* reduction) **21**[7] or the dialkylborane 9-BBN.[8]

MnO$_2$ oxidizes allylic and benzylic alcohols to the corresponding aldehydes.[9] Oxidation of primary saturated alcohols with MnO$_2$ is also possible but much slower. Therefore it is possible to oxidize an allylic or benzylic alcohol in the presence of an unprotected primary alcohol with MnO$_2$. The yields of this procedure strongly depend on the activation grade of MnO$_2$. Other oxidation methods are *Dess-Martin* periodinane,[10] IBX,[11] TPAP[12] or DMSO/C$_2$O$_2$Cl$_2$.[13]

10 Myxalamide A

Problem

[Scheme: tBuS-CH=C(CH₃)-CHO (**6**) converted to **8** via:
1. Oxazolidinone **7** (Bn-substituted, N-propionyl), Bu₂BOTf, NEt₃, CH₂Cl₂, −78 °C, 30 min
2. LiBH₄, Et₂O/H₂O, 0 °C, 1.5 h, 90 % (over two steps)
3. TBSCl, NEt₃, imidazole, CH₂Cl₂, r. t., 18 h, 88 %
4. EtC(O)Cl, pyridine, CH₂Cl₂, r. t., 2.5 h, 92 %]

Hints

- Bu₂BOTf generates the Z-enolate of the oxazolidinone.
- The following aldol reaction proceeds *via* a six-membered transition state.
- LiBH₄ reduces the amide.
- With TBSCl a primary alcohol can be protected selectively in presence of secondary ones.

Solution

[Structure **8**: tBuS-CH=C(CH₃)-CH(OC(O)Et)-CH(CH₃)-CH₂-OTBS]

Discussion

The approach for the enantioselective aldol reaction based on oxazolidinones like **22** and **23** is called *Evans* asymmetric aldol reaction.[14] Conversion of an oxazolidinone amide into the corresponding lithium or boron enolates yields the Z-stereoisomers exclusively. Reaction of the Z-enolate **24** and the carbonyl compound **6** proceeds *via* the cyclic transition state **25**, in which the oxazolidinone carbonyl oxygen and both ring oxygens have an *anti* conformation because of dipole interactions. The back of the enolate is shielded by the benzyl group; thus the aldehyde forms the six-membered transition state **25** by approaching from the front with the larger carbonyl substituent in pseudoequatorial position. The

[Structures **22** and **23**: oxazolidinones with iPr and Ph/Me substituents respectively]

stereoselectivity is higher using boron than it is using lithium because the B-O bond distances are shorter than those in metal enolates. This leads to a more compact structure for the transition state magnifying the steric interactions which control stereoselectivity. Work-up provides oxazolidinone **26** which is converted to **27** without further purification.

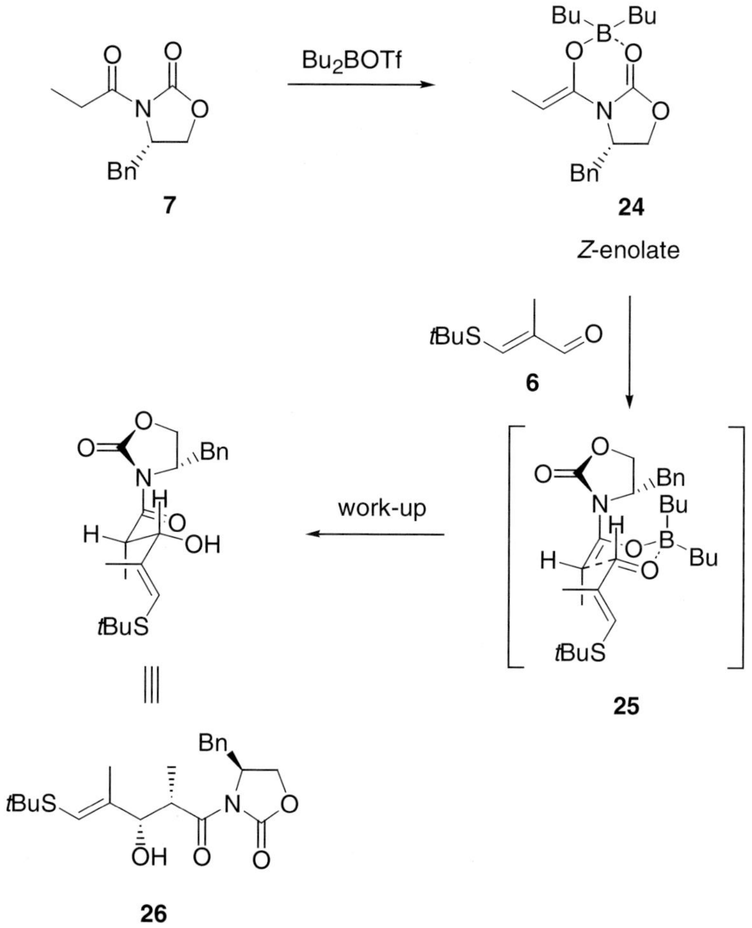

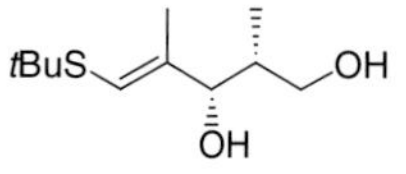

Reductive removal of the chiral auxiliary with $LiBH_4$ and one equivalent of water yields diol **27**. Mechanistically, it is possible that a 1:1 ratio of $LiBH_4$ and water forms a hydroxyborohydride ion acting as reducing agent.[15] Other reductive methods like $LiAlH_4$ or $LiBH_4$ without water are also often employed, but, when substantial steric bulk is introduced in the vicinity of the *N*-acyloxazolidinone moiety, the reduction results in formation of a significant amount of ring-opened byproducts from hydride attack at the oxazolidinone carbonyl group.

The resulting molecule bears a primary and a secondary alcohol function. The primary alcohol can be selectively protected by using TBSCl. This silyl protecting group provides excellent stability towards base, but is relatively sensitive to acid. Lastly the secondary alcohol is acylated using propionyl chloride in a standard procedure to give ester **8**.

Problem

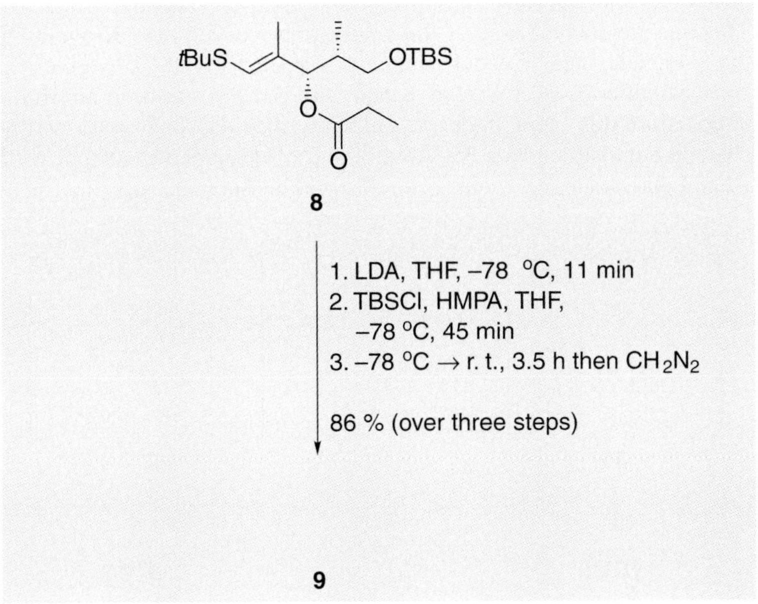

[Structure of **8**: tBuS-CH=CH-CH(OC(=O)Et)-CH(-)-CH₂-OTBS]

8

1. LDA, THF, –78 °C, 11 min
2. TBSCl, HMPA, THF, –78 °C, 45 min
3. –78 °C → r. t., 3.5 h then CH₂N₂

86 % (over three steps)

9

- A silyl enol ether is generated.
- This silyl enol ether undergoes a [3,3]-sigmatropic rearrangement.

Hints

Solution

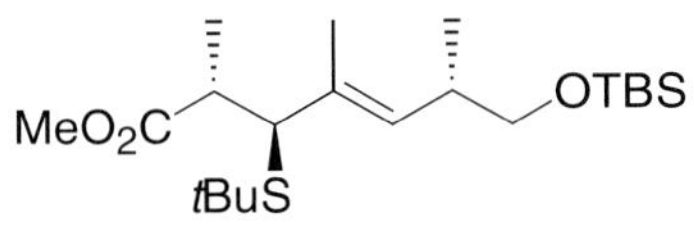

[Structure of **9**: MeO₂C-CH(-)-C(tBuS)=CH-CH(-)-CH₂-OTBS]

9

Treatment of ester **8** with LDA followed by TBSCl produces an *E*-silyl ketene acetal **34**, which undergoes a [3,3]-sigmatropic rearrangement upon warming up to room temperature. This [3,3]-rearrangement is one version of the *Claisen*-rearrangement also called *Claisen-Ireland*-rearrangement.[16] The classical *Claisen* protocol is the rearrangement of an allyl vinyl ether **28** at high temperature under formation of a γ,δ-unsaturated aldehyde **29**.[17] However, the product of

Discussion

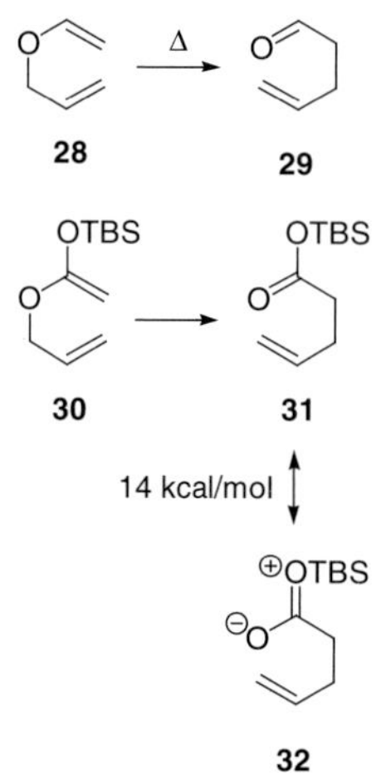

the *Claisen-Ireland* rearrangement is an α-allylated silylether **31**, whose carbonyl double bond is stabilized by mesomeric structure **32** (~14 kcal/mol). This stabilization is an additional driving force compared to the classical reaction whose products cannot be stabilized by resonance structures. The resulting silylester is not stable under hydrolytic conditions and yields the corresponding acid after aqueous work-up.

Enolization of **8** with LDA can proceed *via* the two different transition states **33** and **37**. The steric interactions between the methyl group and the isopropyl group of LDA disfavor transition state **37**. Thus the stereochemistry of the enolization proceeds *via* **33** yielding the *E*-enolate, which is subsequently trapped by TBSCl to give the corresponding *E*-silyl ketene acetal **34**. During warm-up to room temperature this acetal undergoes the described rearrangement *via* the chair-like transition state **35**. After acidic work-up and treatment with etheral diazomethane, methyl ester **9** could be isolated.

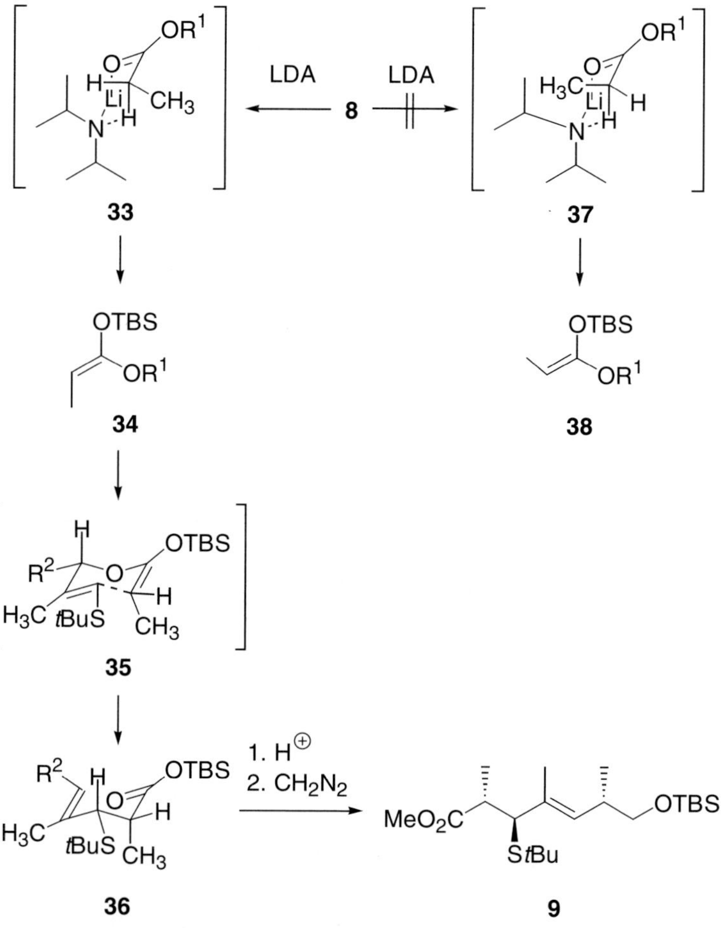

Problem

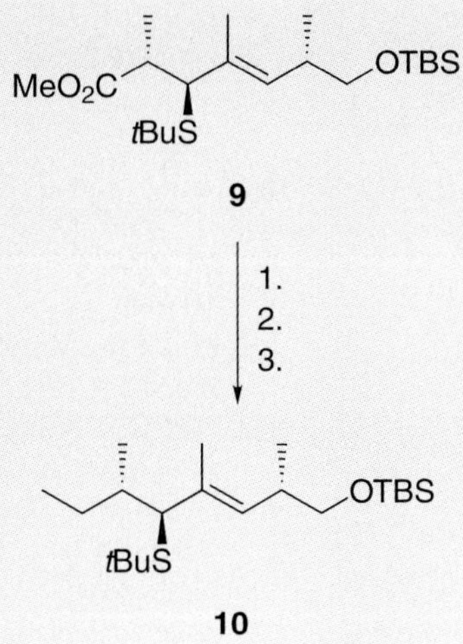

[Structure of compound **9**: MeO₂C-CH(StBu)-C(CH₃)=CH-CH₂-OTBS]

9

1.
2.
3.

[Structure of compound **10**: Et-CH(StBu)-C(CH₃)=CH-CH₂-OTBS]

10

Hints

- During the first two steps the ester is transformed into a leaving group.
- After reduction, the resulting alcohol is treated with tosyl chloride.
- For the nucleophilic substitution a higher-order cyanocuprate is used.

Solution

1. LiAlH$_4$, Et$_2$O, 0 °C, 30 min, r. t., 2 h, 99 %
2. TsCl, DMAP, NEt$_3$, CH$_2$Cl$_2$, r. t., 21 h, 94 %
3. CuCN, MeLi, Et$_2$O, 0 °C, 3.5 h, 94 %

Discussion

These three reactions extend the carbon framework. First methyl ester **9** is reduced with LiAlH$_4$ to the corresponding alcohol **39**, which is then treated with tosyl chloride to yield **40**. The use of cyanocuprates for the following reaction is necessary because they are only a little basic and so nucleophilic that they substitute the tosylate. The addition of two equivalents of methyllithium to CuCN forms cuprate **41**. This compound is called higher-order cyanocuprate.[18] Generation of the reagent at –78 °C followed by warming of the reaction mixture up to room temperature and addition of tosylate **40** yields 76 % of the desired product along with significant amounts of alcohol **39** presumably resulting from attack on the sulfur of the tosylate. When the cuprate is formed at a higher temperature (0 °C) the yield increases to 94 % with little or no formation of alcohol **39**.

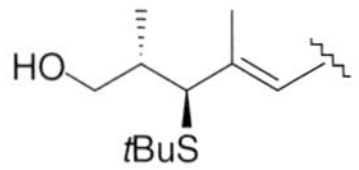

[Structure of **39**: HO-CH₂-CH(StBu)-C(CH₃)=CH-]

39

[Structure of **40**: TsO-CH₂-CH(StBu)-C(CH₃)=CH-]

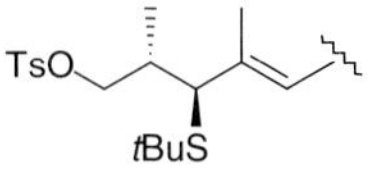

40

2 MeLi + CuCN
↓
[Me$_2$CuCN]Li$_2$

41

10 Myxalamide A

Problem

Compound **10**: structure with tBuS, OTBS groups

Reagents:
1. MCPBA, CH_2Cl_2, –78 °C, 10 min
2. $P(OMe)_3$, MeOH, 55 °C, 8 h, 80 % (over two steps)
3. TIPSOTf, NEt_3, CH_2Cl_2, 0 °C, 20 min, 94 %
4. 5 % H_2SO_4, THF, r. t., 7.5 h, 94 %

Product **11**

Hints

- The thioether is oxidized to the sulfoxide.
- The sulfoxide forms an equilibrium with an other compound, that is trapped with the phosphorus reagent.

Solution

Structure **11** with OTIPS and OH groups

Discussion

Oxidizing reagents like MCPBA can attack the molecule at two different positions: at the double bond and at the thioether functionality. The oxidation of the thioether is faster than the epoxidation of the double bond; therefore oxidation of **10** with one equivalent of MCPBA afforded a mixture of diastereomeric sulfoxides **44**. These allylic sulfoxides tend to undergo a [2,3]-sigmatropic rearrangement, also called the *Evans-Mislow* rearrangement.[19] The allylic sulfoxide structure **45** is strongly favored at equilibrium; therefore the S-O bond has to be cleaved to shift the equilibrium in favor of the sulfenate. Trivalent phosphorus reagents react with the sulfenate to cleave the S-O bond. Presumably they form phosphonium salt **43** via the pentacoordinated phophorane **42**. Finally aqueous work-up yields the allylic alcohol **46**.

$(CH_3O)_3P\begin{smallmatrix}SR\\OR'\end{smallmatrix}$

42

↓

$[(CH_3O)_3POR]^{\oplus}\ SR^{\ominus}$

43

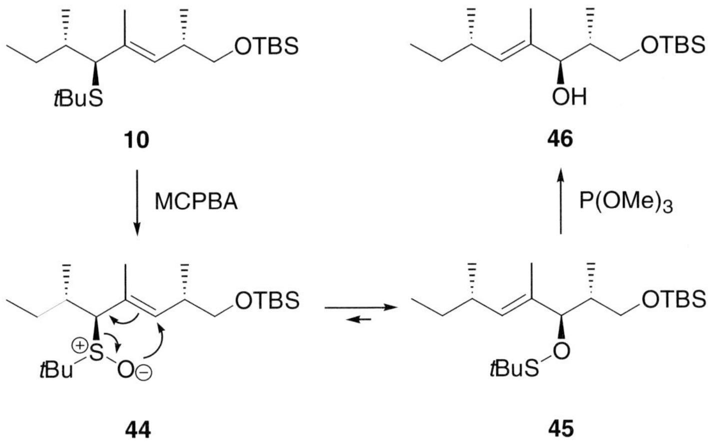

10   Myxalamide A

Next the generated secondary alcohol is protected as triisopropylsilylether using TIPSOTf. The greater bulk of the TIPS group makes it more stable than the TBS group towards acidic hydrolysis; therefore reaction with 5 % H_2SO_4 cleaves the TBS ether selectively to yield **11**.

Problem

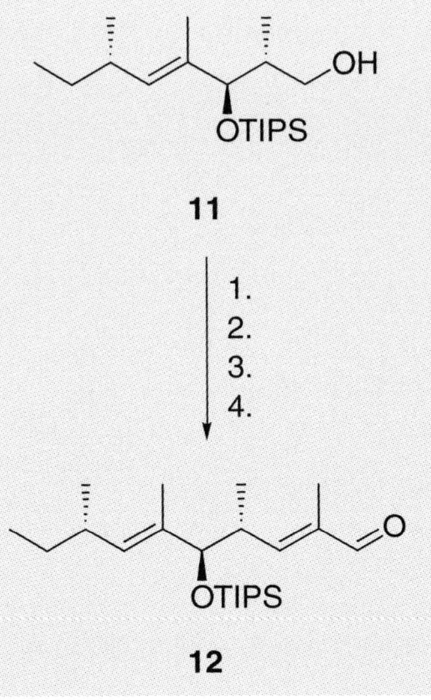

Hints

- A metal-catalyzed oxidation is performed.
- Carbon framework extension is achieved *via* a phosphorus reagent which is generated and used in the second step.
- Which two steps are necessary for the conversion of an ester into an aldehyde?

Solution

1. TPAP, NMO, 4 Å molecular sieves, CH_2Cl_2, r. t., 20 min, 96 %
2. Ethyl-2-bromoprionate, PBu_3, benzene, r. t., 7 h; then NEt_3, **47**, 73 °C, 12.5 h, 92 % (56 % **50** and 26 % Z isomer)
3. DIBAH, CH_2Cl_2, 0 °C, 25 min, 95 %
4. TPAP, NMO, 4 Å molecular sieves, CH_2Cl_2, r. t., 10 min, 100 %

Discussion

These four reactions extend the carbon framework forming an E-α,β-unsaturated aldehyde. This functionalization is achieved using a *Wittig* reaction.[20] The required aldehyde **47** is obtained by oxidation of **11** *via* a standard TPAP oxidation protocol.[12]

The stereoselectivity of the *Wittig* reaction depends strongly on both the structure of the ylide and the reaction conditions. The broadest generalization is that unstabilized ylides give predominantly the Z-alkene while stabilized ylides form mainly the E-alkene (see Chapter 9). In this case the stabilized ylide **49** is generated by reaction of ethyl-2-bromopropionate **48** with tributylphosphine followed by addition of NEt_3. This solution is added to a solution of the aldehyde and yields in a 2:1 (*E*/*Z*) ratio 56 % of **50** together with 26 % of the Z-isomer. The isomers are separable by flash chromatography.

Finally **50** is reduced to the corresponding allylic alcohol (1,2-reduction see above) followed by a further TPAP oxidation to get the corresponding α,β-unsaturated aldehyde **12**.

Problem

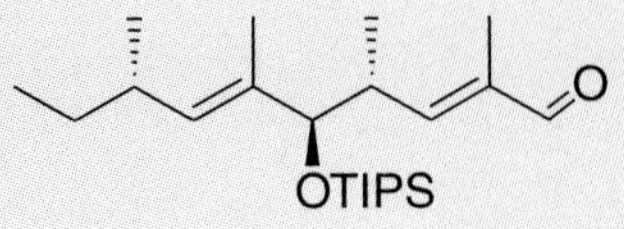

[Structure **12**: aldehyde with OTIPS]

1. (MeO)$_2$P(O)CHN$_2$, KOtBu, THF, −78 °C → r. t., 16 h, 91 %
2. TBAF, THF, r. t., 10 h, 90 %

13

Hints

- The phosphonate reacts in a *Wittig* type reaction.
- The formed diazoethene loses nitrogen and undergoes a hydride migration.
- The last step is a standard deprotection.

Solution

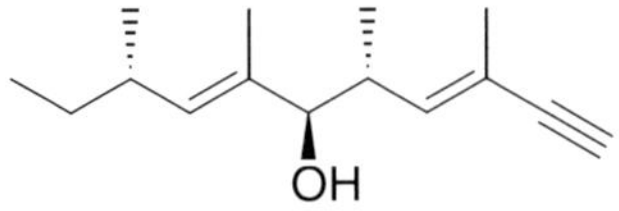

[Structure **13**: alkyne with OH]

Discussion

The phosphonate **51** – also called *Gilbert-Seyferth* phosphonate – is used for the conversion of an aldehyde into an alkyne.[21] One possible mechanistic explanation of this reaction is the following: The nucleophilic carbon of the anion **52** generated by reaction of **51** with KOtBu attacks the electrophilic carbonyl carbon of the aldehyde to form **54**. The driving force of the next step is the formation of a very stable P-O bond. *Via* a four-membered transition state diazoethene **55** is generated and undergoes by loss of molecular nitrogen a 1,2-hydride migration to form alkyne **56**.

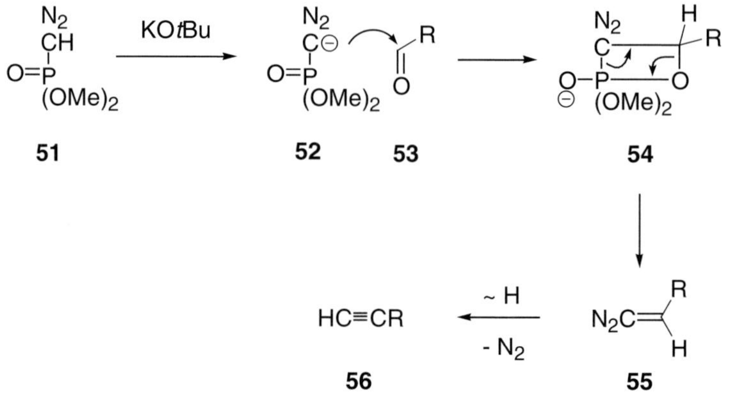

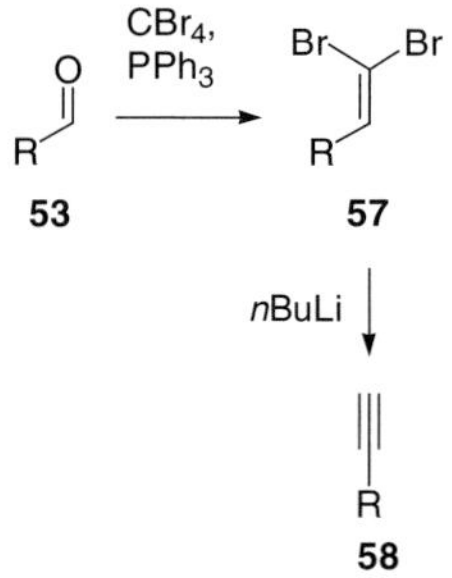

10 Myxalamide A

Another common method for the conversion of an aldehyde into an alkyne is the transformation of the aldehyde *via* a *Corey-Fuchs* reaction into a geminal vinyl dibromide followed by reaction with *n*butyllithium and aqueous work-up.[22]

Next the remaining silyl ether is desilylated to alcohol **13** by reaction with TBAF.

Problem

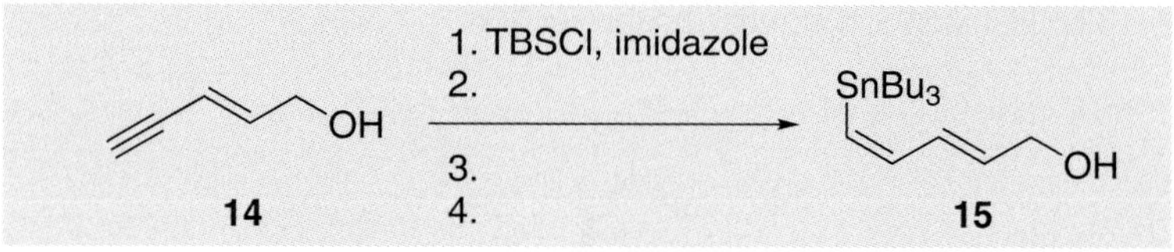

Hints

- After deprotonation, the stannane is introduced.
- Reduction of the triple bond is very uncommon and involves equimolar amounts of a zirconium reagent.

Solution

2. KHMDS, THF, –78 °C, 15 min; 0 °C, 45 min; Bu$_3$SnCl, r. t., 13 h
3. Cp$_2$ZrHCl, THF, r. t., 70 min
4. TBAF, THF, r. t., 30 min

50 % (over three steps)

Discussion

First the enyne alcohol is converted into the TBS ether *via* step 1. For introduction of the stannane moiety the alkyne is deprotonated by a base, in this case by KHMDS, followed by addition of Bu$_3$SnCl. *Cis* selective reduction of the resulting stannylalkyne is achieved *via* hydrozirconation using equimolar amounts of Cp$_2$ZrHCl.[23] This method is not very common and only applied to stannyl alkynes with ether functionalities. Limitation of this method is related to the

10 Myxalamide A

propensity of this hydride source to reduce aldehydes and ketones competitively with hydrozirconation. Catalytic reduction methods using hydrogen are not reported in literature for stannylalkynes. Finally desilylation using TBAF provides allyl alcohol **15**.

Problem

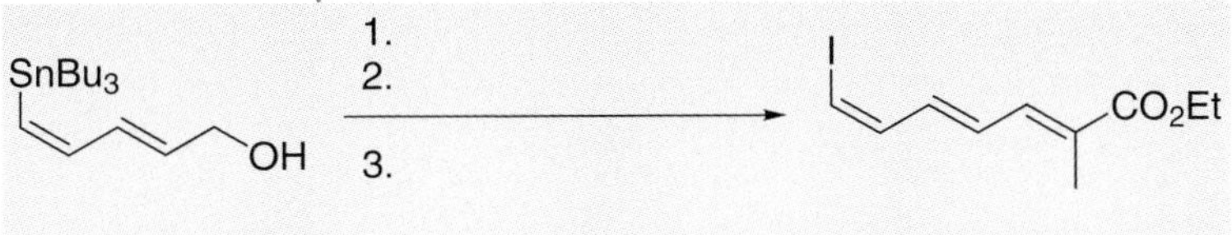

15 → 16 (via steps 1, 2, 3)

Hints

- Which reagent is necessary for the conversion of a vinyl stannane into a vinyl iodide?
- The next two reactions involve manganese and phosphorus.

Solution

1. I_2, CH_2Cl_2, r. t., 5 min
2. MnO_2, CH_2Cl_2, r. t., 24 h
3. Triethyl-2-phosphonopropionate, nBuLi, THF, 0 °C, 10 min, **58**, 20 min

52 % (over three steps)

Discussion

Vinyl stannanes can be very easily converted into the corresponding iodides by reaction with iodine. This reaction proceeds smoothly at room temperature generally with retention of the configuration of the alkene.[24] Without purification the allylic alcohol is oxidized to aldehyde **58** using MnO_2.[9] *Wittig* reaction with the stabilized phosphorus ylide **59** generated from triethyl-2-phosphonopropionate and $n$butyllithium yields the unsaturated ester **16** in 52 % over three steps.[20] However, if the *Wittig* reaction is employed first a 2:1 mixture of *E*- and *Z*-vinyl iodides is obtained.

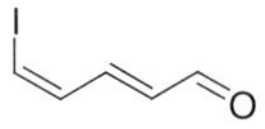

58

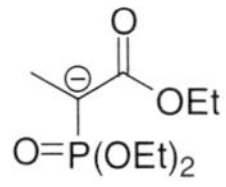

59

Problem

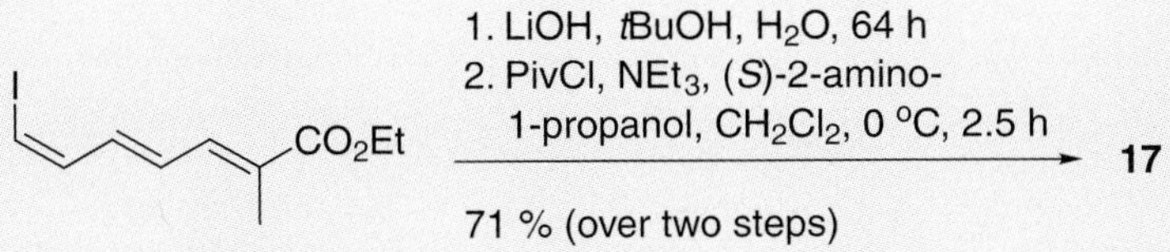

16 → **17**

1. LiOH, *t*BuOH, H_2O, 64 h
2. PivCl, NEt_3, (*S*)-2-amino-1-propanol, CH_2Cl_2, 0 °C, 2.5 h

71 % (over two steps)

10 Myxalamide A

Hints
- Saponification of the ester occurs.
- PivCl activates the resulting acid.
- Which functionality of (S)-2-amino-1-propanol is more basic and reacts with the activated acid?

Solution

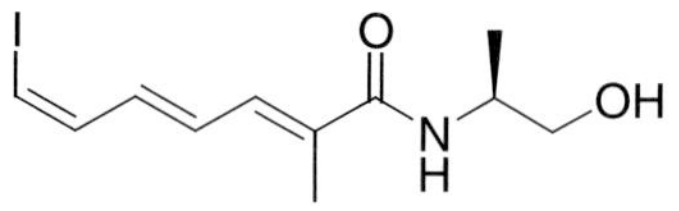

17

Discussion

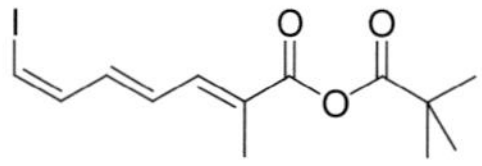

60

Hydrolysis of ester **16** proceeds smoothly. Activation of the resulting acid is achieved *via* conversion into the mixed anhydride **60**. The amino group of (S)-2-amino-1-propanol is more basic than the alcohol function; therefore there is no need for protection. It attacks the anhydride at the carbonyl carbon of the former acid because of the steric interaction with the pivaloyl group and gives amide **17** in 71 % yield.

Problem

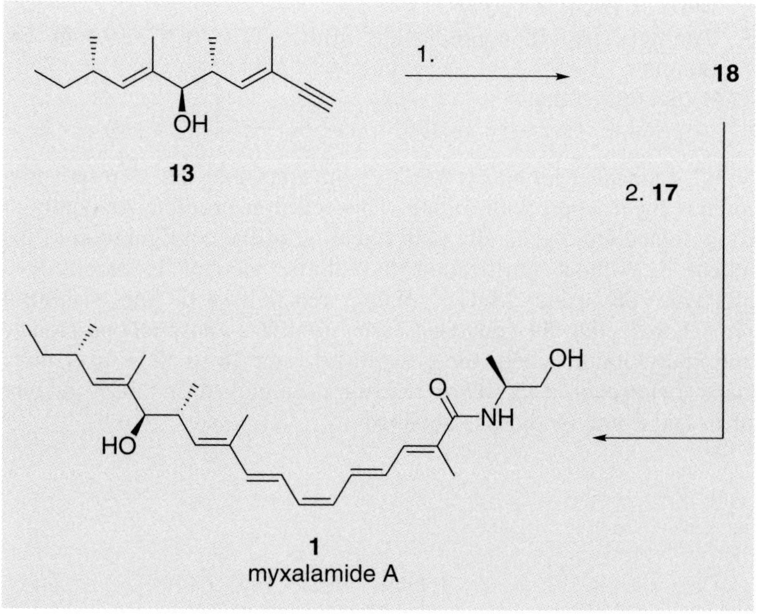

13 → 1. → **18** → 2. **17** → **1** myxalamide A

Hints
- **13** is transformed into a compound that can undergo a metal-catalyzed cross-coupling reaction.
- Which cross-coupling reaction is performed in the second step?

1. Catecholborane, benzene, r. t., 90 min, *N,N*-diethylaniline, 23 h

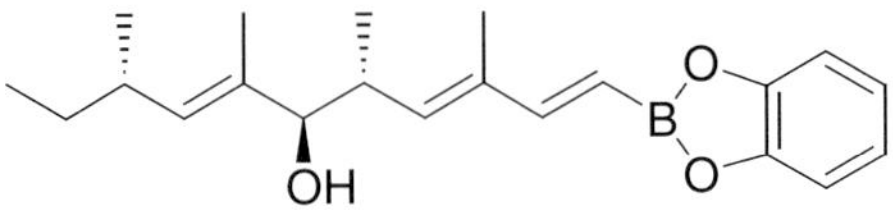

18

2. Pd(OAc)$_2$, TPPTS, *i*Pr$_2$NH, CH$_3$CN, H$_2$O, r. t., 3.25 h
44 % (over two steps)

Solution

Discussion

The key step of this total synthesis is a *Suzuki* reaction[25] forming myxalamide A. In organic synthesis, the *Suzuki* coupling is particularly useful as a method for the construction of conjugated dienes of high stereoisomeric purity. Using a palladium(0) catalyst and a base, the reaction accomplishes a cross-coupling of a 1-alkenylboron compound with an organic electrophile such as vinyl iodide **17**. First the 1-alkenylboron compound is generated *in situ* by reaction of **13** with catecholborane. The *E*-vinylborane **18** is formed exclusively. In the first step of the catalytic cycle a coordinatively unsaturated palladium(0) species **61** – which is formed in situ from Pd(OAc)$_2$ and TPPTS (triphenylphosphine-3,3',3''-trisulfonic acid trisodium salt) **62** – inserts into the alkenyl iodine bond of **17** to give **63**.

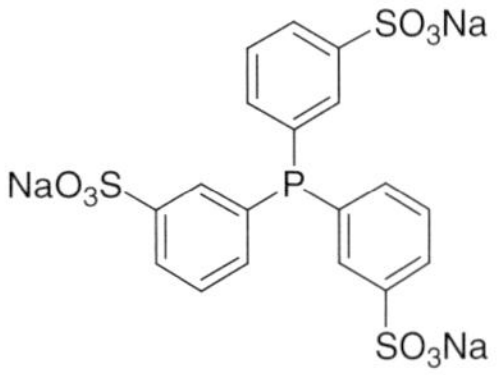

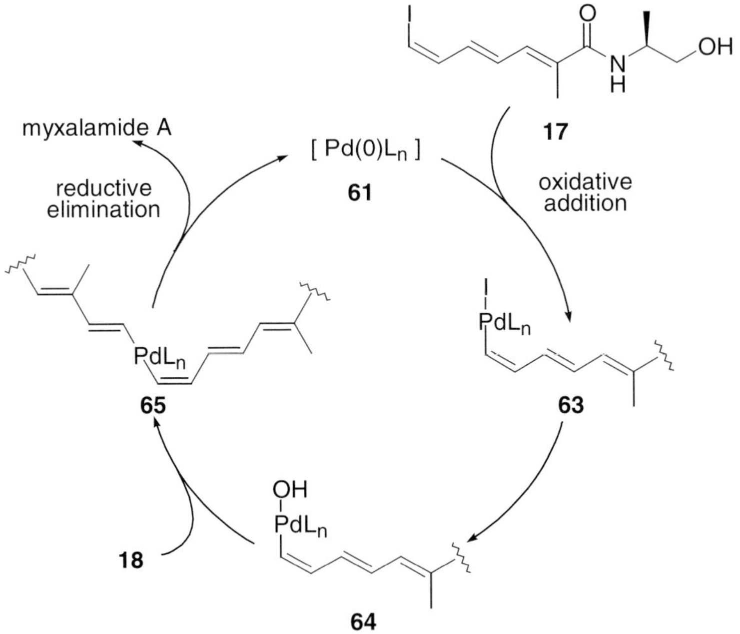

Next it is presumed that a metathetical displacement of the halide substituent in the palladium(II) complex **63** by hydroxide ion (the reaction is carried out in a mixture of water and acetonitrile) takes place to give the hydroxopalladium(II) complex **64**. The latter complex then reacts with the alkenylborane **18**, generating the diorganopalladium complex **65**. Finally reductive elimination of **65** furnishes the cross-coupling product myxalamide A **1** and regenerates the palladium(0)catalyst **61**.

10.4 Conclusion

The total synthesis of myxalamide A (**1**) was accomplished in 22 steps (longest linear sequence) and 4.9 % overall yield from bromide **5**. The completion of the synthesis not only demonstrates the utility of the aldol-*Claisen-Evans-Mislow* strategy but also emphasizes the usefulness of the *Suzuki* coupling for the preparation of polyene-containing natural products. Furthermore the discussed synthetic strategy is also useful for the preparation of other members of the polyene natural product family as well as related diastereoisomers for biological evaluation.

10.5 References

1 a) R. Jansen, G. Reifenstahl, K. Gerth, H. Reichenbach, G. Höfle, *Liebigs Ann. Chem.* **1983**, 1081-1095; b) R. Jansen, W. S. Sheldrick, G. Höfle, *Liebigs Ann. Chem.* **1984**, 78-84.
2 W. Trowitzsch-Kienast, E. Forche, V. Wray, H. Reichenbach, G. Junsmann, G. Höfle, *Liebigs Ann. Chem.* **1992**, 659-664.
3 M. B. Andrus, S. D. Lepore, *J. Am. Chem. Soc.* **1997**, *119*, 12159-12169.
4 C. M. Cox, D. A. Whiting, *J. Chem. Soc., Perkin Trans. 1*, **1991**, 1907-1911.
5 A. K. Mapp, C. H. Heathcock, *J. Org. Chem.* **1999**, *64*, 23-27.
6 A. R. Daniewski, W. Wojceichowska, *J. Org. Chem.* **1982**, *47*, 2993-2995.
7 J.-L. Luche, *J. Am. Chem. Soc.* **1978**, *100*, 2226-2227.
8 K. Krishnamurthy, H. C. Brown, *J. Org. Chem.* **1977**, *42*, 1197-1201.
9 R. E. Counsell, P. D. Klimstra, F. B. Cotton, *J. Org. Chem.* **1962**, *27*, 248-253.
10 D. B. Dess, J. C. Martin, *J. Am. Chem. Soc.* **1991**, *113*, 7277-7287.
11 M. Frigerio, M. Santagostino, S. Sputore, *J. Org. Chem.* **1999**, *64*, 4537-4538.
12 a) S. V. Ley, J. Norman, W. P. Griffith, S. P. Marsden, *Synthesis* **1994**, 639-666; b) W. P. Griffith, S. V. Ley, G. P.

Whitcombe, A. D. White, *J. Chem. Soc., Chem. Comm.* **1987**, 1625-1627.
13 A. J. Mancuso, D. Swern, *Synthesis* **1981**, 165-185.
14 D. A. Evans, *Aldrichimica Acta* **1982**, *15*, 23-32.
15 T. D. Penning, S. W. Djuric, R. A. Haack, V. J. Kalish, J. M. Miyashiro, B. W. Rowell, S. S. Yu, *Synth. Commun.* **1990**, *20*, 307-312.
16 R. E. Ireland, R. H. Mueller, A. K. Willard, *J. Am. Chem. Soc.* **1976**, *98*, 2868-2877.
17 F. W. Schuler, G. W. Murphy, *J. Am. Chem. Soc.* **1950**, *72*, 3155-3159.
18 B. H. Lipshutz, *Synthesis* **1987**, 325-341.
19 a) R. Tang, K. Mislow, *J. Am. Chem. Soc.* **1970**, *92*, 2100-2104; b) D. A. Evans, G. C. Andrews, *J. Am. Chem. Soc.* **1972**, *94*, 3672-3674.
20 B. E. Maryanoff, A. B. Reitz, *Chem. Rev.* **1989**, *89*, 863-927.
21 a) J. C. Gilbert, U. Weerasooriya, *J. Org. Chem.* **1979**, *44*, 4997-4998; b) D. Seyferth, R. S. Marmor, P. Hilbert, *J. Org. Chem.* **1971**, *36*, 1379-1386; c) D. G. Brown, E. J. Velthuisen, J. R. Commerford, R. G. Brisbois, T. R. Hoye, *J. Org. Chem.* **1996**, *61*, 2540-2541.
22 a) E. J. Corey, P. L. Fuchs, *Tetrahedron Lett.* **1972**, *14*, 3769-3772; b) G. J. Hollingworth, J. B. Sweeney, *Synlett* **1993**, 463-465; c) H. Monti, P. Charles, *Synlett* **1995**, 193-194.
23 B. H. Lipshutz, W. Hagen, *Tetrahedron Lett.* **1992**, *33*, 5865-5868.
24 B. Dominguez, B. Iglesias, A. R. de Lera, *Tetrahedron* **1999**, *55*, 15071-15098.
25 N. Miyaura, A. Suzuki, *Chem. Rev.* **1995**, *95*, 2457-2483.

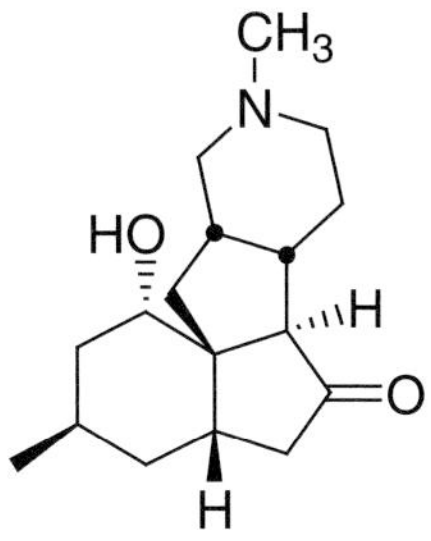

11

(+)-Paniculatine (Sha 1999)

11.1 Introduction

Paniculatine (**1**) was isolated from club moss paniculatum by *Castillo* et al. in the mid of the 1970s and belongs together with magellaninone (**2**) and magellanine (**3**) to a subclass of *Lycopodium* alkaloids.[1] Their tetracyclic framework share in common a diquinane core, that is fused in entirely different ways to a cyclohexanol or cyclohexenone and to a piperidine ring. The structures of **1** and **2** are secured by X-ray crystallography, while absolute configuration was assigned by optical methods. In 1993 and 1994, total syntheses of **2** and **3** were accomplished independently by *Overman* and *Paquette*.[2] This Problem deals with the first total synthesis of **1** finished by *Sha* and co-workers *via* a α-carbonyl radical cyclization as the main step in 1999.[3] The main alkaloid of the $C_{16}N$-type of the *Lycopodium* alkaloids is lycopodin (**4**). It was isolated over 100 years ago, is toxic and shows a curare-like paralysing effect. In China different alkaloids of this type are traditionally used to medicate skin diseases. Although no special features of **1** have so far been published, other *Lycopodium* alkaloids are reported to be potent acetylcholinesterase inhibitors or show promising results in the treatment of Alzheimer's disease.[4]

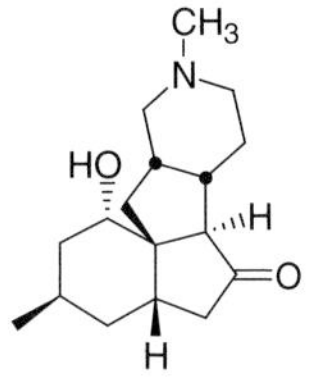

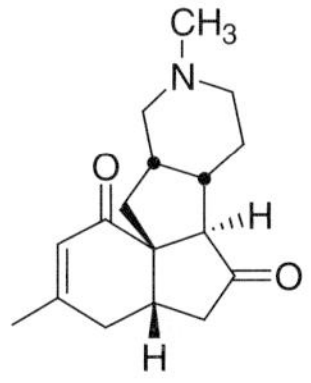

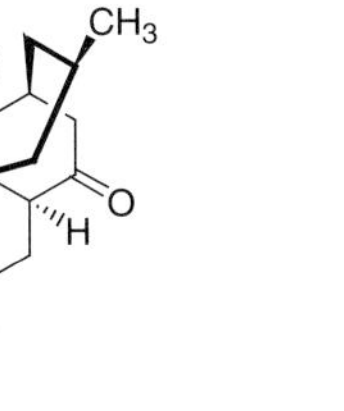

11.2 Overview

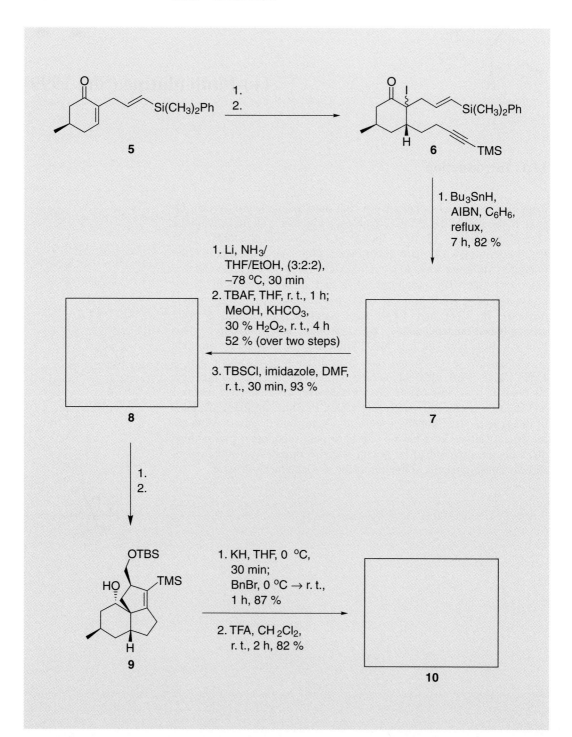

11 (+)-Paniculatine

11.3 Synthesis

Problem

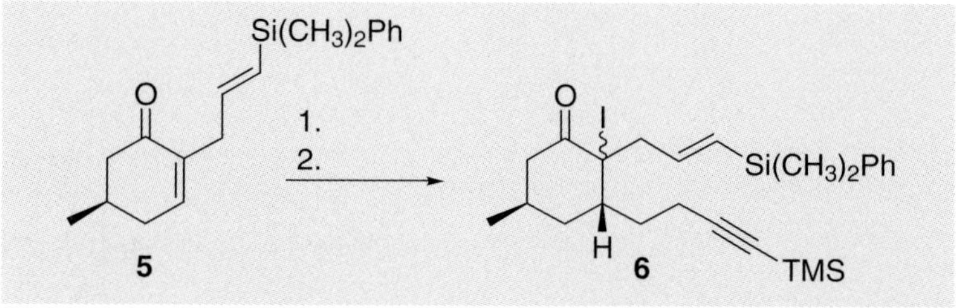

[Scheme: Enone **5** → (1., 2.) → ketone **6** with Si(CH₃)₂Ph allyl group, iodide, and alkynyl-TMS substituent]

Hints

- A *Michael* addition is the preliminary step.
- The reaction proceeds by means of a *Grignard* reagent and a co-metal, which is responsible for exclusive 1,4-addition. Which further additive accelerates such transformation?
- Finally, the iodide is introduced via a iodonium-species.

Solution

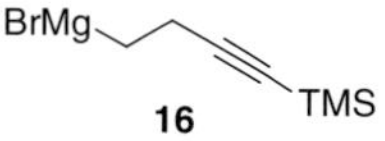

BrMg–CH₂–C≡C–TMS (**16**)

1. **16**, CuI, THF, −78 °C, 30 min; TMSCl, −78 °C → r. t., 15 h
2. NaI, MCPBA, THF, 0 °C → r. t., 30 min; **18**, 0 °C, 30 min

67 % (over two steps)

Discussion

First chiral enantiopure enone **5**, which is prepared from (*R*)-5-methylcyclohexenone, is converted into a trimethylsilyl (TMS) enol ether **18**. This occurs *via* a 1,4-conjugate addition of an organocopper reagent **17** to the α,β-unsaturated ketone system.[5] This cuprate is formed *in situ* by transmetallating the previously synthesized *Grignard* reagent **16** mediated by copper(I) iodide. The exact ratio of Mg to Cu in the active spezies is so far unknown.

To accelerate the addition sequence TMSCl is added, leading to the formation of the TMS enolether **18** at the end.

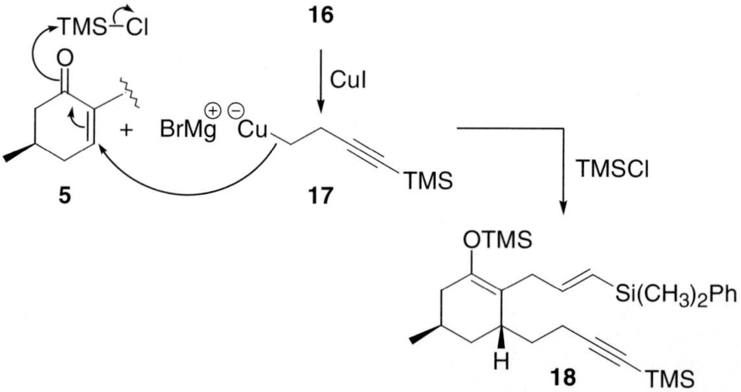

[Mechanism scheme showing enone **5** + cuprate **17** (BrMg⊕ Cu⊖ with alkynyl-TMS), with CuI and TMSCl, leading to TMS enol ether **18** with OTMS, Si(CH₃)₂Ph allyl, and alkynyl-TMS groups]

Overall the attack occurs from the sterically less hindered α-face of the molecule, giving the *R*-configuration at this carbon center.

The positive effect of adding copper salts to *Grignard* reactions was first published by *Normant* in 1971.[6] Thus, today this reagent combination is known as *Normant* cuprate. Furthermore two other copper-metal combinations are common for 1,4-conjugate additions. *Gilman* uses organolithium compounds and *Knochel* established organozinc reagents together with copper salts. Without the copper salt the *Grignard* reaction produces miscellaneous mixtures of 1,2- and 1,4-addition products.

Regarding the mechanistic effect of TMSCl, there has been no direct evidence until today. It is suggested that TMSCl activates the enone by complexing the enone oxygen as *Lewis* acid (**19**). Latest observations on the basis of kinetic isotope effects show that TMSCl accelerates the reaction by interacting with an intermediate π-complex **20**.[7]

After florisil (magnesium-silicate) filtration and concentration, crude **18** was treated with a THF solution of a 1:1 mixture of sodium iodide and *meta*-chloroperoxybenzoic acid (MCPBA) yielding in 67 % (from **5**) the α-iodo ketones **6**.[8] The ratio of diastereomers in this mixture was not described further. Mechanistically MCPBA oxidizes the iodide ion to an iodoniumion-species, which reacts with the double bond, generating intermediate **21**. After TMS is removed the tricyclic iodonium ion collapses to desired **6**. In contrast to this transformation the researchers observed no useful yields by direct treatment of the silyl enol ether with molecular iodine.

Problem

Hints

- In this radical cyclization a five-membered ring is favored over a six-membered one.
- First an iodine atom is abstracted.
- The resulting *exo*-radical reacts to build a second ring.

Solution

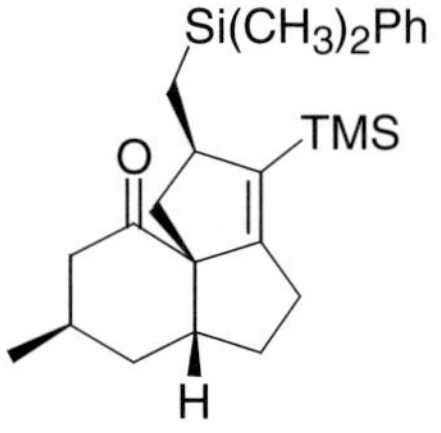

Discussion

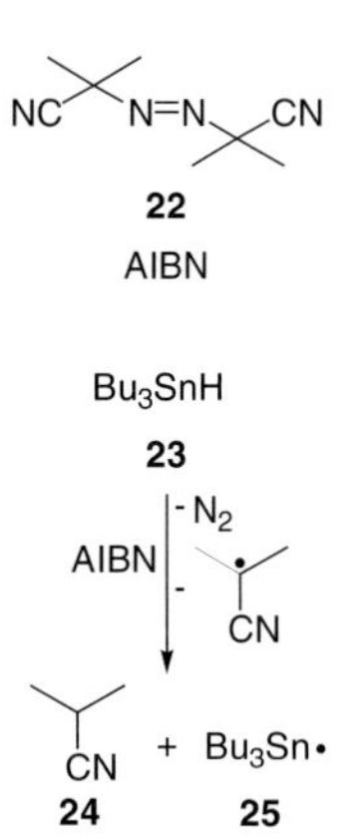

Treatment of **6** with a benzene solution of tributyl tin hydride **23** and azobis*iso*butyronitrile (AIBN) **22**, introduced slowly with a syringe pump at reflux, provided the tricyclic ketone **7** in 82 % yield. Overall only traces of AIBN as radical initiator and not more than a slight excess of Bu$_3$SnH are needed.

This tandem radical cyclization[9] starts with formation of Bu$_3$Sn· **25** initiated by thermal decomposition of unstable AIBN. Next **25** abstracts an iodine atom from the weak C-I bond resulting in alkyl-radical **26**. After *5-exo-dig* cyclization giving **27**, *5-exo-trig* cyclization occurs, affording **7** after hydrogen abstraction from tributyltin hydride. Formed Bu$_3$Sn· then reenters the catalytic cycle. Both cyclizations take place from the α-face of the molecule, according to the stereochemistry of the alkene/alkyne bearing side chains. To use peroxide as a radical starter in this transformation would lead to side reactions as result of uncontrolled hydrogen abstraction through the highly active RO· radical.

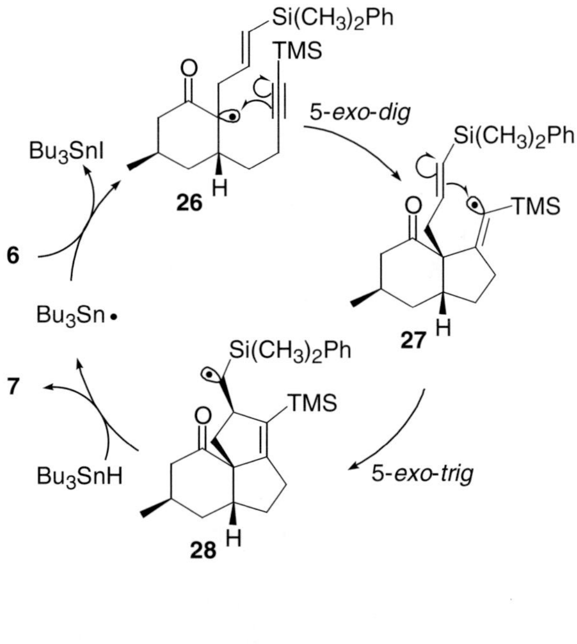

The abbreviations to describe the ring forming processes were established by *Baldwin* in 1976.[10]
- The prefixed number specifies the size of the generated ring.
- The words *exo* or *endo* are used to picture if the breaking bond is *exo*- or *endo*-cyclic to the smallest ring so formed.
- The suffixes *tet*, *trig* or *dig* indicate the geometry of the carbon atom undergoing the ring closure reaction [*tet* = tetragonal (sp^3), *trig* = trigonal (sp^2), *dig* = digonal (sp)].

In this publication the author describes the phenomenon that most times the thermodynamically less stable product (see **29**) of the two possible rings (e.g. 5-*exo* and 6-*endo*) is formed. Today looking at the 5-*exo* cyclization it is known that, although the generated primary radical is less stable than a secondary one, stereoelectronic effects favor reaction to the kinetically controlled product. According to MO-calculations, for a successful cyclization, an angle of 70° of the incoming radical to the plane of the alkene-/alkyne-bond is necessary.[11]

5-*exo-trig*
29

6-*endo-trig*
30

Problem

1. Li, NH$_3$/ THF/ EtOH (3:2:2), −78 °C, 30 min
2. TBAF, THF, r. t., 1 h; MeOH, KHCO$_3$, 30 % H$_2$O$_2$, r. t., 4 h
   52 % (over two steps)
3. TBSCl, imidazole, DMF, r. t., 30 min, 93 %

7 → **8**

Hints

- The phenyldimethylsilyl group is used as surrogate for an alcohol functionality.
- Under *Birch* reaction conditions not only reduction of the carbonyl occurs.
- With TBAF a strong Si-F bond is formed.
- Using TBSCl one alcohol is selectively protected.

Solution

8

Discussion

According to the method of *Taber*[12] compound **7** is first reduced with lithium in a liquid ammonia/THF/EtOH mixture at −78 °C. This leads to **33** with concomitant formation of the alcohol at the six-membered ring and the 1,4-cyclohexadiene from the aromatic ring.

Dissolving elementary lithium or sodium in liquid ammonia gives an intense blue solution, indicating solvated electrons. In a *Birch* reaction these conditions are used for reduction. With time the blue color fades, the electrons reacting slowly with ammonia to form NH_2^- and elemental hydrogen. For this reason a better electron acceptor will get reduced in preference. In this example the electron is transferred *via* SET (single electron transfer) into the LUMO of the benzene ring. The radical anion **32** is very basic and picks up a proton from the ethanol that is present in the reaction mixture. Next a further electron and finally a proton are received, leading to the 1,4-cyclohexadienyl species **33**.

Then **33** is exposed in a one-pot reaction to tetrabutylammonium fluoride (TBAF), forming **34**. Driving force of this reaction is the higher bond energy of an Si-F bond (467 KJ/mol) compared with that of the Si-C bond (251 KJ/mol). The following oxidation with aqueous hydrogen peroxide in the presence of potassium bicarbonate gives exclusively the diol **35**.

The reaction sequence is terminated by protection of the primary alcohol as TBS ether using the standard protocol TBSCl/imidazole.

11 (+)-Paniculatine

Problem

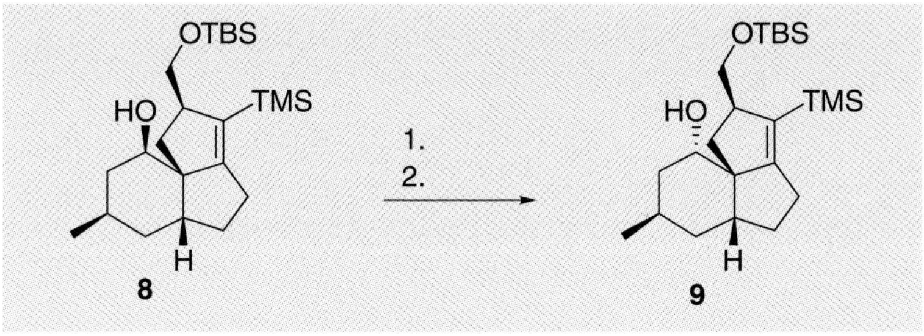

Hints

- Which reaction sequence inverts a stereogenic center?
- Oxidation of the free alcohol is followed by selective reduction.

Solution

1. PCC, CH$_2$Cl$_2$, r. t., 2 h, 88 %
2. L-Selectride® **38**, THF, 0 °C

Discussion

At this stage in order to invert the β-OH to the α-configuration, **8** is first oxidized within 2 h with pyridinium chlorochromate (PCC) at room temperature. PCC is a versatile oxidizing agent for many functional groups and is often used to convert primary or especially secondary alcohols to aldehydes or ketones. Overoxidations are rare. But chromium makes it a suspected cancer agent and thus it should be handled carefully. With compounds bearing acid-sensitive groups milder pyridinium dichromate (PDC) should be used, which is advantageous over PCC in this point.[13]

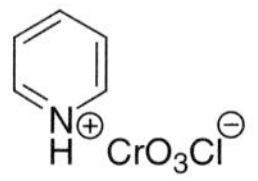

Then lithium tris-sec-butyl-borohydride (L-Selectride®) **38** is added and the ketone is selectively transformed to get **9**.[14]

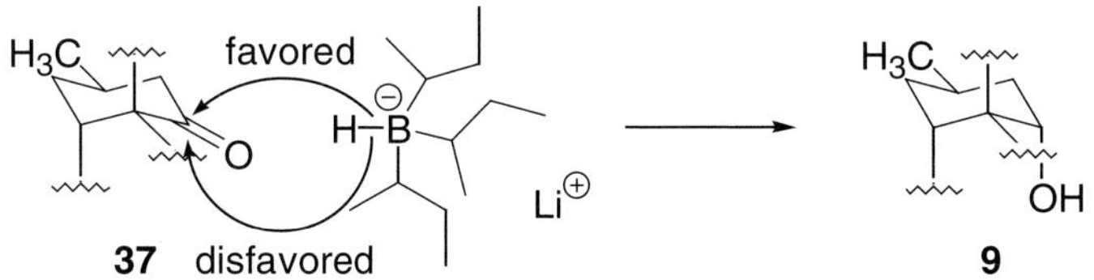

Because of its bulky alkyl substituents this reagent is used as a sterically demanding hydride donor. In the case of this rigid tricyclic ring system the hydride consequently occurs from the sterically less hindered equatorial position of the cyclohexenone ring. That results in an axial position of the OH-group (see conformation of **9a**). This is an example of steric approach control. Reaction with e. g. LiAlH$_4$ would generate the thermodynamically more stable equatorial OH-group.

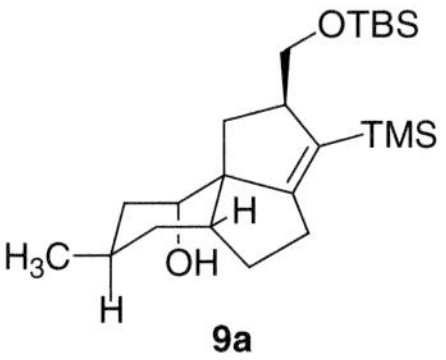

Problem

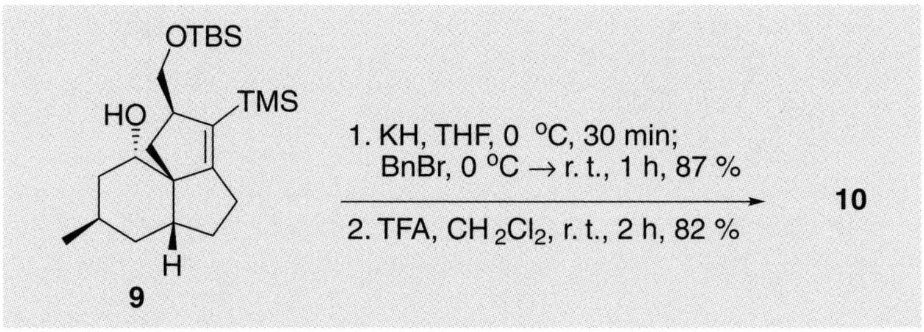

1. KH, THF, 0 °C, 30 min;
BnBr, 0 °C → r.t., 1 h, 87 %

2. TFA, CH$_2$Cl$_2$, r.t., 2 h, 82 %

9 → **10**

Hints

- First an ether is generated.
- Which functionalities are touched under acidic conditions?
- The TBS ether remains under these conditions.

Solution Discussion

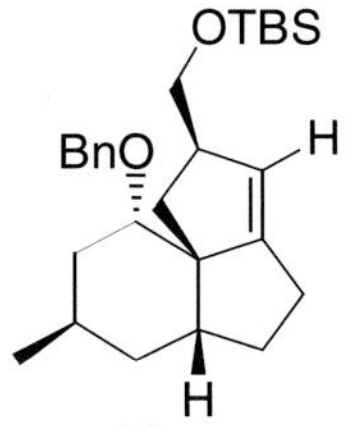

10

Using potassium hydride and benzyl bromide in THF, protection of the secondary alcohol is achieved affording the benzyl ether. *Sha* avoided usage of another TBS group again, because it was essential to deprotect both alcohols at different stages of the synthesis.

Under acidic conditions of trifluoroacetic acid (TFA) in CH$_2$Cl$_2$ at room temperature only the vinylic trimethylsilyl group is removed. Replacement of a C-Si bond with a C-H bond is known as protodesilylation. The TBS ether remains under these conditions.[15]

Problem

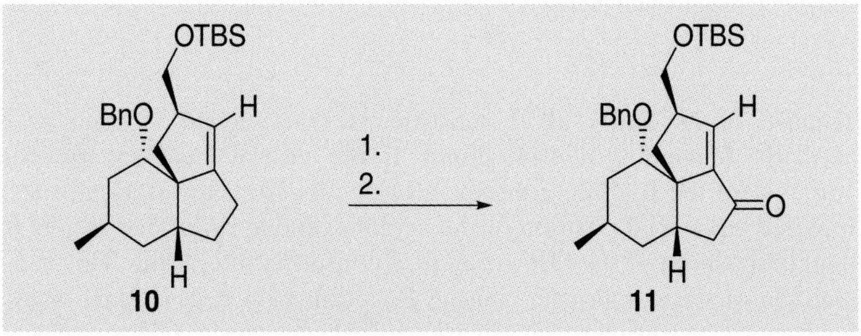

10 → **11**

- Which possibilities do you know to introduce an oxygen atom in the allylic position?
- The element carrying the oxygen is a member of the chalcogens.
- The generated alcohol is oxidized in the second step.

Hints

1. SeO$_2$, dioxane, reflux, 4 h
2. (COCl)$_2$, DMSO, CH$_2$Cl$_2$, −78 °C, 30 min
62 % (over two steps)

Solution

Allylic oxidation is carried out by addition of one equivalent of selenium dioxide. First SeO$_2$ will react with the alkene in a [4 + 2] cycloaddition reminiscent of the ene reaction. The initial product is an allylic selenic acid **40**, which undergoes – like an allylic sulfoxide – allylic rearrangement to give an unstable intermediate, which decomposes rapidly to the allylic alcohol **42**.[16]

To avoid large quantities of selenium(II) byproducts at the end of the reaction *Sharpless* and co-workers suggested the use of only catalytic amounts of SeO$_2$ (1–2 mol%) adding *tert*-butylhydroperoxide to reoxidize the Se(II) species.[17]

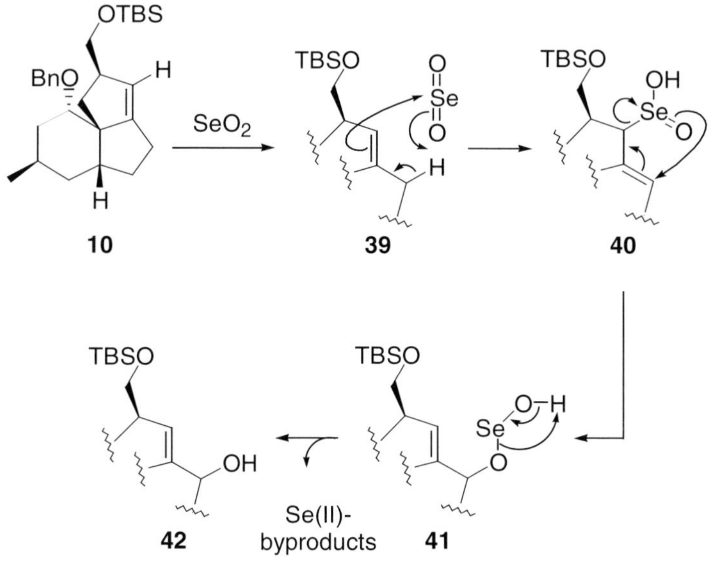

Discussion

The mixture of alcohols generated in this case was then admitted to a *Swern* reaction to get the enone **11**. (For a detailed mechanism see Chapter 2.)

188   *11   (+)-Paniculatine*

Problem

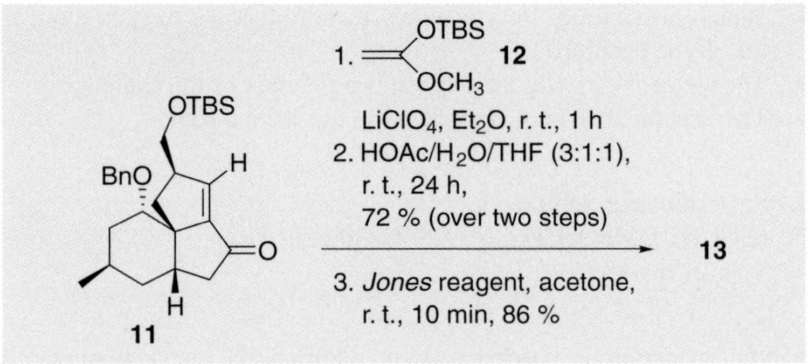

Hints

- In the first step another conjugated 1,4-addition takes place. What is the job of LiClO$_4$?
- Which side of the molecule is the less hindered one?
- Which compounds are hidden behind the *Jones* reagent?
- At the end one acid- and one ester-functionality is present.

Solution

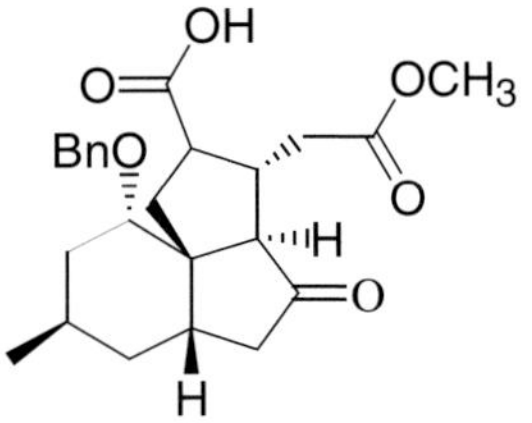

Discussion

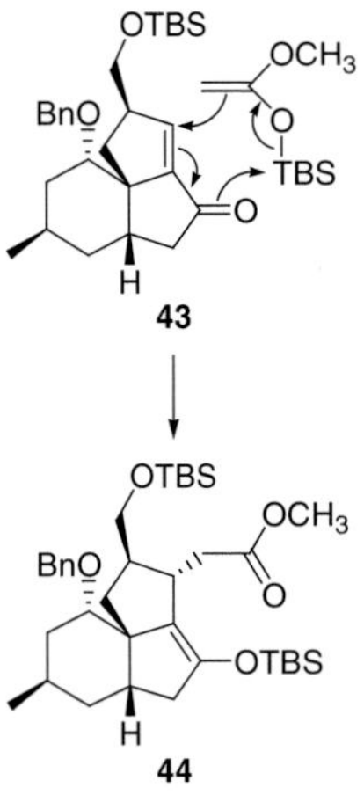

The ester side chain is introduced *via* 1,4-conjugate addition of *tert*-butyldimethylsilyl methyl ketene acetal (**12**) to the α,β-unsaturated ketone. Here **11** is stirred in a 2.0 M solution of lithium perchlorate (LiClO$_4$) in diethyl ether together with **12** for 1 h at room temperature. A large excess of both reagents is necessary (25 equiv. LiClO$_4$ and 10 equiv. **12**).[18]

The advantage of the LiClO$_4$/ether system is its equal or greater polarity compared to that of water. Therefore it promotes solvolysis of **12**. Furthermore the *Lewis* acidity of the lithium cation activates the ketone.[19] As expected, the addition occurred from the sterically less hindered α-face of the molecule. Previously these transformations could be carried out only thermally or under high pressure with the aid of *Lewis* acid catalysis, e. g. TiCl$_4$ or Ti(O*i*Pr)$_4$. *Grieco* also points out that it is very important to keep the exact concentrations of before diluted substrate and added LiClO$_4$ in ether. In disrespect the rate of the formation of the 1,2 addition product is enhanced over that of the 1,4 adduct.

In the second step the TMS enol ether collapses in a 3:1:1 mixture of acetic acid, water and THF to the corresponding ketone. Also the TBS protecting group is removed leading in 72 % yield (from **11**) to the primary alcohol.[15]

The sequence is termintated by oxidation of the free alcohol to the carboxylic acid **13** using *Jones* reagent, namely sodium dichromate ($Na_2Cr_2O_7$) in diluted sulfuric acid. At this stage the stereochemistry of the carboxylic acid bearing side chain was not determined. The reaction is usually run in acetone. During the oxidation process Cr(VI) (orange) is reduced to Cr(III) (green) so that the progress of the reaction is easy to follow by the change of color. Although toxic chromium byproducts are obtained this method is widely used because of its good reaction rates.

Problem

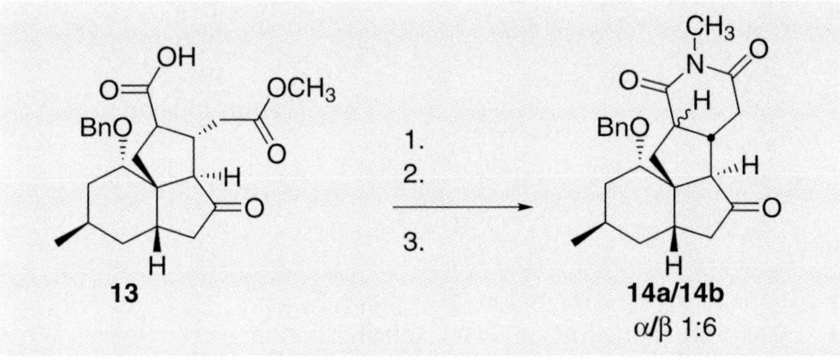

Hints

- Activation of the carboxyl group of **13** is followed by an amide synthesis.
- What pH value is necessary to form the imide?

Solution

1. $(COCl)_2$, 0 °C → r. t., 2 h
2. CH_3NH_2, Et_2O, −20 °C → 0 °C, 1 h
86 % (over two steps)
1. K_2CO_3, xylene, reflux, 15 h, 76 % (α/β 1:6)

Discussion

Activation of **13** with oxalyl chloride, followed by addition of methylamine, afforded amide **45** in 86 % yield. Except for a concentration step after the first reaction to remove excess oxalyl chloride, both reactions were performed in one pot.

11 (+)-Paniculatine

Heating developed **45** in xylene in the presence of potassium carbonate furnishes after cyclization a mixture of imides **14a/14b**. The two diastereomers were formed in a ratio of 1:6 (α/β) in 75 % overall yield. They were separated by silica gel chromatography (EtOAc/hexane 1:3) resulting in the isolation of 65 % of the major isomer. Probably **13** already showed this diastereomeric ratio.

Problem

Hints

- In the first step a reduction takes place.
- After the reduction an extra transformation is necessary. What could be the reason for that?
- Finally the benzyl group is removed.

Solution

1. LiAlH$_4$, THF, reflux, 2 h
2. Oxalyl chloride, DMSO, NEt$_3$, −78 °C, 30 min, 70 % (over two steps)
3. H$_2$, 10 % Pd/C, MeOH, r. t., 3 h, 65 %

Discussion

Reduction of the imide proceeds with lithium aluminium hydride (LiAlH$_4$) in refluxing THF. Even the ketone moiety in the cyclopentanone-ring is reduced generating **50**, which causes subsequent *Swern* oxidation of this secondary alcohol to receive the ketone functionality again.

Using LiAlH$_4$ is a widespread method to generate piperidine rings from lactams. Mechanistically the tetrahedral intermediate **46** is formed first, and this then collapses to the iminium ion **47**. The iminium ion is of course more electrophilic than the carbonyl group. Thus after a hydride transfer to the iminium ion and repetition of the reaction sequence again, the tertiary amine **49** is formed. As a matter of principle parallel to this reduction sequence the ketone in ring B is reduced to form **50**.

Finally the benzyl group is removed by hydrogenolysis using 10 % palladium on charcoal as catalyst. The reaction proceeds in MeOH at atmospheric pressure and is completed after 3 h reaction time. To remove the catalyst the reaction mixture is filtered through celite. Finishing with basic aqueous work-up and silica gel chromatography (MeOH/CHCl$_3$ 1:10), enantiomerically pure (+)-paniculatine (**15**) is afforded in 65 % yield.

11.4 Conclusion

The above described total synthesis shows the application of the α-carbonyl radical-initiated tandem cyclization reaction for the first generation of (+)-paniculatine. With this method starting from 2-substituted-5-(*R*)-cyclohexenone **5** Sha and co-workers obtained **15** (=**1**) in 21 steps. Furthermore the HBr-salt of this alkaloid was prepared and subjected to a single X-ray analysis, which unambiguously confirmed the structure and stereochemistry of this synthetic (+)-paniculatine.

11.5 References

1 a) M. Castillo, G. Morales, L. A. Loyola, I. Singh, C. Calvo, H. L. Holland, D. B. McLean, *Can. J. Chem.* **1975**, *53*, 2513-2514; b) *Can. J. Chem.* **1976**, *54*, 2900-2908.
2 a) G. C. Hirst, T. O. Johnson, L. E. Overman, *J. Am. Chem. Soc.* **1993**, *115*, 2992-2993; b) J. P. Williams, D. R. StLaurent, D. Friedrich, E. Pinard, B. A. Rhoden, L. A. Paquette, *J. Am. Chem. Soc.* **1994**, *116*, 4689-4696.
3 C.-K. Sha, F.-K. Lee, C.-J. Chang, *J. Am. Chem. Soc.* **1999**, *121*, 9875-9876.
4 W. A. Ayer, *J. Nat. Prod. Rep.* **1991**, *8*, 455-463.
5 M. Schlosser, *Organometallics in Synthesis*, Wiley, Chichester, **1994**.
6 J. F. Normant, M. Bourgain, *Tetrahedron Lett.* **1971**, *27*, 2583-2586.

7 D. E. Frantz, D. A. Singleton, *J. Am. Chem. Soc.* **2000**, *122*, 3288-3295.
8 C.-K. Sha, J.-J. Young, T.-S. Jean, *J. Org. Chem* **1987**, *52*, 3919-3920.
9 C.-K. Sha, T.-S. Jean, D.-C. Wang, *Tetrahedron Lett.* **1990**, *31*, 3745-3748.
10 J. E. Baldwin, *J. Chem. Soc., Chem. Comm.* **1976**, 734-736.
11 D. Crich, S. M. Fortt, *Tetrahedron Lett.* **1987**, *28*, 2895-2898.
12 D. F. Taber, L.Yet, R. S. Bhamidipati, *Tetrahedron Lett.* **1995**, *36*, 351-354.
13 S. D. Burke, R. L. Danheiser, *Handbook of Reagents for Organic Synthesis*, John Wiley & Sons, New York, **1999**.
14 W. G. Douben, G. J. Fonken, D. S. Noyce, *J. Am. Chem. Soc.* **1956**, *78*, 2579-2582.
15 a) T. W. Greene, P. G. W. Wuts, *Protective Groups in Organic Synthesis*, 3rd Ed., John Wiley & Sons, New York, **1999**; b) P. J. Kocienski, *Protecting Groups*, Georg Thieme Verlag, Stuttgart, **1994**.
16 K. B. Sharpless, R. F. Lauer, *J. Am. Chem. Soc.* **1972**, *94*, 7154-7155.
17 M. A. Umbreit, K. B. Sharpless, *J. Am. Chem. Soc.* **1977**, *99*, 5526-5528.
18 P. A. Grieco, R. J. Cooke, K. J. Henry, J.-M. VanderRoest, *Tetrahedron Lett.* **1991**, *32*, 4665-4668.
19 H. Waldmann, *Angew. Chem.* **1991**, *103*, 1335-1337; *Angew. Chem. Int. Ed. Engl.* **1991**, *30*, 1306-1309.

12

(+)-Polyoxin J (Gosh 1999)

12.1 Introduction

The polyoxins are an important group of peptidyl nucleosides isolated from the fermentation broth of *Streptomyces cacoi* var *asoensis* by *Isono* et al. in 1969.[1] About 15 compounds showing a closely related structure have been identified and designated with alphabetical letters. The characteristic skeleton of this class of antibiotics includes a peptide linkage between the nucleoside α-amino acid and polyhydroxynorvaline. As main difference among the polyoxins the nucleoside portion can bear different pyrimidine bases. For example polyoxin J is composed of thymine polyoxin C (**1**) and 5-*O*-carbamoyl polyoxamic acid (**2**). The polyoxins are attracting increasing interest as antifungal compounds since they exhibit potent activity against phytopathogenic fungi while being non-toxic to bacteria, plants, or animals.[2] These biological effects are closely related to the inhibition of the enzyme chitin synthetase and therefore the biosynthesis of chitin, an essential component of the fungal cell wall.[3] For example the polyoxins show high inhibitory potencies against isolated chitin synthetase from the human pathogen *Candida albicans*; however, against whole cells the polyoxins are inactive. Moreover, attention has been given to the polyoxins and the structurally related natural products nikkomycins[4] and neopolyoxins[5] because of their inhibitory effects on opportunistic fungal infection by *Candida albicans* in immunocompromised hosts, such as AIDS victims and organ transplant patients.[6]

Since the first synthesis of polyoxin J by *Kuzuhara* in 1973 several other total syntheses have been reported.[7] This problem is based on the stereoselective and convergent total synthesis described by *Gosh* and co-workers in 1999.[8]

12.2 Overview

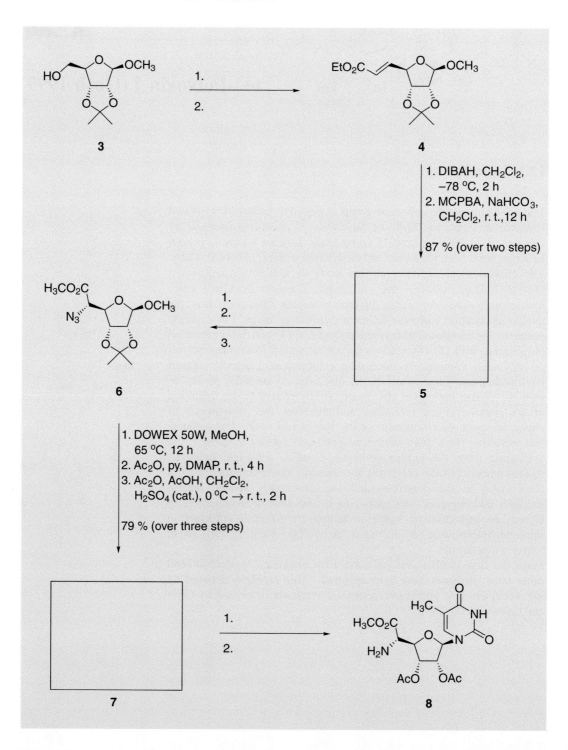

12 (+)-Polyoxin J

12.3 Synthesis

Problem

[Reaction scheme: compound **3** (HO-CH2 furanose with OCH3 and isopropylidene acetal) → compound **4** (EtO2C-CH=CH- furanose with OCH3 and isopropylidene acetal); conditions 1. and 2.]

Hints

- The first step is a *Swern* oxidation.
- Finally a phosphonoacetate is used.

Solution

1. DMSO, (COCl)$_2$, CH$_2$Cl$_2$, −60 → 50 °C, 2 h then Et$_3$N
2. NaH, (EtO)$_2$P(O)CH$_2$CO$_2$Et, THF, 0 °C → r. t., 30 min
72 % (over two steps)

Discussion

The protected methyl glycoside **3** is converted to the corresponding aldehyde by *Swern* oxidation using oxalyl chloride activated DMSO. Further reaction with triethyl phosphonoacetate and sodium hydride – known as the *Horner-Wadsworth-Emmons* reaction – provides selectively the *trans* α,β-unsaturated ester **4** in 72 % yield. This valuable alternative to the *Wittig* olefination protocol uses phosphonate esters as substrates which are readily available from alkyl halides and trialkyl phosphites *via* the *Arbuzov* rearrangement.[9] Reaction of the phosphonate with a suitable base gives the corresponding carbanion **15** which is more nucleophilic than the related phosphorane. **15** reacts with the carbonyl group of the aldehyde to form an alkene and a phosphate ester. Such *Horner-Wadsworth-Emmons* reactions typically occur with *trans* selectivity (see Chapter 9). Generally, the reaction is superior to the analogous *Wittig* olefination. It gives better yields, phosphonate esters are readily available and furthermore the formed phosphate byproduct is water-soluble and thus easily removed from the reaction mixture.

(EtO)$_2$P(O)-C(H)(⊖)-CO$_2$Et

15

Problem

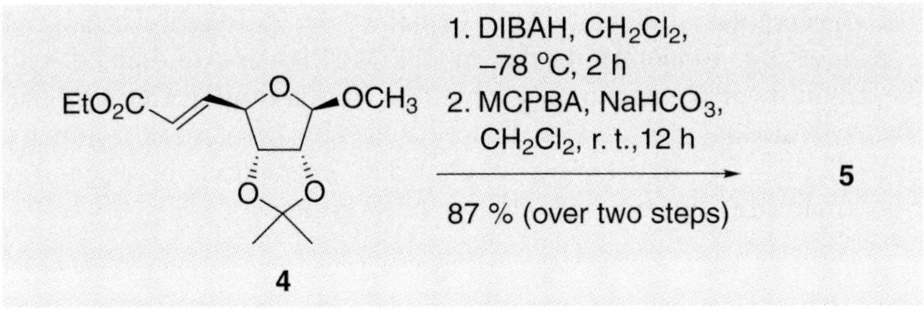

Hints

- DIBAH is a reducing agent.
- Which diastereofacial selectivity do you expect in step 2?

Solution

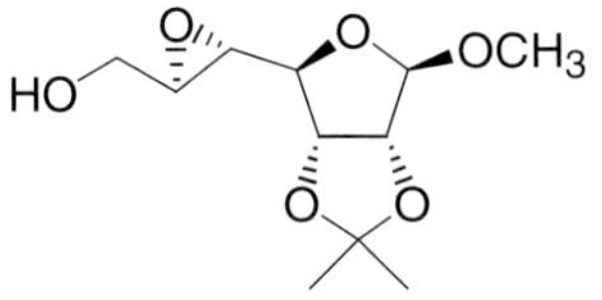

Discussion

DIBAH reduction of **4** at –78 °C provides the corresponding *trans*-allylic alcohol. Successive epoxidation with *meta*-chloroperbenzoic acid (MCPBA) yields a single *syn* epoxide **5**. The stereochemical assignment is proven by a second experiment using the asymmetric *Sharpless* epoxidation protocol. Both MCPBA and the *Sharpless* protocol using (–)-diethyl D-tartrate provided **5**.

Interestingly, MCPBA epoxidation of *cis* alcohol **16** affords a mixture of diastereomeric epoxides (55:45 mixture). Furthermore, protection of the allylic alcohol as TBS ether (**17**) and subsequent epoxidation results as well in hardly any stereochemical selectivity (53:47 mixture). With regard to these results it is suggested that the *trans*-allylic hydroxy group is effectively involved in directing the MCPBA epoxidation event.

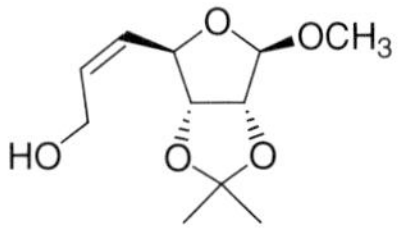

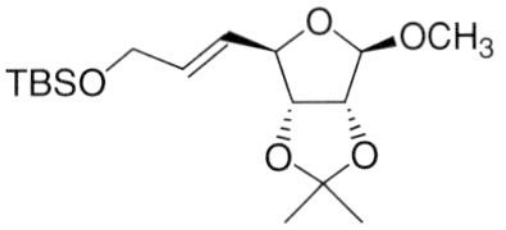

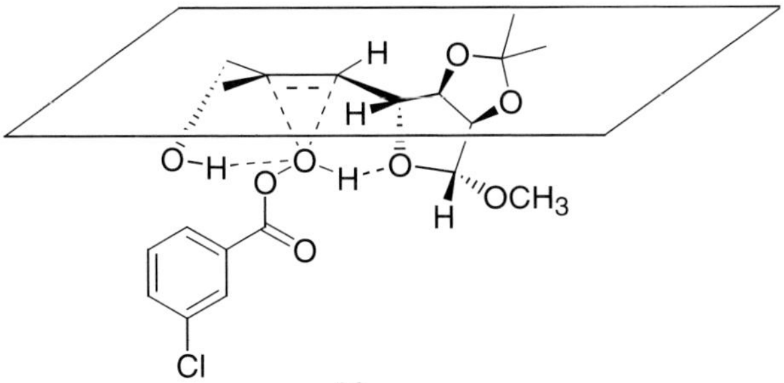

12 (+)-Polyoxin J

The diastereofacial selectivity is explained by the highly organized transition state model **18**. In this model, MCPBA is coordinated with the allylic hydroxy group as well as the ribofuranoside ring oxygen. The *cis* alcohol **16** cannot adopt such transition state geometry because of the developing nonbonded interaction between the ribofuranoside ring and the hydroxymethyl group.

Problem

Hints
- The epoxide is opened by a titanium reagent.
- The resulting diol is cleaved.
- Step 3 generates an ester.

Solution
1. Ti(O*i*Pr)$_2$(N$_3$)$_2$, benzene, 75 °C, 15 min, 79 %
2. RuCl$_3$ (cat.), H$_5$IO$_6$, MeCN/CCl$_4$/H$_2$O, r. t., 2 h
3. MeI, KHCO$_3$, DMF, r. t., 12 h, 80 % (over two steps)

Discussion

The azido group is introduced by a titanium-mediated nucleophilic opening of 2,3-epoxy alcohol **5** invented by *Sharpless*.[10] In principle, nucleophilic attack at C-2' or C-3' is possible.

Interestingly, treatment of **5** with diisopropoxytitanium diazide furnishes **19** and **20** as a nearly 4:1 mixture of regioisomers which are readily separated by silica gel chromatography. The major component of the mixture is azido diol **19** derived from C-3' attack.

Sharpless suggests that this regioselectivity originates from coordination of the epoxy alcohol to the metal center in a bidentate manner (**21**).[11] The bond between C-3' and oxygen appears much better orientated to overlap with an empty d-orbital on titanium than does the bond between C-2' and oxygen which lies nearly in the plane of the five-membered ring. In principle, the magnitude of the selectivity depends on steric and electronic factors. Thus, increasing steric hindrance at C-3' should result in decreased C-3' selectivity as well as the presence of electron-withdrawing groups at C-3'.

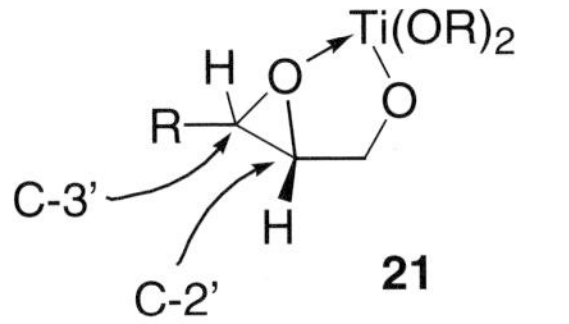

However, regiopure azido 1,2-diol **19** is converted to the corresponding azido carboxylic acid **22** by oxidative glycol cleavage with periodic acid in the presence of catalytic amounts of ruthenium trichloride. Interestingly, the use of sodium periodate instead of periodic acid resulted in a 10–15 % epimerization of the C-5 stereocenter.

The obtained acid **22** is treated with methyl iodide and potassium bicarbonate to afford azido methyl ester **6**.

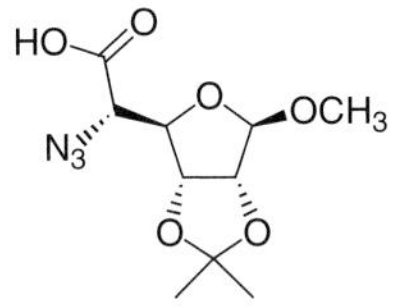

Problem

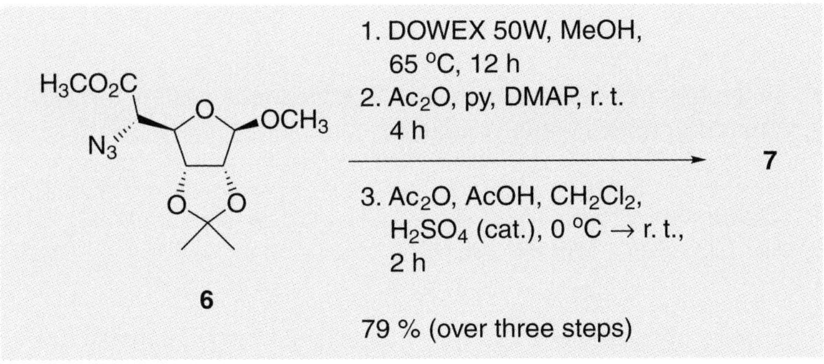

- DOWEX 50W is an acidic ion exchange resin.
- What is the difference between step 2 and 3?

Hints

Solution

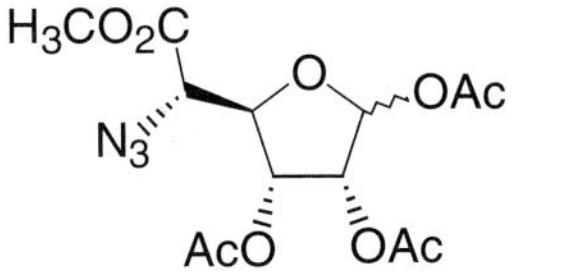

12 (+)-Polyoxin J

Discussion

In principle, acetals are cleaved by acid-catalyzed hydrolysis. In most cases aqueous acetic acid, aqueous trifluoracetic acid, dilute HCl in THF or DOWEX 50W (H⁺) resin are used. Thus, treatment of **6** with DOWEX ion exchange resin in methanol rapidly furnishes the corresponding 1,2-diol without any further chromatographic purification steps. Generally, polymer supported reagents benefit from the ease of removal from the reaction mixture just by filtration of the insoluble resin. The resulting diol is acetylated by addition of acetic anhydride and pyridine. Final acetal exchange is achieved by acetic anhydride and catalytic amounts of concentrated sulfuric acid. A mixture (2:1) of anomers is obtained.

Problem

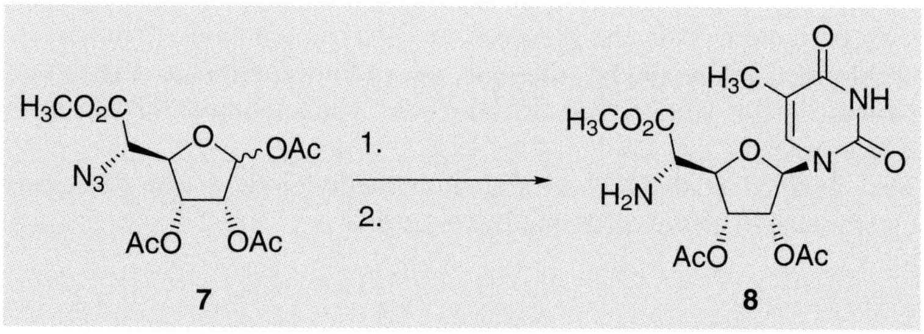

Hints

- In the first step an intermediacy 1,2-acyloxonium salt is formed.
- The azido functionality is reduced.

Solution

1. Thymine-bis-TMS, TMSOTf, Cl(CH$_2$)$_2$Cl, 84 °C, 1 h, 91 %
2. H$_2$, 10 % Pd/C, MeOH, 2 h, 98 %

Discussion

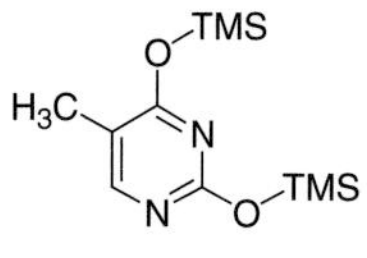

Exposure of triacetate **7** to 5-methyl-2,4-bis(trimethylsilyloxy)-pyrimidine (**23**) in presence of TMSOTf (**24**) provides the protected β-nucleoside **8**. The reaction of silylated heterocyclic bases with peracylated carbohydrates in the presence of *Friedel-Crafts* catalysts yields selectively β-nucleosides.

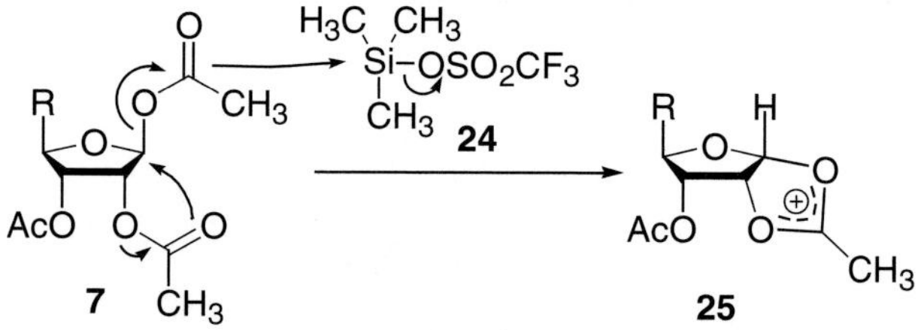

The mechanism is explained *via* the formation of the rather stable 1,2-acyloxonium salt **25** (neighbor group effect).

The activated α-trimethylsilyl group on the pyrimidine moiety reacts with the triflate ion **26** to regenerate the triflate catalyst. Under reversible and thus thermodynamically controlled conditions, the nucleophilic silylated base **23** attacks the carbohydrate cation **25** only from the top (β-side) to afford exclusively the β-nucleoside.

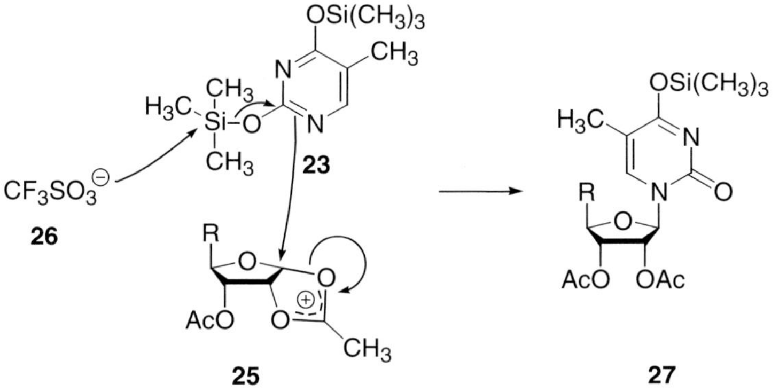

Further hydrolysis yields the protected β-nucleoside **27**.
The azido functionality is finally hydrogenated by means of hydrogen in presence of catalytic amounts of palladium on activated charcoal to furnish **8**.

Problem

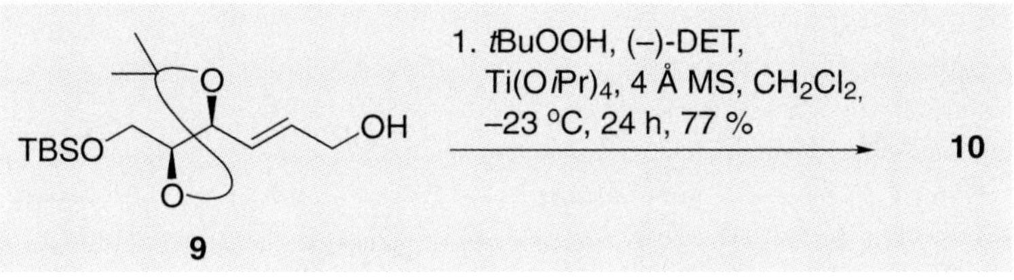

Hints

- The reaction is known as *Sharpless* epoxidation.
- Which stereochemistry do you expect?

Solution

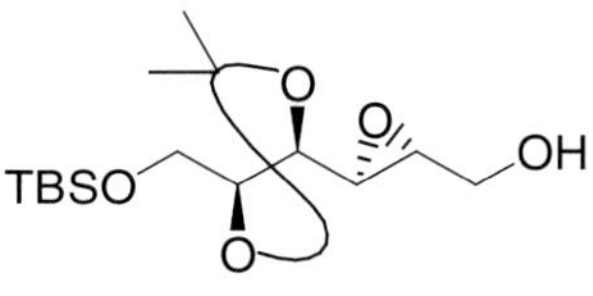

Discussion

The known allylic alcohol **9** derived from protected dimethyl tartrate is exposed to *Sharpless* asymmetric epoxidation conditions with (–)-diethyl D-tartrate. The reaction yields exclusively the *anti* epoxide **10** in 77 % yield. In contrast to the above mentioned epoxidation of the ribose derived allylic alcohol, in this case epoxidation of **9** with MCPBA at 0 °C resulted in a 65:35 mixture of *syn/anti* diastereomers. The *Sharpless* epoxidation of primary and secondary allylic alcohols discovered in 1980 is a powerful reagent-controlled reaction.[12] The use of titanium(IV) tetraisopropoxide as catalyst, *tert*-butylhydroperoxide as oxidant, and an enantiopure dialkyl tartrate as chiral auxiliary accomplishes the epoxidation of allylic alcohols with excellent stereoselectivity. If the reaction is kept absolutely dry, catalytic amounts of the dialkyl tartrate(titanium)(IV) complex are sufficient.

Sharpless and his co-workers studied the mechanism of the oxidation and established the following characteristics: Participation of the free hydroxy group of the substrate is evidently essential for the selectivity. Enantioselectivity is at its optimum at 1:1 rates of Ti(O*i*Pr)$_4$ and tartrate ester. Mixing of the reagents results in formation of a dimeric species. This species was initially believed to be a ten-membered ring, however, because of IR, NMR as well as X-ray studies on the closely related complex titanium (dibenzyltartramide)$_2$(OR)$_4$; structure **28** was suggested to represent the reagent in the absence of substrate and oxidant.[13] Kinetic experiments on the addition of the substrate and the oxidant have shown that the remaining two isopropyloxy ligands are displaced by allyloxy and *tert*-butylhydroperoxy groups as shown in **29**. The reactants are preorganized on the metal in a chiral environment prior to the reaction yielding the corresponding epoxide **30**.

28

29   **30**

Thus, the reaction is very predictable. When a (–)-tartrate ligand such as (–)-DET (diethyl tartrate) or (–)-DIPT (diisopropyl tartrate) is used, the oxygen atom is delivered to the top face of the olefin when the allylic alcohol is depicted as in **31**. The (+)-tartrate ligand, on the other hand, allows the bottom face to be epoxidized.

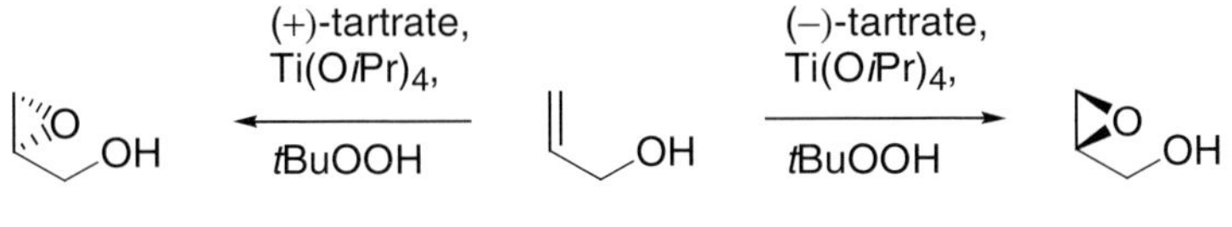

12 (+)-Polyoxin J

Problem

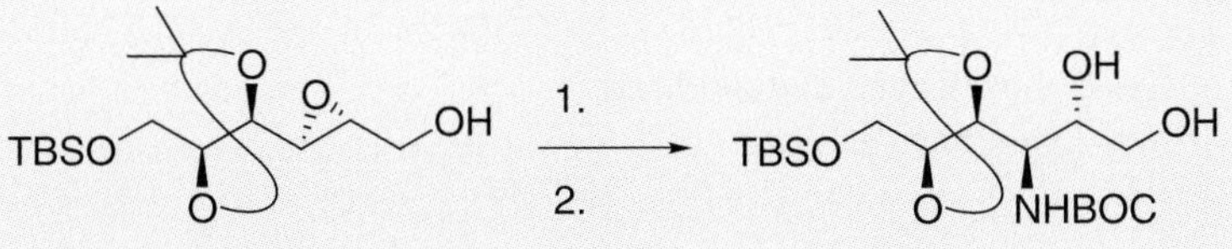

Hints
- The first step has already been mentioned in this chapter.
- Step 2 is a combined reduction/protection operation.

Solution
1. Ti(O*i*Pr)$_2$(N$_3$)$_2$, PhH, 72 °C, 15 min, 96 %, 3:1 mixture
2. H$_2$, 10 % Pd/C, BOC$_2$O, EtOAc, 12 h
60 % (over two steps)

Discussion

The azido group is again introduced by a titanium mediated reaction. The regioselective ring opening of **10** with diisopropoxy-titaniumdiazide in benzene at 72 °C affords the azido diols **34** and **35** as an inseparable 3:1 mixture.
However, the catalytic reduction of the azido functionalities in the presence of BOC$_2$O yields the corresponding BOC-protected hydroxy amine which is easily separated by silica gel chromatography. The variation of the conditions did not improve the mixture ratio of the azido diols. However, **11** could be isolated in 60 % yield in that two step sequence.

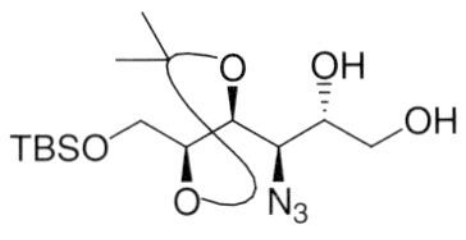

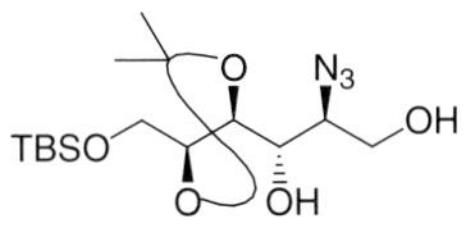

12 (+)-Polyoxin J

Problem

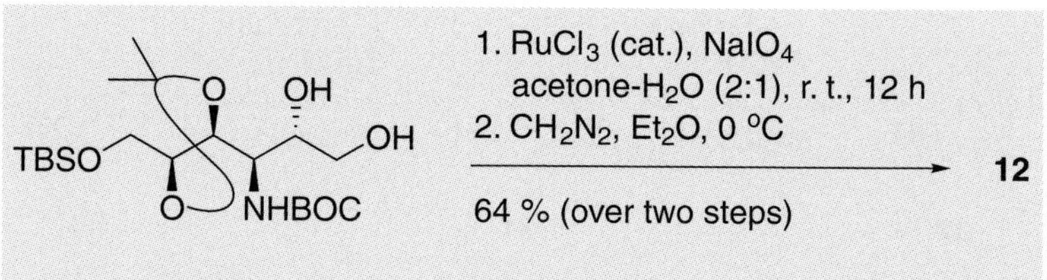

[Structure **11**: TBSO-CH₂ group attached to a quaternary carbon bearing an acetonide (O-C(CH₃)₂-O) linked via CH to NHBOC, with CH(OH)-CH₂OH side chain]

1. RuCl₃ (cat.), NaIO₄ acetone-H₂O (2:1), r. t., 12 h
2. CH₂N₂, Et₂O, 0 °C

64 % (over two steps) → **12**

Hints

- The diol is cleaved and oxidized.
- Step 2 generates an ester.

Solution

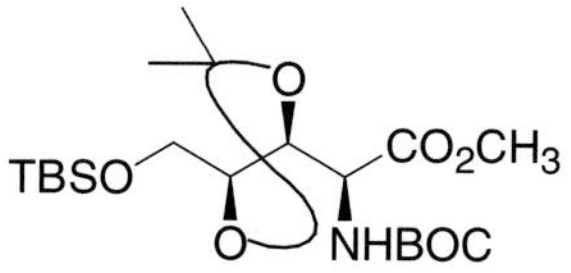

[Structure **12**: same scaffold as 11 but side chain is now CH(NHBOC)-CO₂CH₃]

Discussion

Again the diol is oxidized to the carboxylic acid with RuO₄ generated from RuCl₃ and NaIO₄. Subsequent esterification with diazomethane (see Chapter 13) yields **12** in 64 % over two steps.

Problem

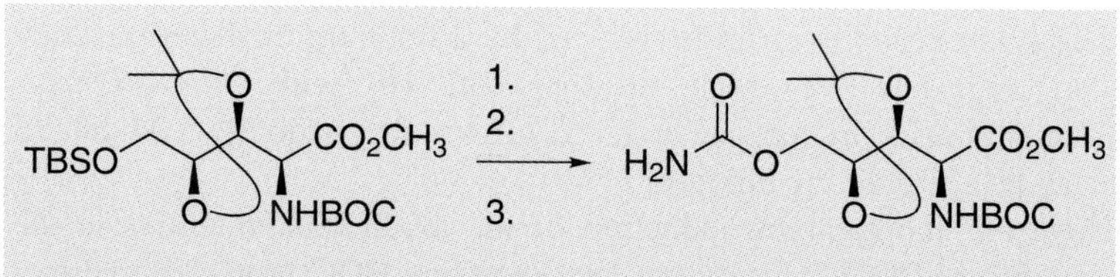

12 → 1. 2. 3. → **13** (H₂N-C(=O)-O-CH₂- replacing TBSO-CH₂-, rest of molecule unchanged)

Hints

- A protecting group is removed.
- The urethane is formed *via* a carbonate.

Solution

1. AcOH-THF-H₂O (3:1:1), r. t., 12 h
2. *p*-NO₂Ph-OCOCl, pyridine, 0 °C, 1 h
3. NH₄OH, THF, 0 °C, 30 min

85 % (over three steps)

Discussion

The TBS-group is removed selectively by treatment of **12** with aqueous acetic acid at room temperature for 12 h.

The carbamoylation of **12** was carried out in a one-pot procedure under standard conditions, i. e. treatment with *para*-nitrophenyl chloroformate followed by ammonolysis of the resulting carbonate **36** to give the urethane in 85 % over three steps.

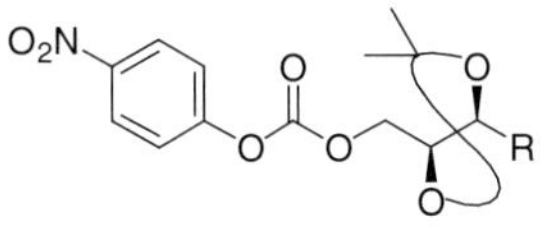

Problem

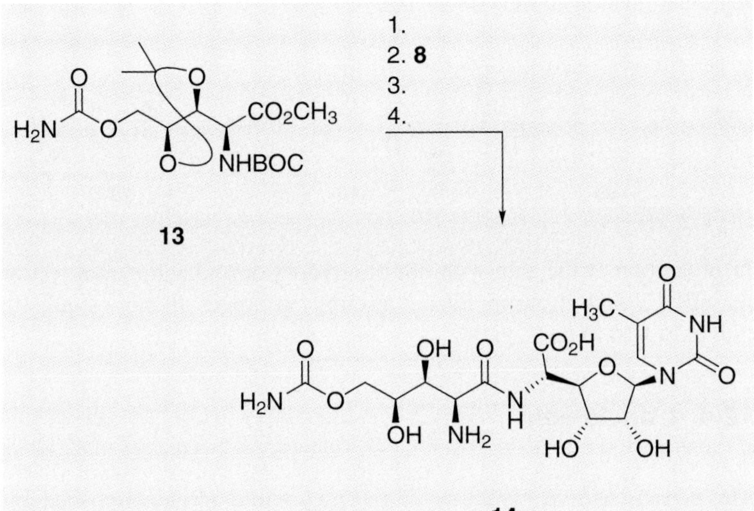

- Saponification of the ester is followed by coupling with **8** using a peptide coupling reagent.
- What peptide coupling reagents do you know?
- Two more deprotections are necessary.

Hints

1. LiOH·H$_2$O, THF-H$_2$O (1:5), 0 °C, 2 h, 90 %
2. **8**, BOP, *i*Pr$_2$NEt, DMF, r. t., 12 h, 63 %
3. LiOH·H$_2$O, THF-H$_2$O (1:5), 0 °C, 2 h
4. CF$_3$CO$_2$H, 0 °C, 2 h, 53 % (over two steps)

Solution

The synthesis of (+)-polyoxin J **14** is completed by selective ester hydrolysis with aqueous lithium hydroxide. The resulting carbamoylpolyoxamic acid was then coupled with the protected thymine polyoxin C **8** with the BOP reagent (*Castro's* reagent)[14] (benzotriazol-1-yloxytris(dimethyl-amino)phosphonium hexafluorophosphate) **37** to furnish the peptide derivative in 63 % yield.

Discussion

The BOP reagent **37** was first prepared to act as a coupling agent (dehydrating agent phosphonium salt) as well as a racemization suppressor (benzotriazole additive). Early investigations have indicated a benzotriazolyl active ester as reactive intermediate like in the activation with the DCC/HOBT reagent system. Now it is believed that the acyloxyphosphonium salt **40** is formed directly presumably through the cyclic complex **38**.[15] No reaction occurs until a tertiary base, *N*-methylmorpholine, triethylamine, or diisopropylethylamine, is added to form salt **39**. Under such conditions the coupling rate is so high that racemization is neglected.

Deprotection of the acid with lithium hydroxide and acidic removal of the BOC and isopropylidene groups afforded finally synthetic polyoxin J **14** in 53 % yield.

12.4 Conclusion

The stereoselective synthesis of (+)-polyoxin J is accomplished by *Gosh* in 24 steps and 3 % overall yield. The key intermediates are protected thymine polyoxin C **8** and the 5-*O*-carbamoyl polyoxamic acid **2**, which were synthesized from D-ribose and dimethyl L-tartrate. Key steps are two different epoxidation reactions, one carried out with MCPBA and the other under *Sharpless* conditions with the D-(–)-tartrate. Both epoxides are opened with diisopropoxytitanium diazide. The coupling of the two fragments was realized with the BOP reagent **37**. This synthesis provides an easy access to the synthesis of various (+)-polyoxin J analogs for biological evaluation.

12.5 References

1. a) K. Isono, K. Asahi, S. Suzuki, *J. Am. Chem. Soc.* **1969**, *91*, 7490-7504; b) K. Isono, S. Suzuki, *Heterocycles* **1979**, *13*, 333-347.
2. K. Isono, *J. Antibiot.* **1988**, *41*, 1711-1739.
3. G. W. Gooday, In *Biochemistry of Cell Walls and Membranes in Fungi* (eds. P. J. Kuhn, A. P. J. Trinci, M. J. Jung, M. W. Goosey, L. G. Copping), Springer, Berlin, **1990**.
4. W. A. König, W. Hass, W. Dehler, H. Fiedler, H. Zähner, *Liebigs Ann. Chem.* **1980**, 622-628.

5   M. Uramoto, K. Kobinata, K. Isono, T. Higashijima, T. Miyazawa, E. E. Jenkins, J. A. McKloskey, *Tetrahedron Lett.* **1980**, *21*, 3395-3398.
6   A. B. Cooper, J. Desai, R. G. Loveley, A. K. Saksena, V. M. Girjiavallabahn, A. K. Ganguly, D. Loebenberg, R. Parmegiani, A. Cacciapuoti, *Bioorg. Med. Chem. Lett.* **1993**, *3*, 1079-1084.
7   a) H. Kuzuhara, H. Ohrui, S. Emoto, *Tetrahedron Lett.* **1973**, *14*, 5055-5058; b) T. Mukaiyama, K. Suzuki, T, Yameda, F. Tabussa, *Tetrahedron* **1990**, *46*, 265-276; c) N. Chida, K. Koizumi, Y. Kitada, C. Yokoyama, S. Ogawa, *J. Chem. Soc., Chem. Commun.* **1994**, 111-114; d) A. Dondoni, F. Junquera, F. L. Merchan, P. Merino, T. Tejero, *J. Chem. Soc., Chem. Commun.* **1995**, 2127-2128; e) A. Dondoni, S. Franco, F. Junquera, F. L. Merchan, P. Merino, T. Tejero, *J. Org. Chem.* **1997**, *62*, 5497-5507; f) H. Akita, K. Uchida, K. Kato, *Heterocycles* **1998**, *47*, 157-162.
8   A. K. Gosh, Y. Wang, *J. Org. Chem.* **1999**, *64*, 2789-2795.
9   B. A. Arbuzov, *Pure Appl. Chem.* **1964**, *9*, 307-335.
10   M. Caron, K. B. Sharpless, *J. Org. Chem.* **1985**, *50*, 1557-1560.
11   J. M. Chong, K. B. Sharpless, *J. Org. Chem.* **1985**, *50*, 1560-1563.
12   T. Katsuki, K. B. Sharpless, *J. Am. Chem. Soc.* **1980**, *102*, 5974-5976.
13   I. D. Williams, S. F. Pedersen, K. B. Sharpless, *J. Am Chem. Soc.* **1984**, *106*, 6430-6431.
14   B. Castro, J. R. Dormoy, G. Evin, G. Selve, *Tetrahedron Lett.* **1975**, *16*, 1219-1221.
15   J. Coste, M. N. Dufour, D. Le Nguyen, B. Castro, In *Peptides, Chem. Struct. Biol.* ESCOM: Leiden, **1990**, 885-888.

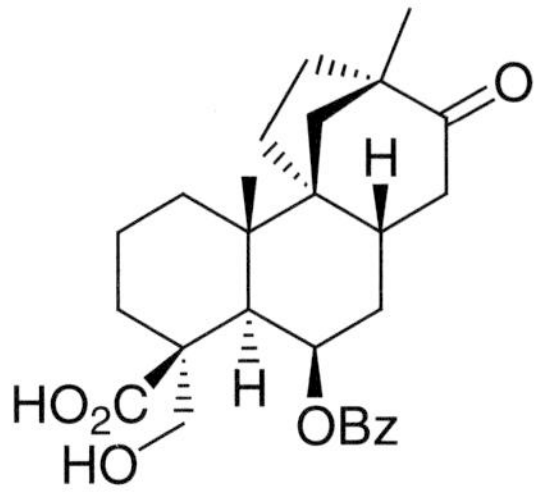

13

(−)-Scopadulcic Acid A (Overman 1999)

13.1 Introduction

1
(−)-scopadulcic acid A

2
scopadulcic acid B

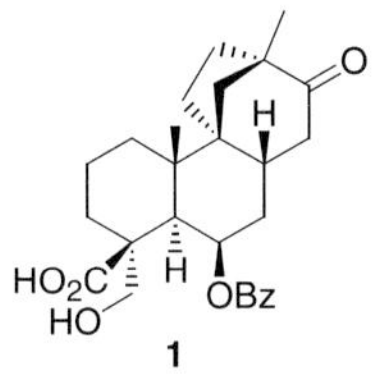

Scoparia dulcis L. (Scrophuraliaceae) is a perennial herb which is found in many tropical countries. It has been used as a medicinal plant in Paraguay to improve digestion and protect the gastrointestinal system. In Taiwan, the same plant is employed as a cure for hypertension, and in India for toothaches, blennorhagia and stomach irritations.[1] From its pharmacologically active extracts *Hayashi* and co-workers identified two structurally unique tetracyclic diterpene acids, scopadulcic acid A (SDA) (**1**) and B (SDB) (**2**).[2] The scopadulcic acids and some of their semisynthetic analogues exhibit a broad pharmacological profile, including *in vitro* antiviral activity against herpes simplex virus type 1,[3] *in vitro* and *in vivo* antitumor activity in various human cell lines,[4] and inhibition of tumor promotion by phorbol esters.[5] The crystal structure of **1** was confirmed by X-ray crystallographic analysis.[6] In 1993 *Overman* and co-workers reported the first total synthesis of (±)-SDA (**1**) and SDB (**2**)[7], subsequently *Ziegler* and *Wallace* described total syntheses following different strategies.[8] This problem is based on the first enantioselective total synthesis of (+)- and (−)-scopadulcic acid A realized by *Overman* in 1999, which for the first time rigorously established the absolute configuration of natural (−)-SDA (**1**).[9]

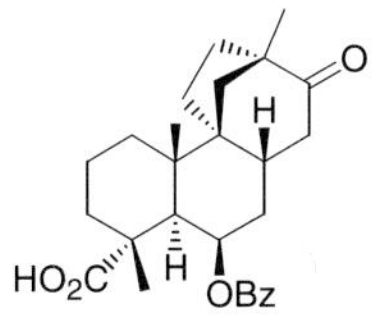

13.2 Overview

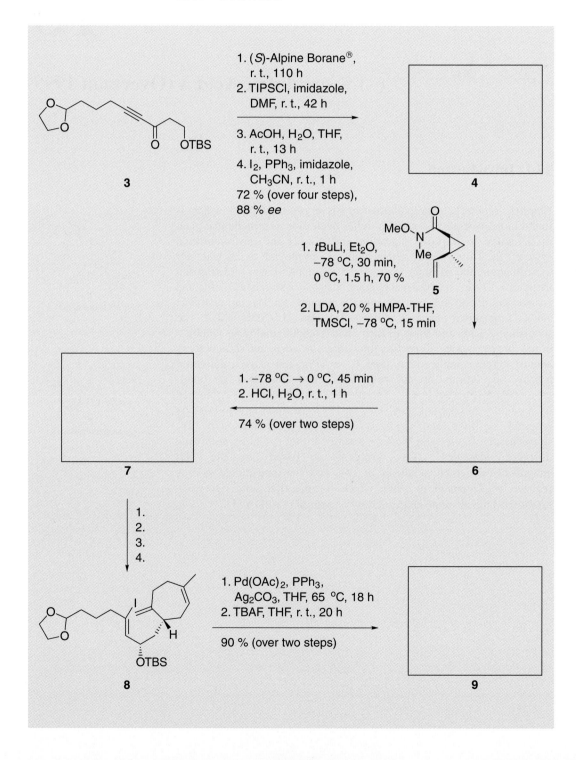

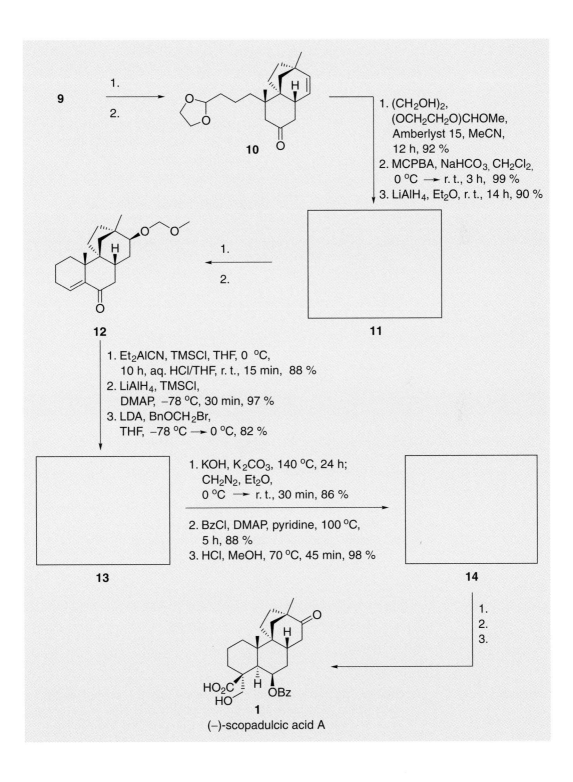

13.3 Synthesis

Problem

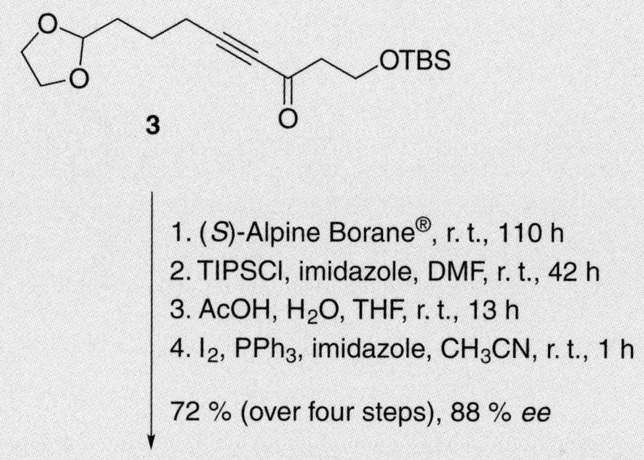

[Reaction scheme: compound **3** (dioxolane-CH₂CH₂CH₂-C≡C-C(O)-CH₂CH₂-OTBS) is converted to **4** by:
1. (S)-Alpine Borane®, r. t., 110 h
2. TIPSCl, imidazole, DMF, r. t., 42 h
3. AcOH, H₂O, THF, r. t., 13 h
4. I₂, PPh₃, imidazole, CH₃CN, r. t., 1 h

72 % (over four steps), 88 % ee]

Hints

- (S)-Alpine Borane®: 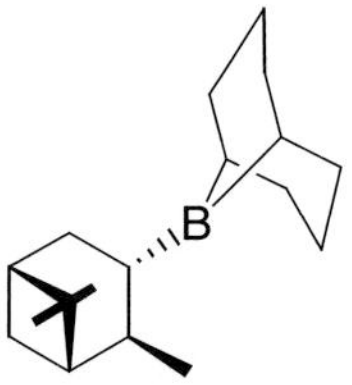 **15**
- **15** reduces ketones stereoselectively.
- Which protecting group is cleaved by AcOH?
- Finally, an alcohol is transformed into a halide.

Solution

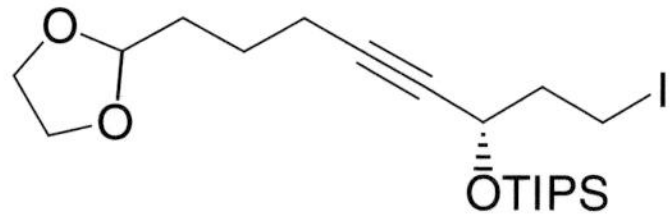

[Structure of **4**: dioxolane-CH₂CH₂CH₂-C≡C-CH(OTIPS)-CH₂CH₂-I]

Discussion

Boranes represent a family of reagents that has proved useful for asymmetric reductions. Here the asymmetric reduction of **3** is realized with *B*-(3-pinanyl)-9-borabicyclo[3.3.1]nonane (**15**) ((*S*)-Alpine

Borane®). This reagent is commercially available or prepared by hydroboration of (−)-α-pinene (**16**) with 9-BBN (**17**).[10]

The stereoselectivity of carbonyl group reduction with (S)-Alpine Borane® is explained *via* six-membered transition state **18**.

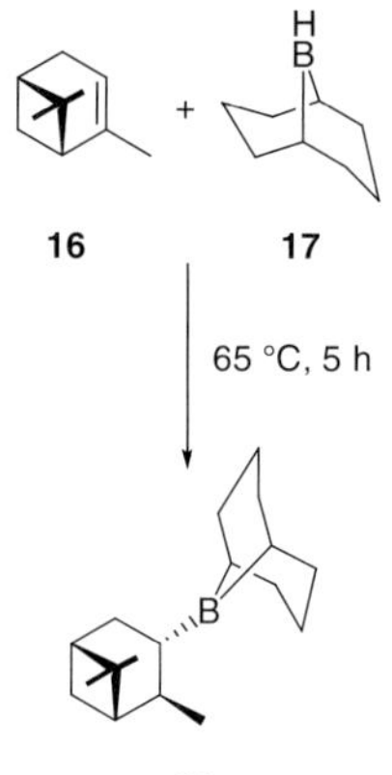

First, the *Lewis* acidic borane reagent coordinates to the carbonyl oxygen. Subsequently the β-hydrogen is transferred to the carbonyl carbon by way of boat type transition state **18**. The preferred orientation of the substituents on the carbonyl group results from the urge of minimization of 1,3-diaxial interactions with the β-methyl group of the chiral borane reagent. Alternatively various other chiral agents were used to effect the reduction. However, only inferior results were obtained using (S)-BINAL-H **20**[11] or the borane reduction catalyzed by oxazaborolidine **21**.[12]

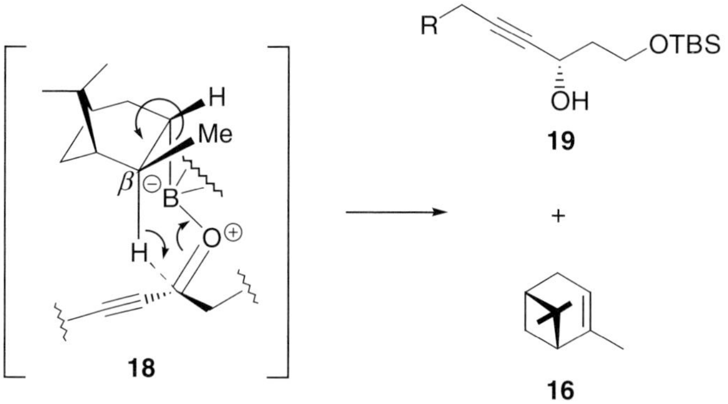

The resulting propargylic alcohol is protected as TIPS ether by a standard procedure using the corresponding silyl chloride and imidazole in DMF. Optionally the more reactive silyl triflate and 2,6-lutidine may be employed in order to shorten the reaction time. Under acidic conditions TIPS and TPS are nearly stable protecting groups. Therefore the TBS ether is selectively cleaved with acetic acid even in presence of the acetal moiety.[13] Subsequent reaction with iodine and triphenylphosphine, known as the *Appel* reaction[14], provides the desired iodide **4**.

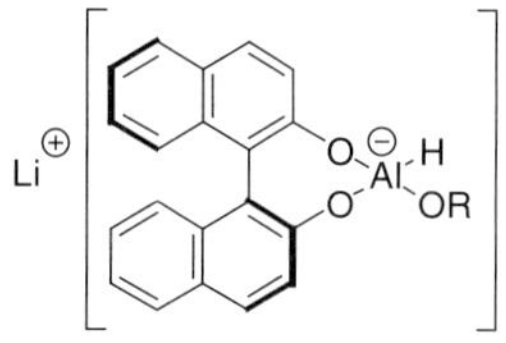

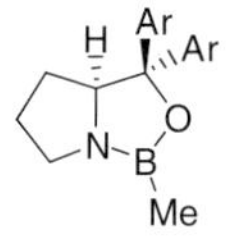

13 (−)-Scopadulcic Acid A

Problem

[Scheme showing compound **4** (dioxolane-protected aldehyde with alkyne, OTIPS group, and terminal iodide) reacting via:
1. *t*BuLi, Et₂O, −78 °C, 30 min, 0 °C, 1.5 h, 70 %; with Weinreb amide **5** (cyclopropyl vinyl N-methoxy-N-methylamide)
2. LDA, 20 % HMPA-THF, TMSCl, −78 °C, 15 min

to give compound **6**]

Hints

- *t*BuLi reacts with the iodide.
- The resulting species reacts as nucleophile.
- Carbonyl compounds are easily silylated.

Solution

[Structure of **6**: dioxolane-CH₂CH₂CH₂-C≡C-CH(OTIPS)-CH=C(OTMS)-cyclopropyl-vinyl]

Discussion

The organolithium species is generated by reaction of **4** with 2 eq. of *t*BuLi at −78 °C in ether. Acylation with *N*-methoxy-*N*-methylamide (*Weinreb* amide)[15] **5** forms the intermediate complex **22**.

Since this complex is stable at the reaction conditions a multiple nucleophilic addition to the ketone is inhibited. Aqueous work-up finally provides cyclopropyl ketone **23**.

[Structure **22**: tetrahedral intermediate with Li··O, OMe, NMe, cyclopropyl vinyl]

The following kinetically controlled enolization using LDA in 20 % HMPA-THF, known as *Ireland* conditions[16], can proceed *via* two different transition states **24** and **25**.

Unhindered aliphatic ketones usually react with LDA to form *E*-enolates. However, in this case the steric demand of the alkyl side chain provokes strong 1,2-interactions disfavoring transition state **25**. Thus the stereochemistry of the enolization proceeds preferred *via* **24** yielding *Z*-enolate **26**, which is subsequently trapped by TMSCl to give the corresponding silyl enol ether **6**.

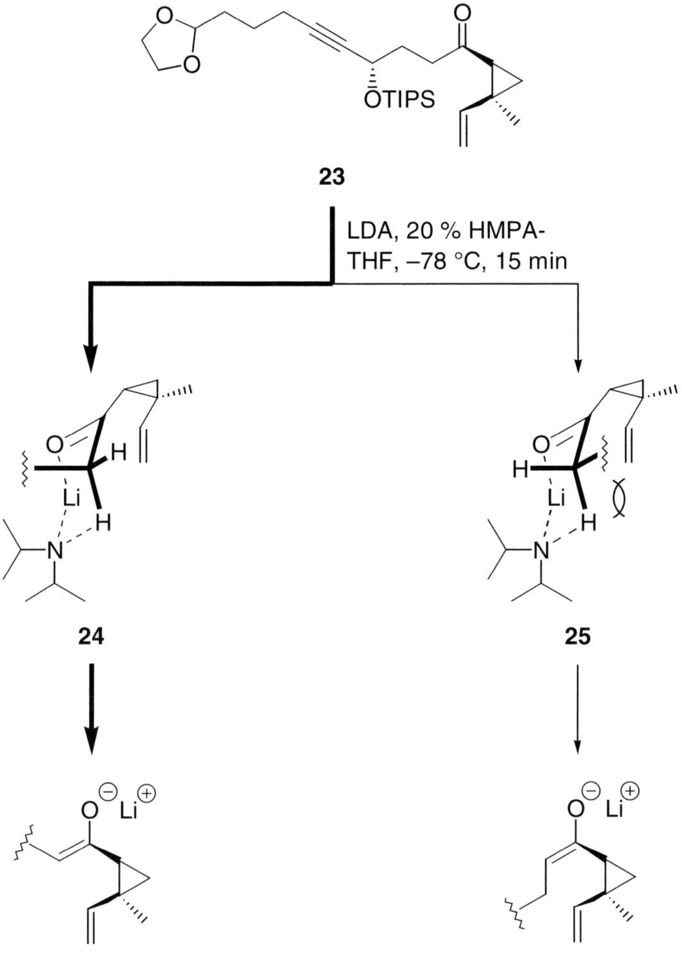

In contrast, using TMS triflate and triethylamine as base the reaction follows the thermodynamically controlled pathway and the isomeric *E*-enoxysilane is predominantly obtained (*E*:*Z* 4:1).

Problem

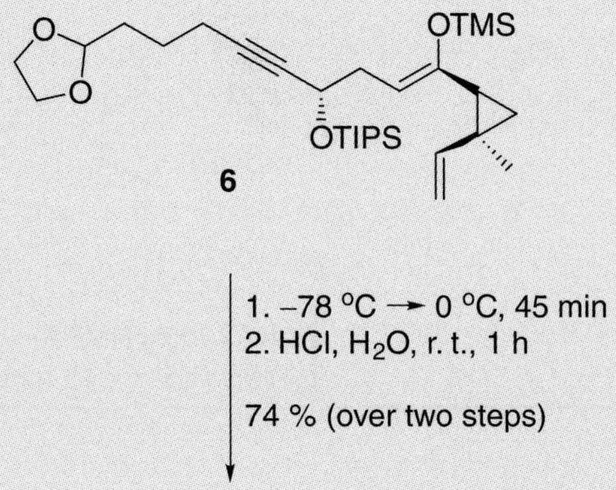

[Structure of compound **6**: dioxolane-CH₂CH₂CH₂-C≡C-CH(OTIPS)-CH₂-C(OTMS)=CH-cyclopropyl with vinyl substituent]

1. −78 °C → 0 °C, 45 min
2. HCl, H₂O, r. t., 1 h

74 % (over two steps)

7

Hints

- The TMS enol ether is only stable below −78 °C.
- Warming to 0 °C induces a rearrangement of the 1,5-diene.

Solution

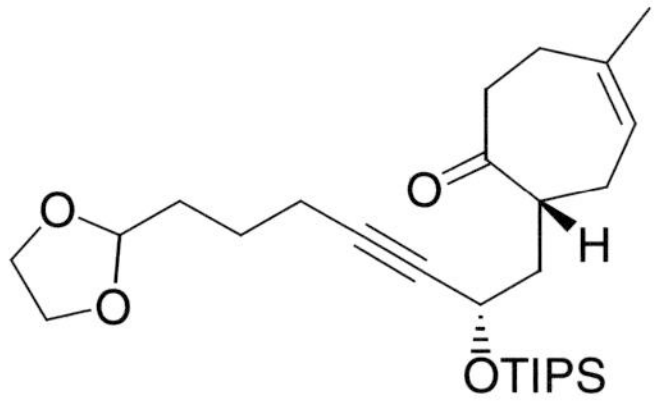

[Structure of compound **7**: cycloheptenone with methyl substituent, connected via CH₂-C≡C-CH(OTIPS)-CH₂CH₂CH₂-dioxolane]

7

Discussion

Generally, when 1,5-dienes are heated, they isomerize in a [3,3]-sigmatropic rearrangement known as a *Cope* rearrangement[17] (not to be confused with the *Cope* elimination process). The mechanism is a simple six-membered pericyclic process, the chair form being the usual transition state. However, in this case the cyclopropyl moiety forces the geometry of the transition state into boat form **28**.

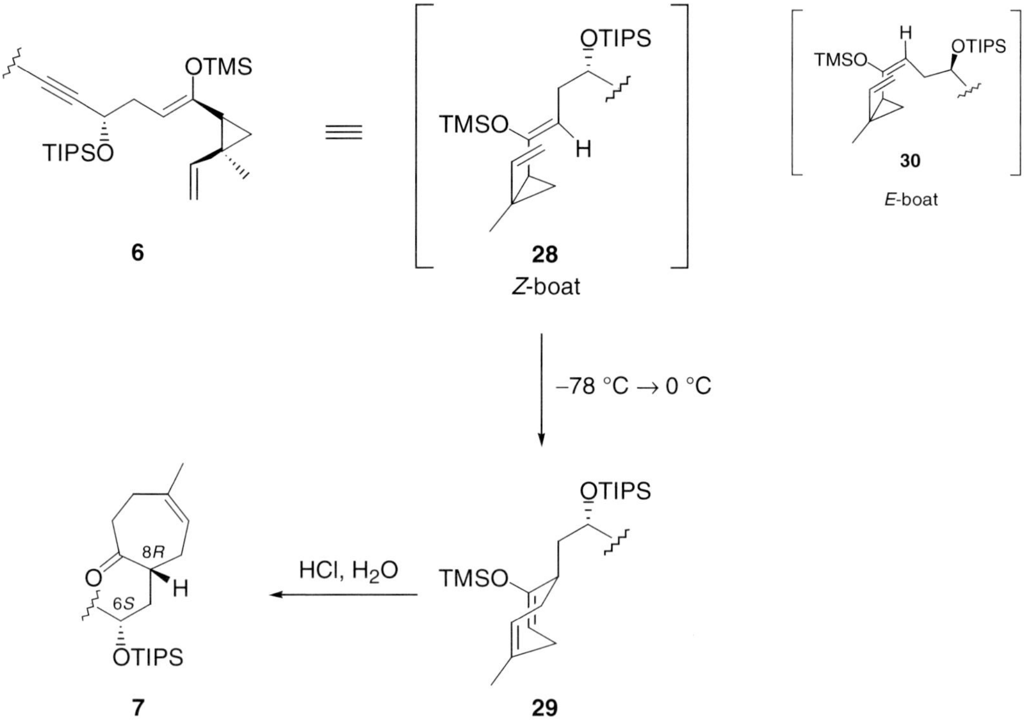

The presumed Z-enoxysilane **28** undergoes rearrangement prior to warming to room temperature furnishing siloxy cycloheptadiene **29**. Hydrolysis of silyl enol ether **29** yields (6S, 8R)-cycloheptenone **7**.

It is noteworthy that *Cope* rearrangement of the Z-enoxysilane intermediate **28** takes place at a temperature at least 80 °C lower than that required for rearrangement of the corresponding E-enoxysilane intermediate, since boat conformer **28** has a hydrogen (rather than the alkyl chain in conformer **30**) thrust over the cyclopropane ring.

The resulting stereoisomeric (6R, 8S)-cycloheptenone **31** is used for the synthesis of (+)-scopadulcic acid A **32** as an enantiodivergent approach.

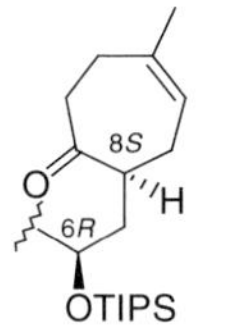

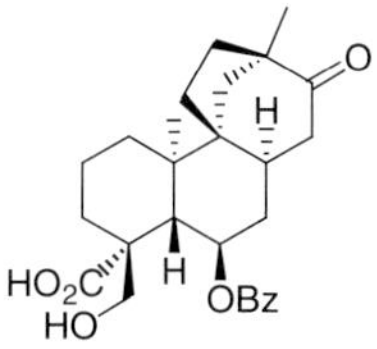

Problem

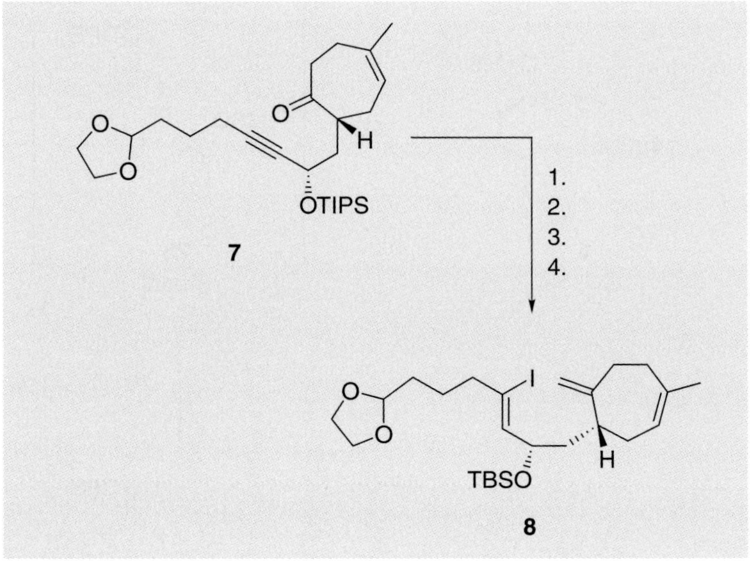

Hints
- Initially the carbonyl group is transformed into an alkene.
- Reduction of the triple bond is followed by iodination.
- The alcohol is protected in a standard procedure.

Solution
1. Ph$_3$P=CH$_2$, THF, −78 °C → 0 °C, 1 h, 85 %
2. TBAF, THF, r. t., 1 h, 100 %
3. Red-Al®, Et$_2$O, r. t., 15 h then NIS, −78 °C → r. t., 5 min, 97 %
4. TBSCl, imidazole, DMF, r. t., 8.5 h, 97 %

Discussion

phosphorane form

Ph$_3$P=CH$_2$

↕

⊕ ⊖
Ph$_3$P-CH$_2$

ylide form

The first step is a *Wittig* reaction[18] in which the ketone is converted to the terminal olefin by reaction with a phosphorus ylide (also called a phosphorane). Phosphoranes are resonance-stabilized by overlap between the carbon p-orbital and one of the d-orbitals of the phosphorus.

These reagents are usually prepared by treatment of a phosphine with an alkyl halide and successive addition of base, like BuLi or NaHMDS. Reaction with the ketone takes place by attack of the carbanionoid carbon of the ylide form on the electrophilic carbon of the carbonyl group *via* a four-membered heterocyclic transition state, the oxaphosphetane **35**. The driving force for this transformation is provided by formation of the very strong phosphorus-oxygen bond. Subsequent collapse of the oxaphosphetane furnishes the desired exomethylene group under retention of geometry.

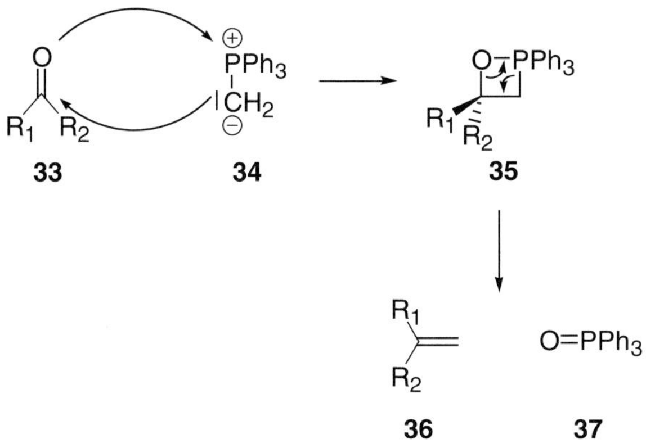

In this case the *E/Z*-selectivity does not matter because a terminal double bond is generated. In principle the *E/Z*-selectivity depends on the stability of the ylide employed (see chapter 9). The most valuable feature of the *Wittig* procedure is that, in contrast to elimination and pyrolytic reactions, it gives rise to alkenes with unambiguous position of the double bond.

In the second step a fluoride-induced desilylation furnishes the propargylic alcohol **38**.

Next, the propargylic alcohol is *trans*-selectively reduced to an allyl alcohol with sodium bis(2-methoxyethoxy)aluminum hydride (Red-Al®) **39**. This selectivity arises through the five-membered ring **40** formed by coordination of the aluminum to the former triple bond and the alcohol.

Reductive iodination of this vinylalane intermediate with *N*-iodosuccinimide (NIS) yields the *Z*-trienyl iodide. Finally the free secondary alcohol is protected as TBS ether in a standard procedure to provide **8**.

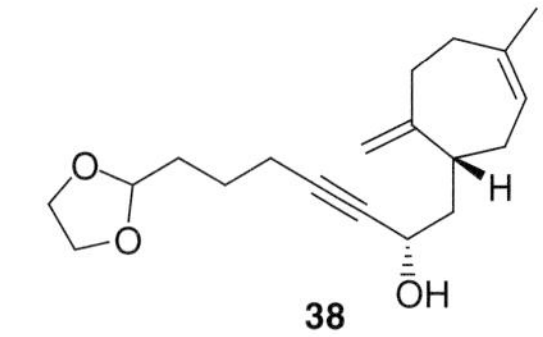

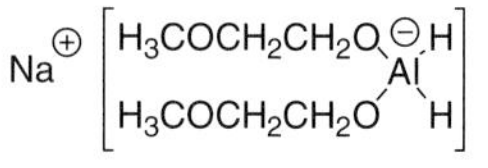

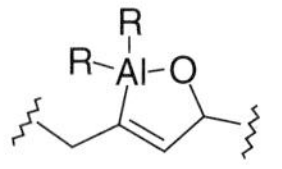

Problem

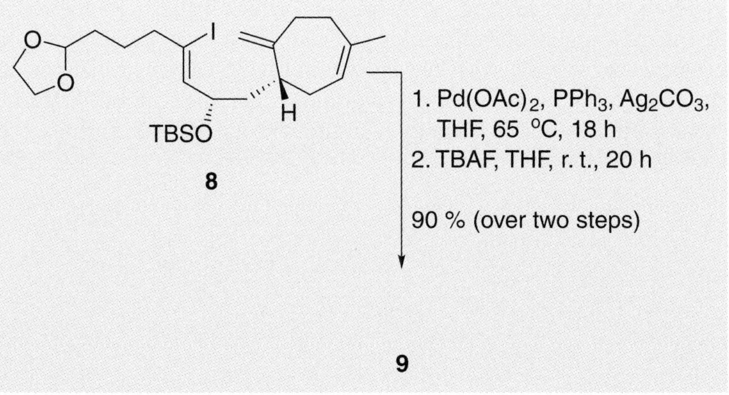

Hints

- What kind of palladium-mediated reactions are known?
- The first step is a domino-*Heck* reaction.
- One five-membered and two six-membered rings are formed with complete stereo- and regiocontrol.
- Step 2 is a deprotection operation.

Solution

Discussion

One of the key steps in this synthesis is the domino-*Heck* reaction forming **45**.[19] The Pd-catalyzed coupling of alkenes with vinyl- and aryliodides was discovered in 1968 by R. F. *Heck*; nowadays also triflates or nonaflates are used as substrates.[20]

In the first step of the catalytic cycle a coordinatively unsaturated Pd(0) species – which is formed *in situ* from Pd(OAc)$_2$ and PPh$_3$ – inserts into the alkenyl-I bond of **8** to give **42** (*syn* addition). Next an insertion of the terminal olefin into the σ-alkenyl-C-Pd bond forms the six-membered ring in **43**. The stereochemistry can be explained by **41**: reaction of the *si*-face of the exomethylene group involves a nearly coplanar orientation of the Pd-C bond and the C-C-π bond. The siloxy substituent is placed pseudo-equatorially.

β-Hydrogen elimination is usually faster than another insertion step. However, because of a missing β-hydrogen **43** is only able to undergo a second insertion. The regioselectivity is assumed to be influenced by the decreasing stability of metal-C-σ bonds as the degree of substitution at carbon increases. Product **45** arises from insertion of the double bond in **43** into the alkenyl-Pd bond to generate **44**, having palladium attached to the secondary and not the tertiary carbon center. The catalytic cycle is completed by a reversible *syn* elimination yielding alkene **45** and a hydridopalladium halide. Reductive HI elimination regenerates the catalyst from the hydridopalladium halide promoted by a base. Ag$_2$CO$_3$ as base in this bis-*Heck* cyclization was paramount in suppressing migration of the initially formed 1,2-substituted double bond by readdition of a palladium species.

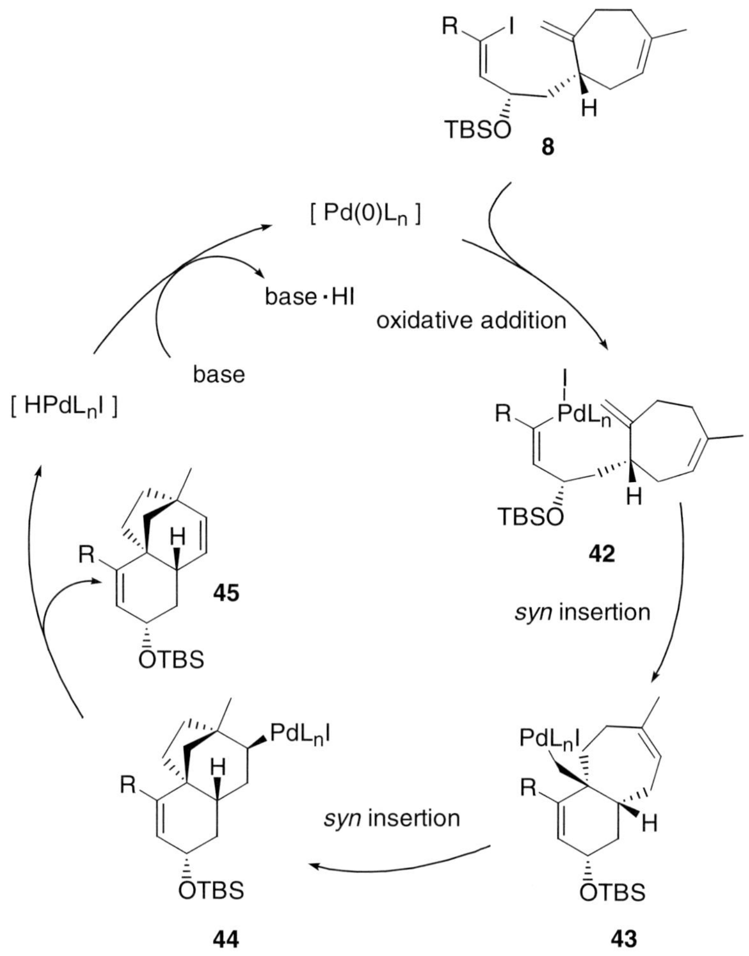

Step 2 cleaves the silyl protecting group using a standard protocol with tetrabutylammonium fluoride (TBAF) in THF.

Problem

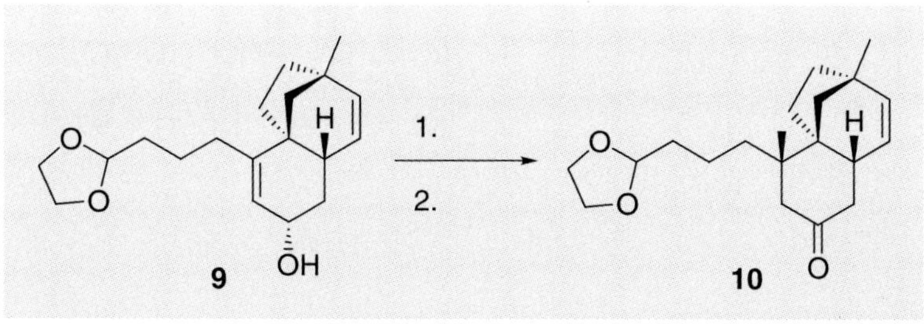

13 (−)-Scopadulcic Acid A

Hints
- Which oxidation methods for alcohols do you know?
- *Overman* used TPAP for oxidation; which co-reagent is necessary?
- The angular methyl group is introduced by a Ni-catalyzed reaction.

Solution
1. TPAP, NMO, CH$_2$Cl$_2$/MeCN, r. t., 1 h, 95 %
2. Ni(acac)$_2$, Me$_2$Zn, Et$_2$O, 0 °C → r. t., 17 h, 88 %

Discussion

The oxidation to the enone was realized with catalytic amounts of tetra-*n*-propylammonium perruthenate (TPAP)[21] (**46**), which is a mild oxidant for conversion of multifunctionalized alcohols to aldehydes or ketones. Catalytic TPAP oxidations are carried out in the presence of stoichiometric or excess *N*-methylmorpholine-*N*-oxide (NMO)[22] (**47**) as cooxidant. Other common reagents for oxidation of alcohols are e.g. DMSO/C$_2$O$_2$Cl$_2$[23], *Dess-Martin* periodinane[24], PCC[25], PDC[26] or the *Jones* reagent[27].

The second step is a *Michael* addition, that was not easy to perform because of the steric demand of the quaternary center nearby. Organocopper compounds have been the most extensively studied reagents for conjugate 1,4-additions to α-enones.[28] In this case, however, the addition of a cuprate failed to give the desired product. Especially with β,β-disubstituted enones allylic alcohols are often produced rather than the normally desired conjugate-addition adducts. In contrast, the Ni-catalyzed reaction of ZnMe$_2$[29] is highly regio- and stereoselective and occurs from the β-side opposite to the ethano-bridge in 88 % yield.

Problem

1. (CH$_2$OH)$_2$, (OCH$_2$CH$_2$O)CHOMe, Amberlyst 15, MeCN, 12 h, 92 %
2. MCPBA, NaHCO$_3$, CH$_2$Cl$_2$, 0 °C → r. t., 3 h, 99 %
3. LiAlH$_4$, Et$_2$O, r. t., 14 h, 90 %

13 (−)-Scopadulcic Acid A

Hints

- Amberlyst 15 is an acidic ion exchange resin.
- The first step introduces a protecting group.
- The second step is a substrate controlled epoxidation. What stereochemistry do you expect?
- LiAlH₄ is able to open epoxides.

Solution

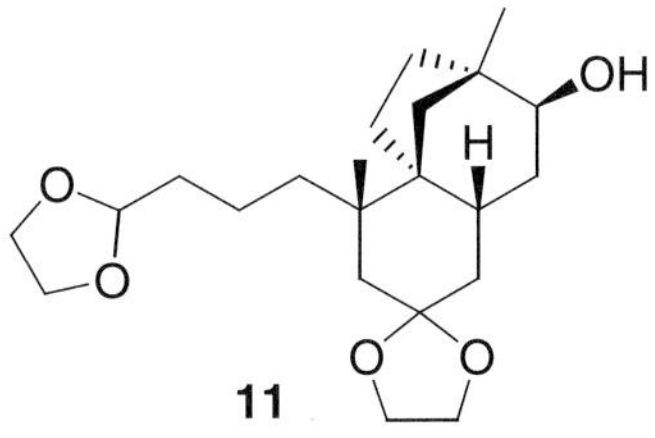

11

Discussion

First, the carbonyl group is protected as an acetal under acidic conditions. The most common acid-catalysts are PTSA, CSA, PPTS or, as used in this case, acid exchange resins like Amberlyst 15. An advantage of ion exchange resins is that aqueous work-up is not necessary since the insoluble catalyst can be easily removed by filtration and then recycled.

In the second step *meta*-chloroperbenzoic acid (MCPBA) epoxidizes the resulting bis-acetal from the β-face. The weak O-O bond of MCPBA undergoes attack by electron rich substrates like alkenes. This reaction is *syn* stereospecific and believed to take place *via* transition state **48**.[30]

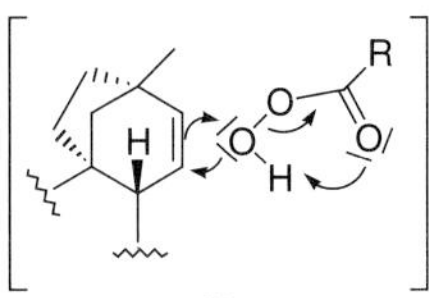

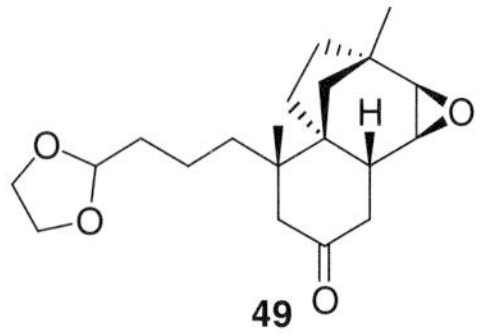

Reduction of **49** with LiAlH₄ provides **11** by an axial hydride transfer to the sterically less hindered carbon atom. Generally, LiAlH₄ attacks at the less substituted carbon and in addition 1,2-epoxycyclohexanes exhibit a strong preference for axial directed reactions.[31]

48

49

13 (−)-Scopadulcic Acid A

Problem

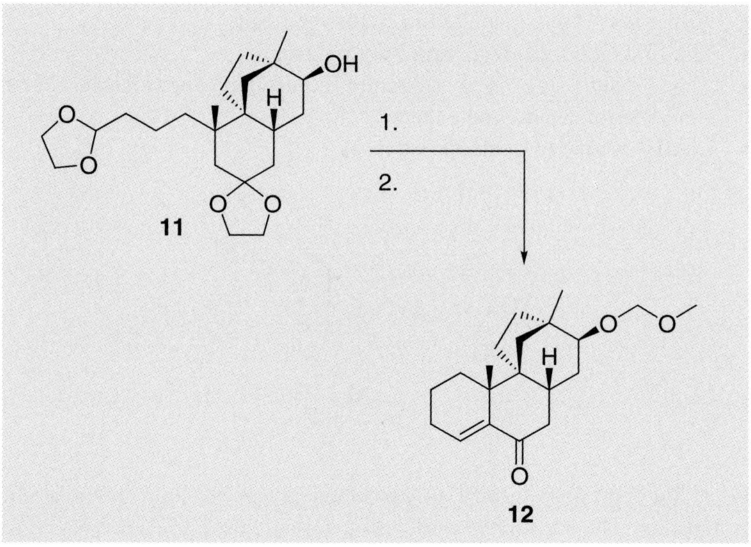

Hints
- Deprotection of **11** is followed by a cyclization.
- The second step introduces a MOM protecting group.

Solution
1. 20 % HCl, THF, r. t. → 50 °C, 24 h, 74 %
2. MOMCl, *i*Pr$_2$EtN, CH$_2$Cl$_2$, 0 °C, 7 h, 92 %

Discussion
Acetals can be removed under acidic conditions. In this case the two dioxolane groups are cleaved, followed by acid-catalyzed cyclization of the keto-aldehyde to form the A-ring. It is the enol tautomer of the ketone that functions as nucleophile while the aldehyde is activated towards nucleophilic attack by oxygen protonation (**51**). Cyclization is completed by water elimination to furnish the enone system in **53** (acid-catalyzed aldol-condensation).

The MOM ether is produced under standard conditions with MOMCl and *Hünig's* base in CH_2Cl_2.

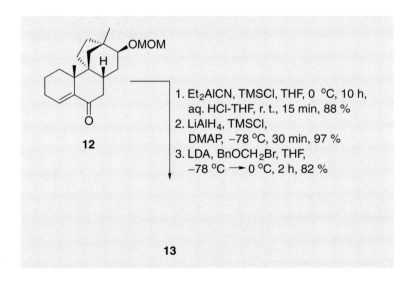

Problem

13 (−)-Scopadulcic Acid A

Hints

- First a conjugate addition to the enone takes place.
- The keto-function is reduced and subsequently protected as TMS ether.
- A benzyloxy-methyl substituent is introduced in the last step.

Solution

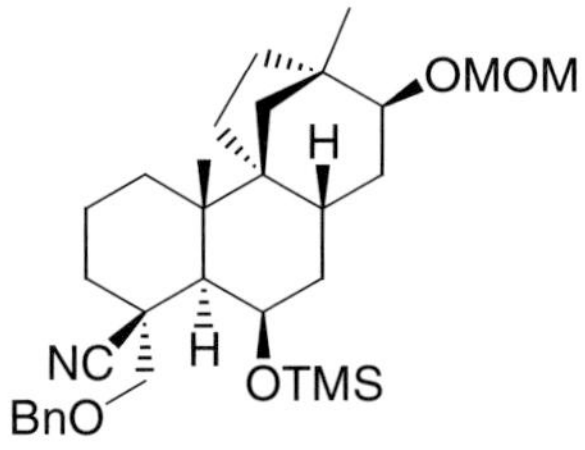

13

Discussion

1,4-addition of $Et_2Al^+ CN^-$ takes place selectively from the β-face. However, quenching of the resulting aluminum enolate **54** with acid provides some *cis* product **56**. To overcome this problem the enolate was quenched with TMSCl. The resulting enoxysilane **57** was hydrolyzed with diluted HCl to provide only the *trans*-fused ketone **55** in 88 % yield.

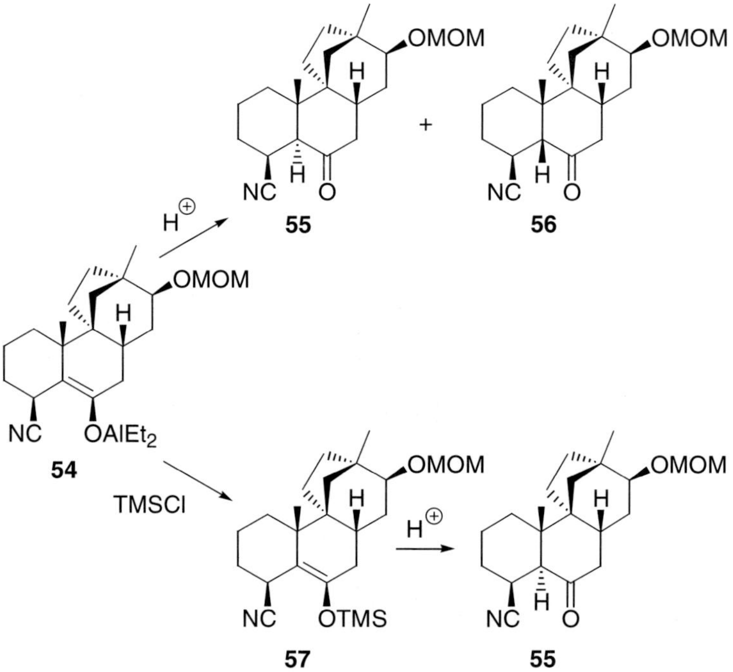

Reaction with LiAlH$_4$ at −78 °C selectively reduces the keto-function from the face opposite to the angular methyl group. Subsequently, the resulting alcohol is protected as TMS ether **58**.

The carbinyl nitrile acidifies the α-proton, which is deprotonated with LDA. The nucleophilic lithium salt is quenched with benzyloxymethyl bromide as electrophile to give **13** in 82 % yield. Again, the angular methyl group controls the facial selectivity: the benzyloxymethyl group is introduced from the α-face.

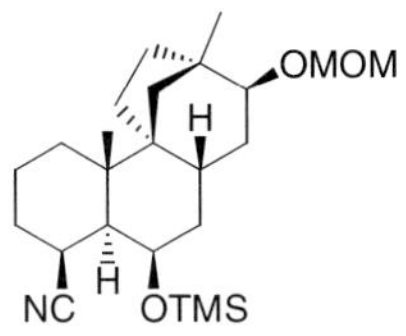

Problem

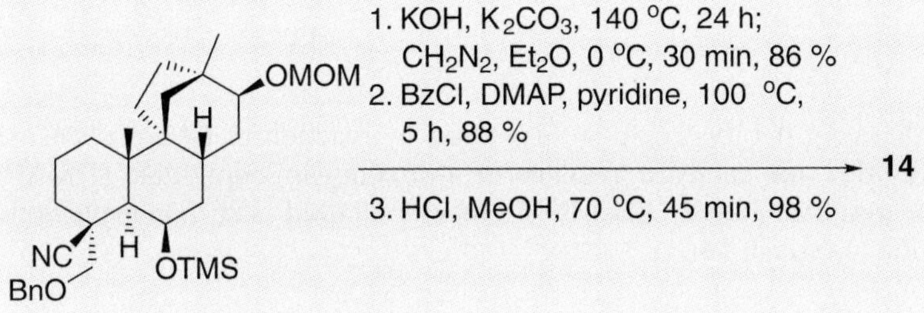

- First, a carboxylic acid is generated and subsequently esterified.
- Methoxymethyl ethers are stable to bases.

Hints

Solution

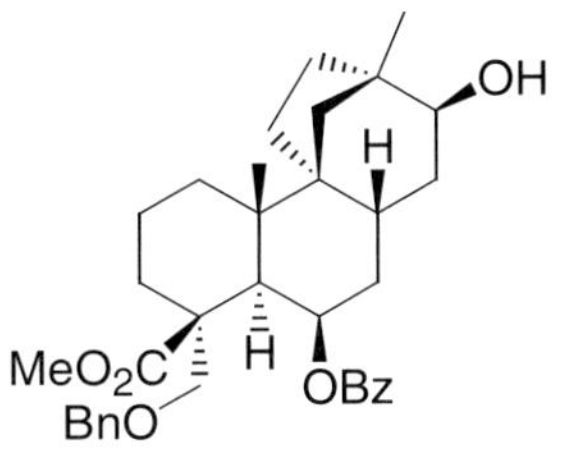

Discussion

Hydrolysis of the axial tertiary carbinyl nitrile as well as cleavage of the TMS group is accomplished under forcing conditions by heating up to 140 °C with KOH. Acidification of this product and esterification of the resulting carboxylic acid with diazomethane provides **64**. Diazomethane represents a mild and feasible reagent for preparing methyl esters especially in presence of acid or base sensitive functional groups. The mechanism can be explained as a two-step sequence. First the acid **59** is deprotonated in order to attack the protonated electrophilic diazomethane **62** in the second step. Finally methyl ester **63** is formed under loss of nitrogen.

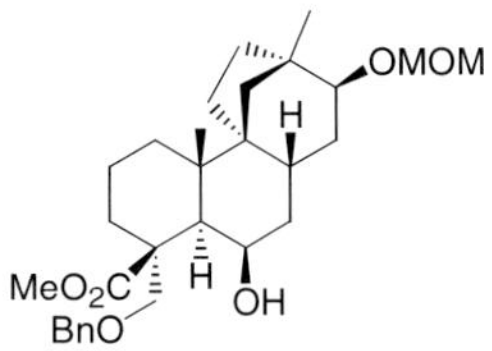

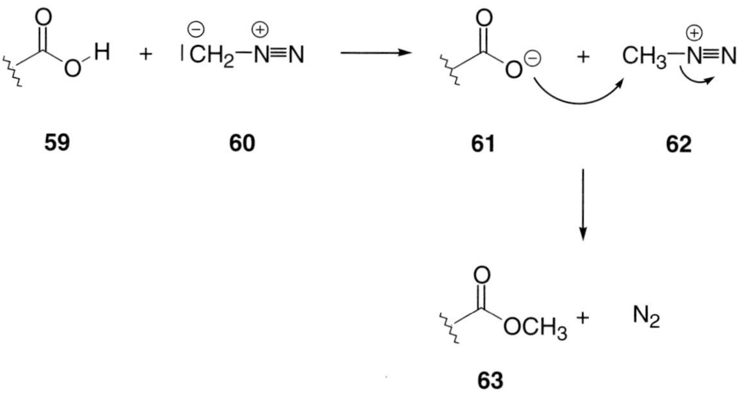

The axial C-6 benzoate is introduced by reaction of **64** with benzoyl chloride and catalytic amounts of 4-dimethylaminopyridine (DMAP) in pyridine. Finally the MOM ether is cleaved with hot methanolic HCl to furnish **14**.

Problem

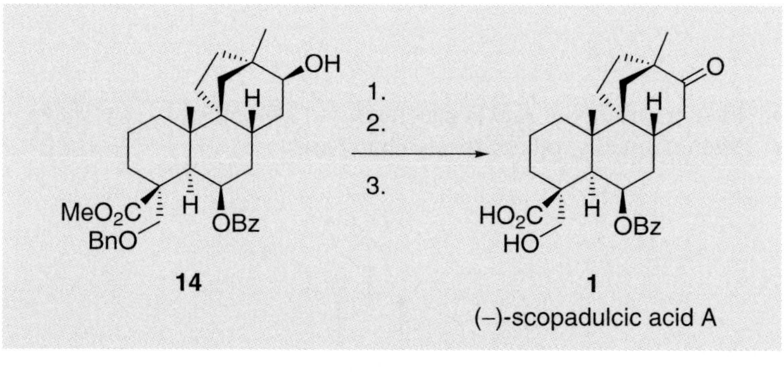

Hints

- The first step is an oxidation.
- Afterwards two protecting groups are cleaved.

Solution

1. PCC, r. t., 2.5 h, 90 %
2. nPrSLi, HMPA, r. t., 8 h, 72 %
3. H_2, 10 % Pd/C, MeOH, r. t., 9 h, 90 %

The secondary alcohol is readily oxidized with pyridinium chlorochromate (PCC)[25] (**65**) which is commercially available or easily prepared by addition of pyridine to a solution of chromium(VI)-oxide in hydrochloric acid.

Finally the methyl ester is cleaved with lithium *n*-propyl-mercaptide by nucleophilic displacement and the benzyl ether by conventional catalytic hydrogenolysis to furnish (−)-scopadulcic acid (**1**).

Discussion

65

13.4 Conclusion

The above described total synthesis features the first enantiodivergent approach to (+)- and (−)-scopadulcic acid A. The central transformations are the stereoselective carbonyl group reduction with (*S*)-Alpine Borane®, the use of enolization stereoselection to dictate which enantiomer is produced, and the palladium-catalyzed bis-*Heck* cyclization which occurs with complete stereo- and regiocontrol to establish the scopadulan scaffold.

This synthesis is completed in 29 steps with an overall yield of approximately 4 % from ketone **20**.

13.5 References

1   T. Hayashi, K. Okamura, M. Kakemi, S. Asano, M. Mizutani, N. Takeguchi, M. Kawasaki, Y. Tezuka, T. Kikuchi, N. Morita, *Chem. Pharm. Bull.* **1990**, *38*, 2740-2745.
2   T. Hayashi, M. Kishi, M. Kawasaki, M. Arisawa, M. Shimizu, S. Suzuki, M. Yoshizaki, N. Morita, Y. Tezuka, T. Kikuchi, L. H. Berganza, E. Ferro, I. Basualdo, *Tetrahedron Lett.* **1987**, *28*, 3693-3696.
3   T. Hayashi, K. Hayashi, K. Uchida, S. Niwayama, N. Morita, *Chem. Pharm. Bull.* **1990**, *38*, 239-242.
4   T. Hayashi, M. Kawasaki, Y. Miwa, T. Taga, N. Morita, *Chem. Pharm. Bull.* **1990**, *38*, 945-947.
5   H. Nishino, T. Hayashi, M. Arisawa, Y. Satomi, A. Iwashima, *Oncology* **1993**, *50*, 100-103.
6   T. Hayashi, M. Kishi, M. Kawasaki, M. Arisawa, N. Morita, *J. Nat. Prod.* **1988**, *51*, 360-363.
7   L. E. Overman, D. J. Ricca, V. D. Tran, *J. Am. Chem. Soc.* **1993**, *115*, 2042-2044.
8   F. E. Ziegler, O. B. Wallace, *J. Org. Chem.* **1995**, *60*, 3626-3636.
9   M. E. Fox, C. Li, J. P. Marino, Jr., L. E. Overman, *J. Am. Chem. Soc.* **1999**, *121*, 5467-5480.
10 H. C. Brown, G. G. Pai, *J. Org. Chem.* **1985**, *50*, 1384-1394.
11 R. Noyori, I. Tomino, Y. Tanimoto, M. Nishizawa, *J. Am. Chem. Soc.* **1984**, *106*, 6709-6716.

12   E. J. Corey, C. J. Helal, *Angew. Chem.* **1998**, *110*, 2092-2118; *Angew. Chem. Int. Ed. Engl.* **1998**, *37*, 1986-2012.
13   P. J. Kocienski, *Protecting Groups*, Thieme, Stuttgart, **1994**, 33.
14   R. Appel, *Angew. Chem.* **1975**, *87*, 863-874; *Angew. Chem. Int. Ed. Engl.* **1975**, *14*, 801-812.
15   H. M. R. Hoffmann, M. Mentzel, *J. Prakt. Chem.* **1997**, *339*, 517-524.
16   R. E. Ireland, R. H. Müller, A. K. Willard, *J. Am. Chem. Soc.* **1976**, *98*, 2868-2877.
17   A. C. Cope, E. M. Hardy, *J. Am. Chem. Soc.*, **1940**, *62*, 441-444.
18   B. E. Maryanoff, A. B. Reitz, *Chem. Rev.* **1989**, *89*, 863-927.
19   L. F. Tietze, *Chem. Rev.* **1996**, *96*, 115-136.
20   A. de Meijere, F. E. Meyer, *Angew. Chem.* **1994**, *106*, 2473-2506; *Angew. Chem. Int. Ed. Engl.* **1994**, *33*, 2379-2411.
21   S. V. Ley, J. Norman, W. P. Griffith, S. P. Marsden, *Synthesis* **1994**, 639-666.
22   W. P. Griffith, S. V. Ley, G. P. Whitcombe, A. D. White, *Chem. Comm.* **1987**, 1625-1627.
23   A. J. Mancuso, D. Swern, *Synthesis* **1981**, 165-185.
24   D. B. Dess, J. C. Martin, *J. Am. Chem. Soc.* **1991**, *113*, 7277-7287.
25   G. Piancatelli, A. Scettri, M. D'Auria, *Synthesis* **1982**, 245-258.
26   E. J. Corey, G. Schmidt, *Tetrahedron Lett.* **1979**, *5*, 399-402.
27   K. Bowden, I. M. Heilbron, E. R. Jones, B. C. L. Weedon, *J. Chem. Soc.* **1946**, 39-45.
28   G. H. Posner, *An Introduction to Synthesis Using Organocopper Reagents*, Wiley-Interscience; New York, **1980**.
29   J. L. Luche, C. Petrier, J. P. Lansard, A. E. Greene. *J. Org. Chem.* **1983**, *48*, 3837-3839.
30   K. W. Woods, P. Beak, *J. Am. Chem. Soc.* **1991**, *113*, 6281-6283.
31   D. K. Murphy, R. Alumbaugh, B. Rickborn, *J. Am. Chem. Soc.* **1969**, *91*, 2649-2653.

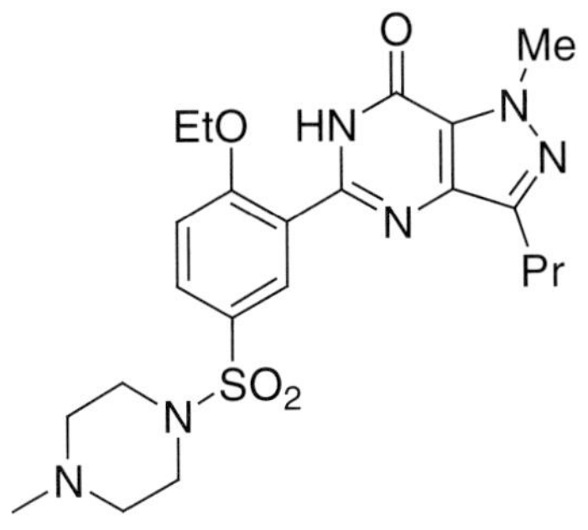

Sildenafil (VIAGRA™) (Pfizer 2000)

14

14.1 Introduction

Sildenafil is a selective inhibitor of phosphodiesterase 5 (PDE5) and is the first agent with this mode of action for the treatment of male erectile dysfunction. Pfizer produces this drug under the trademark VIAGRA and worldwide sales in its first year after FDA approval in 1998 totaled $788 million, increasing ever since. Before Viagra there was no orally active therapy for male erectile dysfunction and sildenafil was tested against hypertension and later against angina with little success. Rumor has it that the male patients involved in the clinical studies were reluctant to discontinue the tests without apparent reason.[1] The mode of action is described to proceed *via* inhibition of the phosphodiesterase enzyme PDE5 which is found in the human corpus cavernosum. This enzyme regulates the level of cyclic guanosine monophosphate (cGMP) by its conversion to guanosine monophosphate (GMP). The inhibition of PDE5 leads to higher levels of cGMP which in turn causes improved smooth muscle relaxation, increasing blood flow in the relevant body parts.

The chemical development of the commercial route to sildenafil also serves as an excellent example of the different issues that need to be considered when moving from drug discovery to commercial quantities. This problem is therefore based on the commercial synthesis of sildenafil as published by the Pfizer research group around *P. J. Dunn*.[2]

14.2 Overview

14 Sildenafil

14.3 Synthesis

Problem

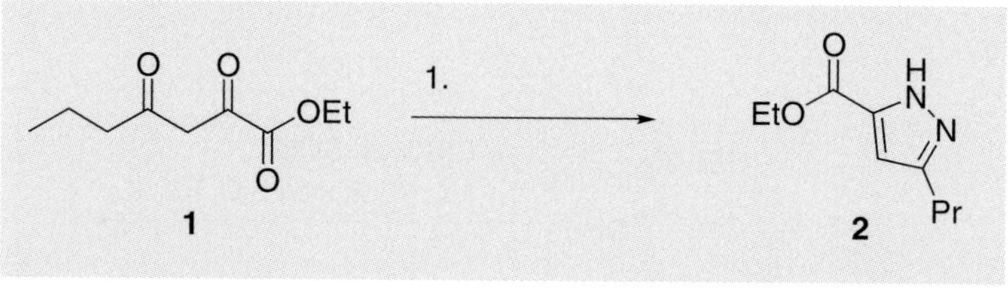

[Structure 1: CH3CH2CH2-C(=O)-CH2-C(=O)-C(=O)-OEt] → [Structure 2: ethyl 5-propyl-1H-pyrazole-3-carboxylate, with EtO-C(=O)- group on pyrazole and Pr group]

Hints
- You need to introduce two nitrogens in exchange for the ketones.
- An amine and a carbonyl group can form an imine.

Solution

1. $NH_2NH_2 \cdot H_2O$, EtOH, reflux, 2 h[3]

Discussion

Pyrazoles are very common parts of commercially available pharmaceuticals, agrochemicals and dyestuffs. The reaction of β-diketones with hydrazines is the most widely used method to synthesize pyrazoles. The reaction proceeds *via* the formation of hydrazone **13** which on subsequent cyclization and dehydration produces the corresponding pyrazole **2**. This method usually has the disadvantage that with unsymmetrical diketones generally a mixture of isomeric pyrazoles is formed. Here the authors report no other isomer probably because of the great electronic differences of the two ketones being substituted with an ester vs. an alkyl group.

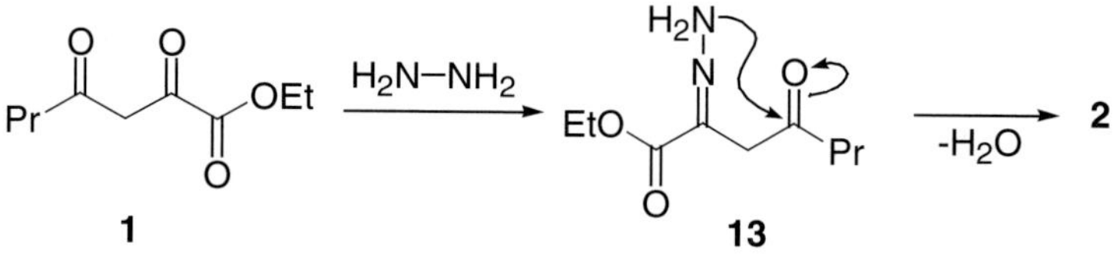

1 + H_2N-NH_2 → **13** → **2** (−H_2O)

Problem

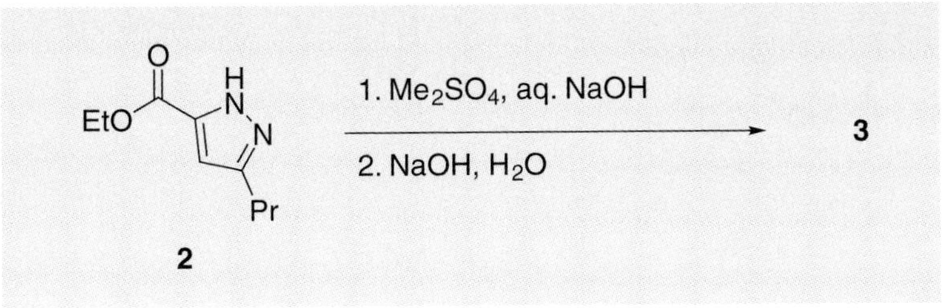

[Reaction: compound **2** (ethyl ester of 3-propyl-1H-pyrazole-5-carboxylate) → **3** via 1. Me$_2$SO$_4$, aq. NaOH; 2. NaOH, H$_2$O]

Hints

- Me$_2$SO$_4$ is a methylating agent.
- Basic conditions can cleave carboxylic esters.

Solution

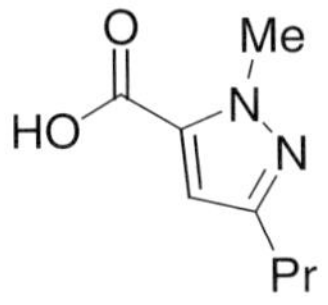

[Structure **3**: 1-methyl-3-propyl-1H-pyrazole-5-carboxylic acid]

Discussion

Me$_2$SO$_4$ is used under basic conditions as source of a methyl nucleophile. It can thus transform alcohols into methyl ethers or transform amines into the methylamine. Usually a phase transfer catalyst such as tetrabutylammonium iodide is added to the aqueous basic solution.[4] Other reagents to create methyl ethers include iodomethane or trimethoxonium tetrafluoroborate (Meerwein's reagent).[5]

Ester hydrolysis can proceed under acidic or basic (saponification) conditions to the carboxylic acid. In this case aqueous NaOH is used. The order of reaction, methylation before saponification, both under basic conditions, is presumably important since the free acid would be again methylated by Me$_2$SO$_4$.

Problem

Hints

- This is a standard nitration protocol.

Solution

Discussion

Nitration is one of the classical examples of electrophilic aromatic substitution as taught in introductory organic chemistry courses. When (hetero)aromatic compounds are treated with a mixture of nitric and sulfuric acid, nitronium ions (NO_2^+) are the electrophilic species. Aromatic heterocycles are divided into two general groups by their electronic properties and the resulting reactivity. Six-membered (pyridine-type) heteroaromats are electron poor (π-deficient) and thus their reactivity is affected by the electron-withdrawing effect of the heteroatom.[6] Six-membered heteroaromats therefore react fast with nucleophiles and often not in an electrophilic substitution. Five-membered (pyrrole-type) hetereoaromatic compounds are electron rich (π-excessive) and the lone electron pair of the heteroatom is part of the delocalized aromatic system. These compounds therefore react easily with electrophiles. Pyrazole belongs to the group of five-membered electron rich heterocycles and contains a pyrrole type nitrogen which releases electrons into the aromatic system. However, its second nitrogen is of the pyridine-type and deactivates the ring by its electron-withdrawing properties. Usually pyrazole reacts with electrophiles very well in the 4 position. In this case only the 4 position is sterically available making the intrinsic selectivity irrelevant.

This reaction is the most dangerous in the commercial synthesis since it requires at least 50 °C to start, and once the reaction starts it is very exothermic. This would increase the heat under adiabatic conditions still further until at 120 °C the carboxylic acid would decarboxylate in another exothermic reaction creating CO_2 gas thus increasing the pressure in the reactor. The research group at Pfizer designed the process to minimize the risk by dividing the procedure into three steps: First the pyrazole **3** was dissolved in conc. H_2SO_4, next the fuming nitric acid was mixed with conc. H_2SO_4 and added to the pyrazole depending on the conversion.

Problem

[Scheme: compound **4** (HO-C(O)- pyrazole with Me, N-N, O_2N, Pr substituents) →(1.) **5** →(2.) **6** (H_2N-C(O)- pyrazole with Me, N-N, H_2N, Pr)]

Hints

- The acid is transformed into the amide first.
- How can you activate the acid to make it reactive towards nucleophilic substitution?
- Ammonia is the nucleophile that is attacking the carbonyl group.
- In the second step the nitro group is reduced.
- Nitro groups can be reduced by H_2 under metal catalysis.

1. $SOCl_2$, DMF, toluene, 55 °C, 6 h; NH_3 (aq.), 20 °C, 2 h, 92 %

Solution

[Structure of compound **5**: H_2N-C(O)- pyrazole with Me, N-N, O_2N, Pr]

2. H_2 (50 psi), 5 % Pd/C, EtOAc, 50 °C, 4 h, directly used in the next step

The formation of amides is one of the most important reactions in organic chemistry, especially since peptides contain amide bonds between the amino acids. Since carboxylic acids are not reactive towards nucleophilic substitution they have to be activated first. One of the simplest methods is the transformation of the acid into the acid halide. This can be afforded by thionyl chloride ($SOCl_2$), or thionyl bromide ($SOBr_2$), phosphorus trichloride (PCl_3), phosphorus oxychloride ($POCl_3$), phosphorus pentachloride (PCl_5) or oxalyl chloride [$(COCl)_2$]. The addition of catalytic amounts of DMF enhances the reactivity by forming a *Vilsmeyer* intermediate such as

Discussion

14, which can also be isolated and used in the formation of acid chlorides.[7] Another mild and simple method to generate acid halides involves the use of triphenylphosphine and CCl_4 creating the acid chloride and triphenylphosphine oxide.[8]

In the described transformation the created acid chloride is attacked by aqueous ammonia as nucleophile to create the amide **5** in excellent yield.

In the medicinal chemistry route to sildenafil, the nitro group in **5** was reduced using $SnCl_2$ and hydrochloric acid in ethanol; in the commercial route this was changed to the application of H_2 under palladium catalysis in ethyl acetate. Since stannous compounds are highly toxic the aqueous waste stream from the reduction would have been hard to purify. The heterogeneous catalysis of palladium on charcoal in ethyl acetate leaves no residues and therefore the solution containing **6** could be used directly in the following coupling step.

Problem

Hints

- An electrophilic aromatic substitution takes place.
- A sulfonic acid is formed as intermediate.
- The sulfonic acid is transformed into a sulfonyl chloride.

Solution

Discussion

Molten 2-ethoxybenzoic acid (**7**) was added to a mixture of chlorosulfonic acid and thionyl chloride while keeping the reaction temperature below 25 °C. In this straightforward electrophilic aromatic substitution the ethoxy group directs the electrophile towards the *ortho* and *para* position whereas the carboxylic acid directs *meta* giving an overall selectivity for the attack at C-5. It was necessary to add thionyl chloride to transform the intermediate sulfonic acid into the sulfonyl chloride. After quenching with icewater the product **8** could simply be precipitated and washed with water. The benzoic acid chloride, which was probably formed intermediately by thionyl chloride, was hydrolyzed in the work-up, too.

Problem

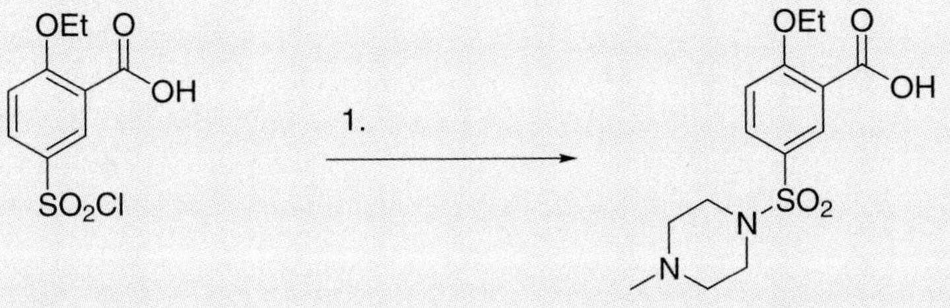

8 → **9**

Hints

- The sulfonyl chloride is reactive towards nucleophilic substitution at sulfur.
- What amine would be the necessary nucleophile?

Solution

1. N-methylpiperazine, H$_2$O, 10 °C, 2 h, 86 %

Discussion

In this simple step, the sulfonyl chloride **8** can be slurried in H$_2$O directly as the wet filter cake from the previous reaction. N-methylpiperazine is added and the nucleophilic substitution takes place below 20 °C. The title compound again can be crystallized out of the water solution and isolated by simple filtration.

14 Sildenafil

Problem

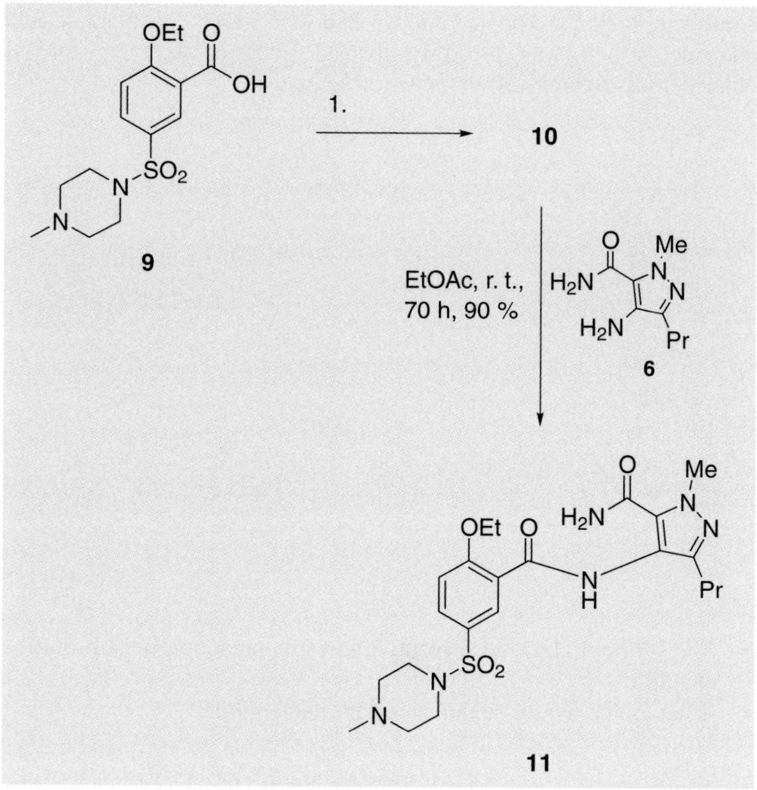

[Scheme: compound **9** (2-ethoxy-5-(4-methylpiperazin-1-ylsulfonyl)benzoic acid) → **10** via step 1; then reacted with amine **6** (4-amino-1-methyl-3-propyl-1H-pyrazole-5-carboxamide) in EtOAc, r.t., 70 h, 90% → **11**]

Hints

- The benzoic acid needs to be activated before amide bond formation can take place.
- **10** is the activated species which will react with the amine **6**.
- A five-membered heterocycle is used to activate the acid.

Solution

1. N'-carbonyldiimidazole, EtOAc, reflux, 2 h

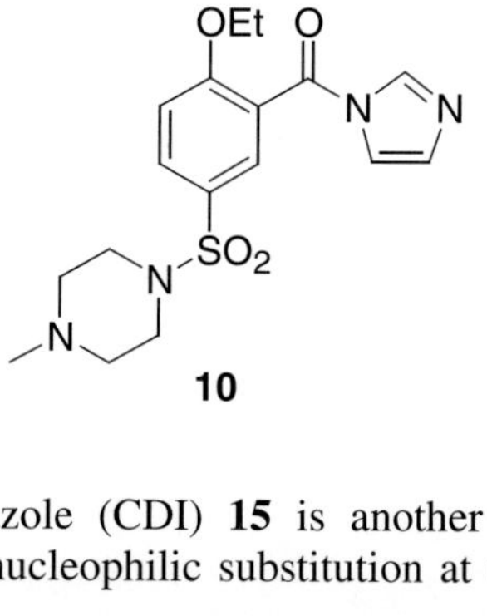

[Structure of **10**: the benzoyl imidazolide]

Discussion

N,N'-Carbonyldiimidazole (CDI) **15** is another reagent to activate carboxylic acids for nucleophilic substitution at the carbonyl group.[9]

14 Sildenafil

In aprotic media a 1-(acyloxycarbonyl)imidazole such as **16** is formed primarily which reacts to the acylimidazole and carbon dioxide. Imidazole now serves as a good leaving group and so the previously synthesized amine **6** could be added and the desired amide was formed *via* the usual addition elimination mechanism. One of the advantages of using this more expensive way of activation is the possibility to run the nitro reduction, acid activation and acylation in the same solvent (ethyl acetate); thus all three reactions could be telescoped into a single step during production.

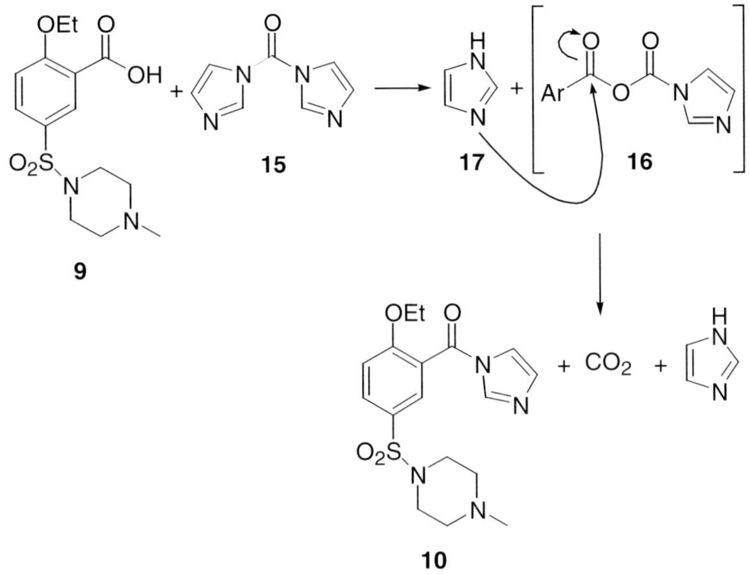

The acylimidazoles could be considered as amides themselves; they do not have a free electron pair to engage in amide mesomerism, however. Thus, the bond between imidazole nitrogen and carbonyl carbon is not as strong as an amide bond, making it easy to cleave. Amide bonds are one of the strongest connections in organic molecules and necessary for the stability of peptides. Interestingly, nature uses enzymes that contain histidine (the imidazole amino acid) in the active site to cleave amide bonds by an analogous intermediate as in the formation of amides with imidazole.

Other ways to form amide bonds are explained in Chapters 5 and 12 or can be found in excellent monographs.[10]

Problem

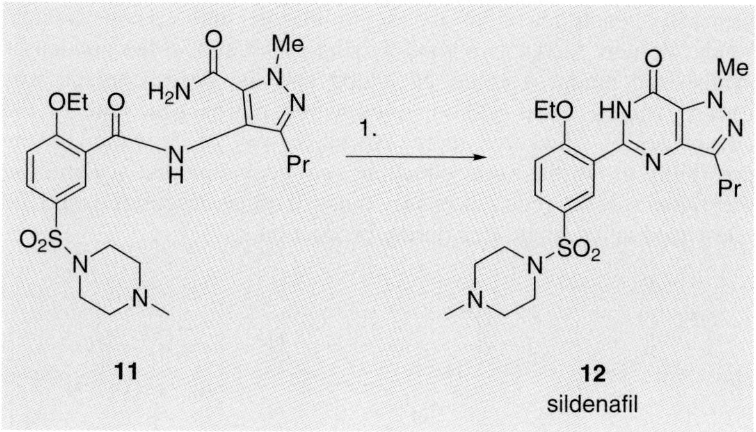

Hints
- Ketones and amines can form imines; why shouldn't two amides do the same?
- The amide needs to be activated by a base in order to be more nucleophilic.

Solution

1. KO*t*Bu, *t*BuOH, reflux, 8 h; HCl/H$_2$O, 10 °C, 1 h, 90 %

Discussion

The primary amide is deprotonated by the base *tert*-butoxide making it more nucleophilic. The nitrogen will then act as nucleophile and attack the other amide carbon closing the ring. Isomerization will lead to the pyrimidone ring in sildenafil concluding this synthesis. This last step involves only water soluble solvents and reagents and the final product again precipitates out of the aqueous solution upon reaching pH 7.5. No further purification is necessary and clinical quality Viagra™ is obtained directly from the filtration.

14.4 Conclusion

The synthesis of sildenafil serves as an excellent example of the demands of commercial chemistry. The route described contains all of the desired attributes required in chemical development, namely a safe, robust route, a convergent synthesis and a high yielding process. The authors managed to improve the yield from 7.5 % in the medicinal chemistry to 75.8 % overall from pyrazole **3**. The synthesis also has an exceptionally low environmental impact. Only toluene and ethyl acetate are organic waste while the other solvents (ethanol and *tert*-butanol) can be treated in the water plants. The synthesis has been reassembled to make it more convergent and to put clean steps at the end of the process.

This synthesis also gives a small glimpse at the chemistry of heterocyclic compounds. Most active compounds in today's pharmaceuticals or agrochemicals include heterocycles, as well as most vitamins and natural products. The chemistry of heterocycles is thus very important and lectures or textbooks should be consulted.[6]

Formation of amide bonds also plays a large role in this problem. It was demonstrated that the strong amide bond can be formed from an amine and a carboxylic acid only after the acid has been activated. This can be done by transformation into the carboxylic halide or imidazolide or by application of an activating agent developed for peptide synthesis.

14.5 References

1. J. Kling, *Modern Drug Disc.* **1998**, *1*, 31-38.
2. D. J. Dale, P. J. Dunn, C. Golightly, M. L. Hughes, P. C. Levett, A. K. Pearce, P. M. Searle, G. Ward, A. S. Wood, *Org. Proc. Res. Develop.* **2000**, *4*, 17-22.
3. For the synthesis of compound **3** see: N. K. Terrett, A. S. Bell, D. Brown, P. Ellis, *Bioorg. Med. Chem. Lett.* **1996**, *6*, 1819-1824. However no reaction details and yields are given. Whenever possible details are taken from analogous reactions published.
4. A. Merz, *Angew. Chem. Int. Ed. Engl.* **1973**, *12*, 846-847.
5. H. Meerwein, G. Hinz, P. Hofmann, E. Kroning, E. Pfeil, *J. Prakt. Chem.* **1937**, *147*, 257-261.
6. A) T. Eicher, S. Hauptmann, *The Chemistry of Heterocycles: Structure, Reactions, Syntheses and Applications*, Thieme, Stuttgart **1994**; b) R. R. Gupta, M. Kumar, V. Gupta, *Heterocyclic Chemistry I and II*, Springer Verlag, Berlin **1999**.
7. H. H. Bosshard, R. Mory, M. Schmid, H. Zollinger, *Helv. Chim. Act.* **1959**, *42*, 1653-1658.
8. L. E. Barstow, V. J. Hruby, *J. Org. Chem.* **1971**, *9*, 1305-1306.
9. H. A. Staab, M. Lücking, F. H. Dürr, *Chem. Ber.* **1962**, *95*, 1275-1283.
10. A. J. Pearson, W. J. Roush (ed.) *Handbook of Reagents for Organic Synthesis – Activating Agents and Protecting Groups*, John Wiley & Sons, Chichester **1999**.

15

GM2 (Schmidt 1997)

15.1 Introduction

Gangliosides are sialic acid-containing glycosphingolipids, which are widely expressed in mammalian tissues. They can be tumor-associated antigens, and furthermore efficient receptors for the adhesion of bacteria and viruses to cells, a prerequisite for infection. In general, glycoconjugates play important roles in many cellular and physiological processes, for example development and differentiation of cells, recognition of microorganisms and intracellular transport of proteins.[1]

The GM2 ganglioside (**14**) is an example of a carbohydrate antigen, specific for melanoma, sarcoma and kidney carcinoma. It could suit the purpose of an immuno therapy with monoclonal antibodies. In previous examinations the used GM2 was extracted from human and animal tissues. In this, one ran the risk of biological contamination of the GM2, which could influence the results of an examination. For this reason *R. R. Schmidt* et al. developed a total synthesis of the GM2 ganglioside to sustain sufficient substance for further immunological examinations without faking artifacts.[2]

GM2 can be separated into a hydrophile oligosaccharide part and a lipophile term. With its oligosaccharid moiety it is exposed on the cell surface of eukaryotic cells as part of the glycocalix. The GM2 carbohydrate structure consists of D-(+)-lactose (**1**), *N*-acetyl-D-(+)-galactosamine (**2**) and *N*-acetyl-neuraminic acid (**3**). The lipophile ceramide term (**4**) of GM2, serves as membrane anchor on a cell. Ceramide consists of a fatty acid with C_{16}-C_{34} carbon atoms length and an amino alcohol with a chain length of C_{12}-C_{20} carbon atoms, two to three hydroxyl groups and zero to two double bonds. The amino alcohols are homologs of the most widespread C_{18}-aminodiol base sphingosine.

1
D-(+)-lactose
(β-Gal-(1→4)-β-Glc)

2
N-acetyl-D-(+)-galactoseamine
(GalNAc)

3
N-acetyl-neuraminic acid
(Neu5Ac)

4
ceramide
(Cer)

15.2 Overview

1. Sn(OTf)$_2$, MeCN, −40 °C
61 %, α : β 9:1

7 + 10 → **11**

15
D-(+)-galactose

α-15
α-D-(+)-glucopyranose

↕ H⊕ or OH⊖

16
aldehyde form

↕ H⊕ or OH⊖

β-15
β-D-(+)-glucopyranose

If you already have some experience in carbohydrate chemistry, you can directly step over to *15.3 Synthesis* on page 249.

D-(+)-galactose (**15**) is an example of the consecutive numbering of the carbon ring atoms in a monosaccharide (disaccharide see p. 253). Carbohydrates can exist in a cyclic and an acyclic structure. For this reason there is a special position in the structure of a monosaccharide, the carbon atom C-1 and so called anomeric center. You can see that there is an equilibrium between α-anomer α-**15** and β-anomer β-**15** of D-(+)-glucopyranose over the acyclic aldehyde structure **16**. Both are cyclic *hemi*-acetals. The β-anomer is the preferred conformation, but there are a few effects, like sterical or stereoelectronical effects (anomeric effect, inverse anomeric effect), which have influence on the α : β rate.

Often the synthesis of the demanded compound is linear over most of the steps. Therefore it is very important to build a new C-O bond with high stereoselectivity and only one anomer in the ideal case. To build a new C-O bond you need a so-called glycosyl donor and a glycosyl acceptor. The donor possesses a leaving group at the carbon atom which should be the "donated" C-atom for the new bond. There are many different leaving groups in use, but a few like the trichloroacetimidate[3,4] are used more frequently.

The acceptor should have only one unprotected OH-group, which "accepts" another carbon atom to build a new bond. If there is more than one unprotected OH-group it is necessary to take advantage of the different reactivity of these groups under optimized conditions. The unique reactivities depend on the sugar itself and for example the neighboring groups to this reaction center.

Carbohydrates are polyfunctional molecules; therefore you have to turn your attention to the protecting group strategy during the synthesis.

Activation or deactivation of a desired position in a carbohydrate compound is achieved by the choose of protecting groups. Another possibility is the implementation of an activating neighbor group. The less one has to utilize protecting groups, the better it is in the face of the yield and the length of a synthesis. Each protecting group stands for two synthetic steps which can be associated with a reduction in yield and an increase in costs.

Another important influence is the selection of the solvent for every reaction step. The α : β ratio of a glycosylation can be completely reversed by changing the solvent.

For further more detailed information you may have a look at the literature referred to or some textbooks about carbohydrate chemistry and protecting groups.[5,6]

15.3 Synthesis

Problem

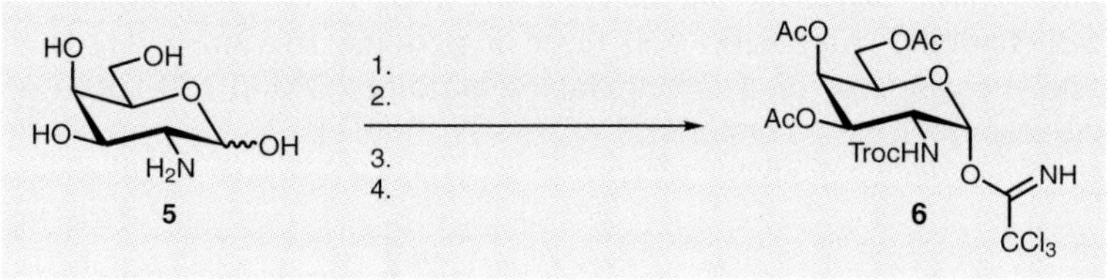

[Scheme: compound **5** (aminosugar with HO, OH, H₂N, OH groups) → compound **6** (AcO, OAc, AcO, TrocHN, anomeric O–C(=NH)CCl₃) via steps 1., 2., 3., 4.]

- First the amino functionality is protected and the carbohydrate is peracetylated.
- How can the protecting group at the anomeric center be cleaved selectively?
- Cl$_3$CCN is used to activate the anomeric center. (trichloroacetimidate method)

Hints

1. TrocCl, NaHCO$_3$, H$_2$O, 0 °C, 12 h
2. Ac$_2$O, pyridine, DMAP, r. t., 24 h, 91 % (over two steps)
3. N$_2$H$_4$·HOAc, DMF, 45 °C, 30 min, 93 %
4. Cl$_3$CCN, DBU, CH$_2$Cl$_2$, –20 °C, 2 h, 80 %

Solution

The first step, protection of the NH$_2$ group, is performed in order to deactivate this reaction center for later glycosylation to the tetrasaccharide. Using 2,2,2-trichloroethoxycarbonyl chloride (TrocCl) **17** is a standard procedure to protect amino groups.[7] The acetyl group is not suitable because of its strong neighbor group effect, stabilizing the intermediate oxonium ion **20**. During the subsequent glycosylation the stable oxazoline **21** would be formed.

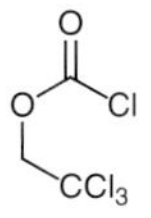

Discussion

17
TrocCl
[structure: Cl–C(=O)–O–CH$_2$–CCl$_3$]

[Scheme showing compounds **18** → **19** (−X⁻), **19** ⇌ **20**, **20** → **21** (−H⁺)]

18 **19**
21 **20**

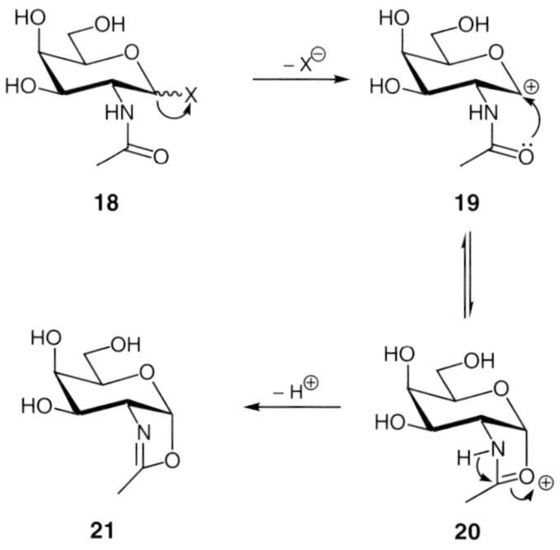

The Troc group has a weaker neighbor group effect than the acetyl group. (Ether protected trichloroacetimidates are more reactive than ester protected ones.) Furthermore the Troc group can be removed easily and stereoselectively.

There are different methodologies known to peracetylate a carbohydrate. An acetic anhydride / pyridine mixture (2:1) with catalytic amounts of 4-dimethylaminopyridine (DMAP) to speed up the reaction is very common.[8,9]

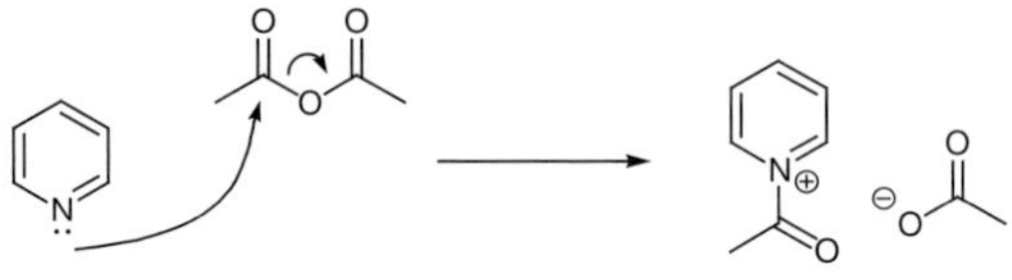

First *N*-acetylpyridinium acetate (**24**) is formed, which is the acetylating species during this step. This is a relatively mild method and leads to **25**.

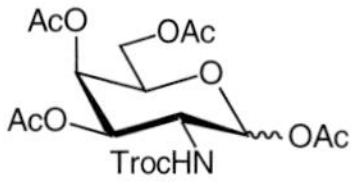

To get the desired trichloroacetimidate at C-1 you have to deprotect this position selectively. With hydrazine acetate ($N_2H_4 \cdot HOAc$) in DMF there is a possibility to do this.[10] This reaction is non-catalytic; you have to take equivalent amounts of hydrazine acetate. Generally, trichloroacetimidates are stable and represent good glycosyl donors.[11]

Non-participating protecting groups like benzyl ethers in non-polar solvents, with weak *Lewis* acids as catalysts, lead to the opposite stereochemistry of the utilized trichloroacetimidate. This is comparable with S_N2 reactions. Participating neighbor groups with a strong catalyst in a polar solvent lead to 1,2-*trans* glycosides in a subsequent glycosylation.[12]

One can manage the formation of the kinetical *β*- or the thermodynamical *α*-anomer by means of different reaction protocols.

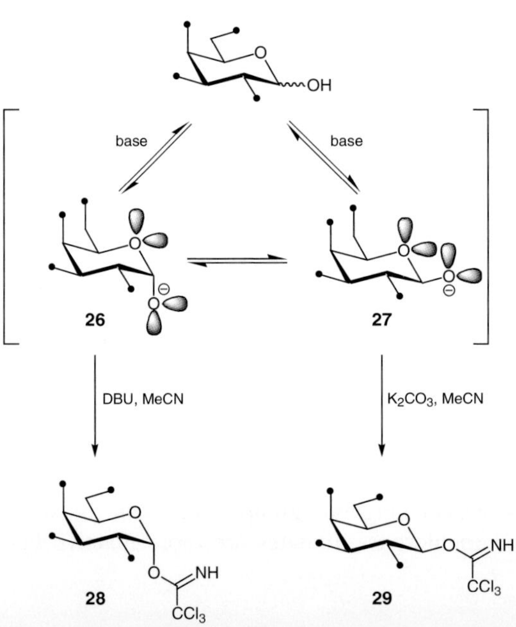

Application of 1,8-diazabicyclo[5.4.0]undec-7-ene (DBU) (**30**) as base will exclusively yield the α formation; using K_2CO_3 you will get the β-anomer. After abstraction of a proton at the anomeric center you get an equilibrium **26** ↔ **27**. The interactions of the electron pairs of **26** lead to a bigger nucleophilicity than **27**. Compound **26** is stabilized by the anomeric effect.

The anomeric effect, a stereoelectronic effect, is explained in terms of lone pair-lone pair repulsion, dipole-dipole interactions and by M.O. theory. The equatorial positions of a carbohydrate are favored by sterically demanding substituents. However, electronegative groups at the anomeric center prefer the axial position because of the stereoelectronic effects. This fact is known as the anomeric effect.[13]

If there is a positive charge at the anomer substituent of a carbohydrate, the equatorial conformation is preferred. To explain this result a reverse anomeric effect was proposed and first detected at N-(α-glycopyranosyl)pyridinium ions **31** and **32**.[14]

30
1,8-diazabicyclo[5.4.0]undec-7-ene (DBU)

31   **32**

The existence of the reverse anomeric effect is controversial and seems to be rebuttet for pyranoses by NMR titration studies with N-(D-glucopyranosyl)imidazole (**33**). Nitrogen-protonation of the imidazole should increase the proportion of the β-anomer if there is a reverse anomeric effect. During and after titration no shift of the α : β- equilibrium to the β-anomer was observed.[15]

R = H, Ac

33

Problem

Neu5Ac **3** → (1. 2. 3.) → **7**

Hints

- This time the COOH group is protected first.
- Selective protection of all other functionalities except the OH group at C-2 of the neuraminic acid is performed.
- Finally sialic acid is transformed into the shown glycosyl donor.

Solution

1. CH_2N_2, Et_2O, 0 °C → r. t., 12 h
2. Ac_2O, $HClO_4$, 20 °C, 2 h
3. $ClP(OEt)_2$, CH_2Cl_2, $EtNiPr_2$, r. t., 10 min

97 % (over three steps)

Discussion

The methylester **38** of the carboxylic acid **34** is formed by using diazomethane **35**.[16]

The 1,3 dipole diazomethane is a mild reagent to furnish methyl esters (see Chapter 13), but it has some disadvantages, too: it is volatile, toxic and furthermore explosive. For this reason it has to be prepared by reaction of KOH with N-methyl-N-nitroso-*para*-toluenesulfonamide (carcinogenic!) or *in situ*.[17] Another simple method to protect the COOH functionality of the neuraminic acid is the esterification with methanol as solvent and reactand under H^+ catalysis e. g. ion exchanger.

In the following, all alcohol functionalities besides the C-2 of the neuraminic acid have to be protected. *Kuhn* et al. describe a simple method to get product **39** in good yields.[18] By means of the different reactivities of the OH-groups the peracetylated neuraminic acid or **39** is obtained by carrying out the reaction at 20 °C for 2 h in acetic anhydride with catalytic amounts of perchloric acid ($HClO_4$). If the temperature or the reaction time is rising you get the completely peracetylated product. The advantage of perchloric acid is the low basicity of the anion ClO_4^-. Instead, the use of sulfuric acid as catalyst yields the peracetylated glycal **40**.

Finally a good leaving group has to be introduced to transform **39** into the glycosyl donor **7**. It has been shown that glycosyl phosphite donors can easily be synthesized and are more stable than some other common glycosyl donors. Within a few minutes stirring at room temperature in the presence of N-ethyldiisopropylamine **41** (*Hünig*'s base) you get **7** in good yields.[19]

Problem

Lactose **1** →(1., 2.) **8**

Hints

- How can an acetal be sustained?
- What kinds of silylating reagent are used to protect a secondary alcohol as TDS ether?

Solution

1. Me$_2$CO, H$^+$ (ion-exchange resin), r. t., 24 h
2. TDSCl, imidazole, DMF, r. t., 3 h

Discussion

Lactose is transformed into a glycosyl acceptor finally. A view on the oligosaccharide moiety of the GM2 shows the connection of the two already built glycosyl donors to the hydroxy groups at C-3b and C-4b of the lactose. To get the free OH functionalities later, in due time they have to be deprotected selectively.

First the isopropylidene ketal at C-3b and C-4b is generated using acetone and an ion-exchange resin at first. There is no further ketal formation at other positions of the lactose. It is also possible to use 2,2-dimethoxypropane (Me)$_2$C(OMe)$_2$ instead of acetone. The problem in this step is the partial formation of *hemi*-ketals with the other hydroxy groups which lower the yield. They are destroyed by heating to 100 °C in glacial acetic acid.[20]

Hereafter the anomeric center is protected. When the ceramide building block has to be connected to the oligosaccharide, normally selectively deprotection of the C-1a position of the lactose is necessary. The thexyldimethylsilyl (TDS) group is a less common trialkylsilyl protecting group in carbohydrate synthesis. Advantageously it survives some rather harsh conditions. The liquid TDSCl **42** with imidazole **43** or DMAP as basic activator in a dipolar aprotic solvent like DMF forms the C-1 protected compound **8** as a mixture of anomers.[21]

1

42
TDSCl

43
imidazole

Problem

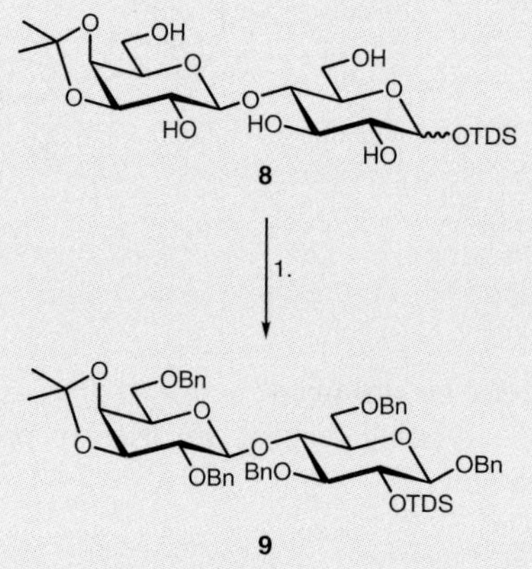

[Structure 8: disaccharide with isopropylidene, free OH groups, and OTDS]

↓ 1.

[Structure 9: same disaccharide with OBn groups replacing OH, retaining isopropylidene and OTDS]

Hints

- Which method is used to benzylate an alcohol?
- The exocyclic *O*-atoms of the sugar have to be activated before the benzyl ether is formed.

Solution

1. NaH, BnBr, DMF, r. t., 15 h
65 % (over three steps)

Discussion

Deprotection of the hydroxy groups at C-3b and C-4b will take place under acidic conditions later in the synthesis. Therefore a protecting group, which is stable during the cleavage of the isopropylidene ketal, is needed. The benzyl group is an often used protecting group and is suitable for this purpose. Also it increases the reactivity of the lactose acceptor. It can be removed by hydrogenolysis for instance.

During the benzylation of an alcohol the hydroxy functionalities are turned into alkoxides. Therefore NaH is necessary. Next benzyl bromide is added. For an efficient alkylation using sodium hydride as base dipolar aprotic solvents like DMF or DMSO are necessary.[22]

During the benzylation a reversible TDS rearrangement between C-1a and C-2a of the lactose occurs.[23] First deprotonation with NaH leads to C-2a-oxide **44** formation and then to the rearrangement.

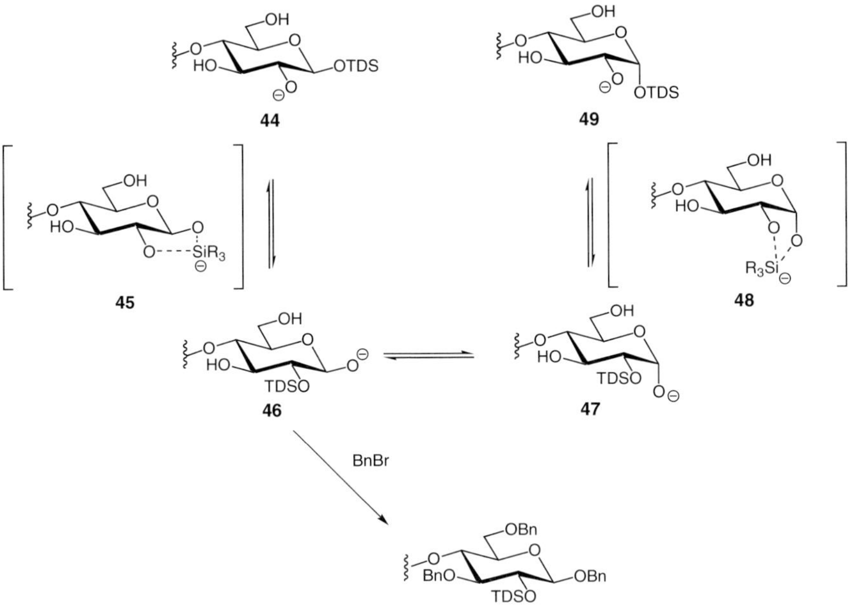

As displayed above there is an anomerization of **44** to **49** due to *trans*-(2→1)-*O*- (**44**→**46**) and ensuring *cis*-(2→1)-*O*- (**47**→**49**) silyl group migration, presumably *via* pentacovalent silicon intermediates or transition states **45** and **48**.

Because of the high nucleophilicity **46** is trapped by means of fast irreversible anomeric *O*-alkylation and leads to **50**.[3,23] In general the equatorial oxygen bears the highest nucleophilicity under these reaction conditions. Accompanying the *O*-silyl group rearrangement and the anomeric benzylation excess NaH and benzyl bromide provides fully benzyl protected lactose derivative **9**.

Problem

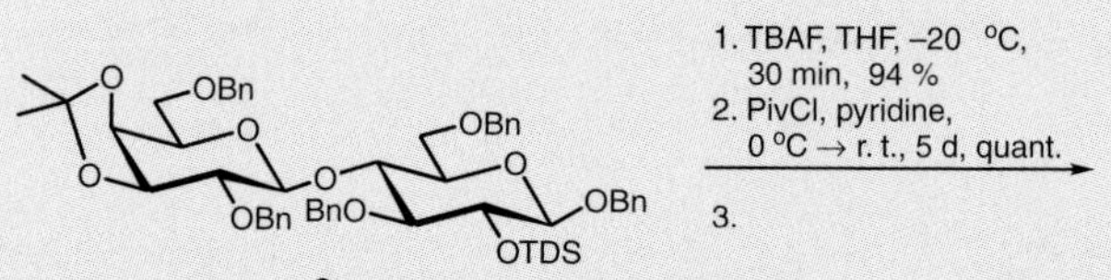

[Scheme: compound **9** → **10** with reagents:
1. TBAF, THF, −20 °C, 30 min, 94 %
2. PivCl, pyridine, 0 °C → r. t., 5 d, quant.
3.]

Hints

- Protecting groups are changed.

Solution

3. TFA/ H$_2$O, CH$_2$Cl$_2$, 100 °C, 1 h, 85 %

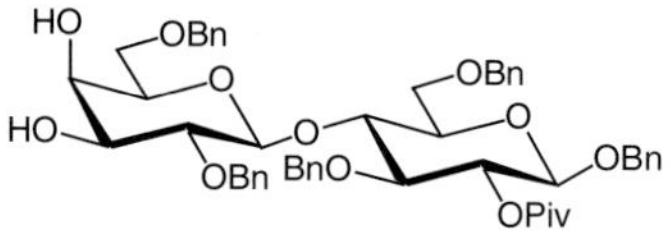

[Structure **10**]

Discussion

In the first step tetrabutylammonium fluoride (TBAF) cleaves the silyl protecting group at C-2a (**51**). Secondly the pivaloate ester **53** is formed.[24] In the last step the ketal was removed under acidic conditions by using trifluoroacetic acid (TFA) in water and CH$_2$Cl$_2$.

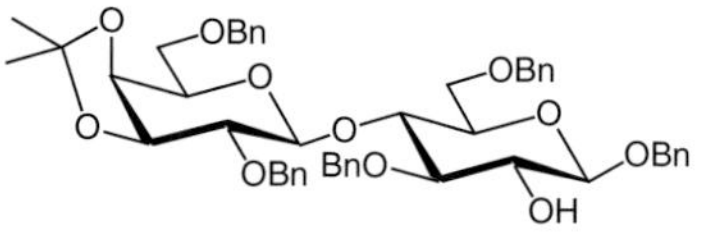

[Structures: **51**, **52** pivaloyl chloride, **53**]

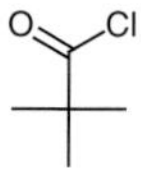

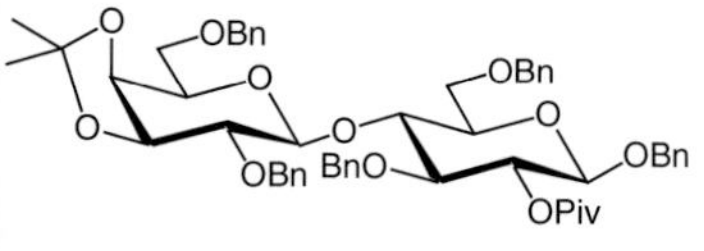

[Mechanism scheme: After protonation of the ketal **54** and addition of water hemi-ketal **56** is formed via intermediate **55** (+H$_2$O, −H$^⊕$), then proton migration gives **57** + acetone **58**.]

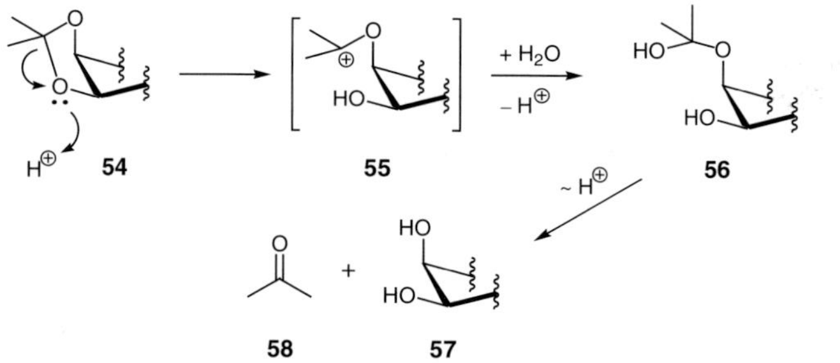

water *hemi*-ketal **56** is formed and leads after proton migration to the deprotected diol **57** and acetone **58**.

Although acetic acid is less acidic than TFA, with acetic acid (80 %) at 100 °C the same result is reached.

Problem

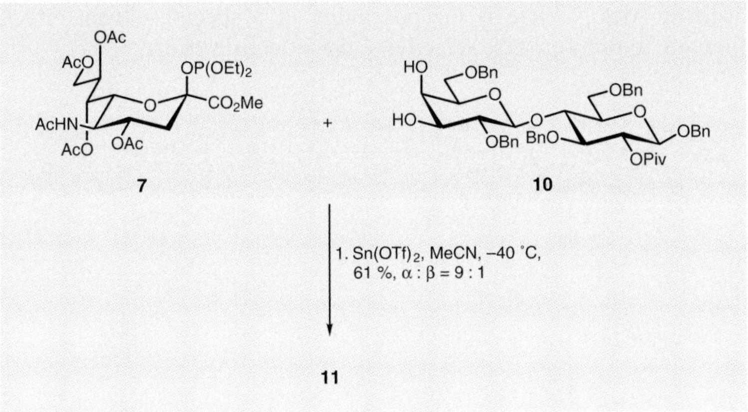

[Scheme: Compound **7** (sugar with OAc, AcO, AcHN, OAc, OP(OEt)₂, CO₂Me groups) + Compound **10** (trisaccharide with HO, HO, OBn, OBn, BnO, OBn, OPiv groups) → via 1. Sn(OTf)₂, MeCN, –40 °C, 61 %, α : β = 9 : 1 → Compound **11**]

- A new glycosidic bond is formed.
- Which hydroxy group of your acceptor is more reactive?

Hints

Solution

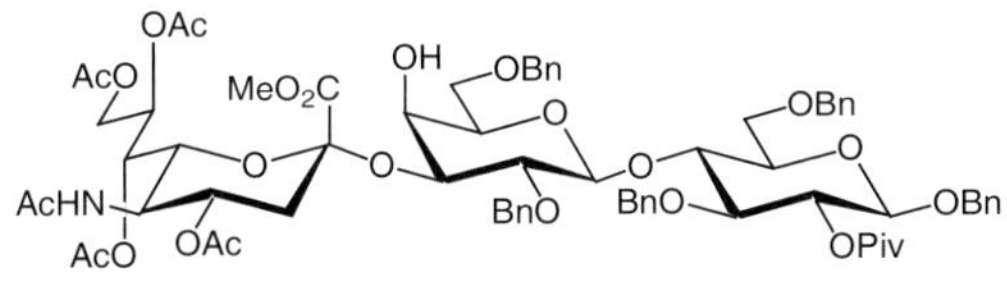

[Structure of compound **11**: tetrasaccharide with OAc, AcO, AcHN, AcO, OAc, MeO₂C, OH, OBn, BnO, BnO, OBn, OPiv groups]

Stannous triflate in acetonitrile as solvent at –40 °C catalyzes the glycosylation of the phosphite donor **7** with the acceptor **10**. As result an α-selective linkage takes place.

This is the first glycosylation in this synthesis. There are several methods to build a new glycosydic C-O bond. Very important methods are the *Koenigs-Knorr*[25] and related reactions and the trichloroacetimidate method.[26] It depends on the carbohydrates themselves, what kind of method is useful and which donor and acceptor you have to use.

The solvent plays an important role for the stereoselectivity of a glycosylation. For this reaction protocol acetonitrile is used.

Stannous triflate originally is used for the selective formation of β-glucosides.[27] In this example *Schmidt* et al. get an α : β ratio of 9 : 1.

Discussion

Utilizing MeCN there is the possibility of a special solvent effect, directing to this converse selectivity, named nitrile effect.[28,29,30]

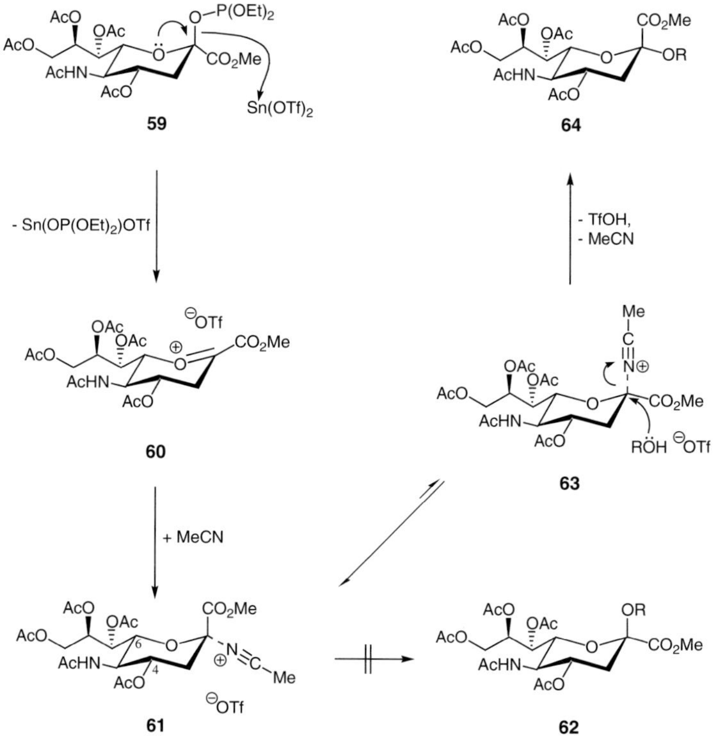

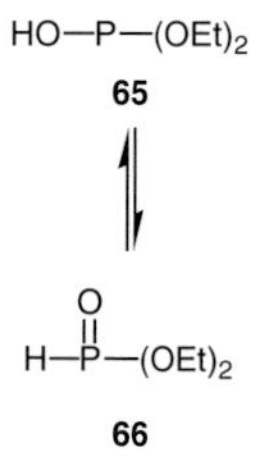

Starting with phosphite **59** stannous triflate catalyzes the formation of an oxenium ion **60**, which is stabilized by the solvent (**61**). The complex **61** is more favored than **63**, justified by the reverse anomeric effect until now. But as it is shown there is no possibility for **61** to react in an S_N2 reaction to **62**. The intermediate **63** can be attacked by the alcohol functionality of the glycosyl acceptor, and with MeCN as leaving group you get **64**. Ultimately the equilibrium between **61** and **63** shifts completely to **63**. This kind of stereocontrol during a glycosylation is known as the nitrile effect.[28]

The released TfOH reacts with the trivalent stannoyl phosphite and gives the hydrogen phosphite **65**.[31] Stannous triflate acts as catalyst.

Problem

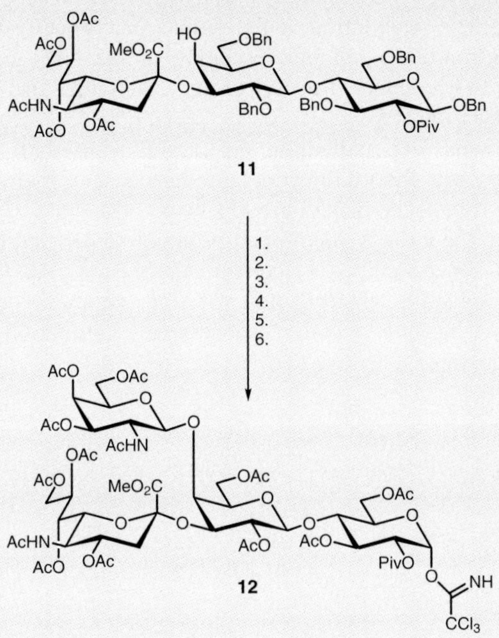

11

1.
2.
3.
4.
5.
6.

12

Hints

- Building block **6** and an Si-reagent as catalyst are used in the first step.
- A transition metal for reductive elimination is needed to cleave Troc.
- Benzyl groups can be removed hydrogenolytically with a transition metal as catalyst.
- The last step to build the donor is an already known method.

Solution

1. TMSOTf, CH_2Cl_2, **6**, −40 °C, 1 h, 89 %
2. Zn, Ac_2O, r. t., 5 h, 86 %
3. Pd/C, H_2 (5 bar), MeOH/ HOAc, r. t., 24 h
4. Ac_2O, pyridine, DMAP, r. t., 12 h, 84 % (over two steps)
5. $N_2H_4 \cdot$HOAc, DMF, 45 °C, 30 min, 91 %
6. Cl_3CCN, DBU, CH_2Cl_2, r. t., 3 h, 95 %

Discussion

TMSOTf is a strong *Lewis* acid and can be used as catalyst for selective synthesis of α-gluco- and galactosamines − respectively pyranosides, both 1,2-*cis* configurated, too. Here β-connection to C-4b of the lactose is obtained in high yields.

Next, the reductive elimination of the Troc protecting group can be done by employing zinc in acetic anhydride with the positive side effect of a single acetylation at the amino functionality in one step. Under these conditions no bisacetylation of the NH_2 group occurs, because the first connected acetyl group deactivates the nitrogen for the next acetylation step.

In the following step palladium on carbon is used to remove the benzyl ethers only. Platinum is not suited because of hydrogenation of the aromatic ring of the benzyl group (see p. 148).

Subsequently the already mentioned protocol of peracetylation of the free alcohol functionalities is performed. In step five you have to repeat the selective deprotection of the anomeric center with hydrazinium acetate with successive formation of the α-trichloroacetimidate in step six.

Problem

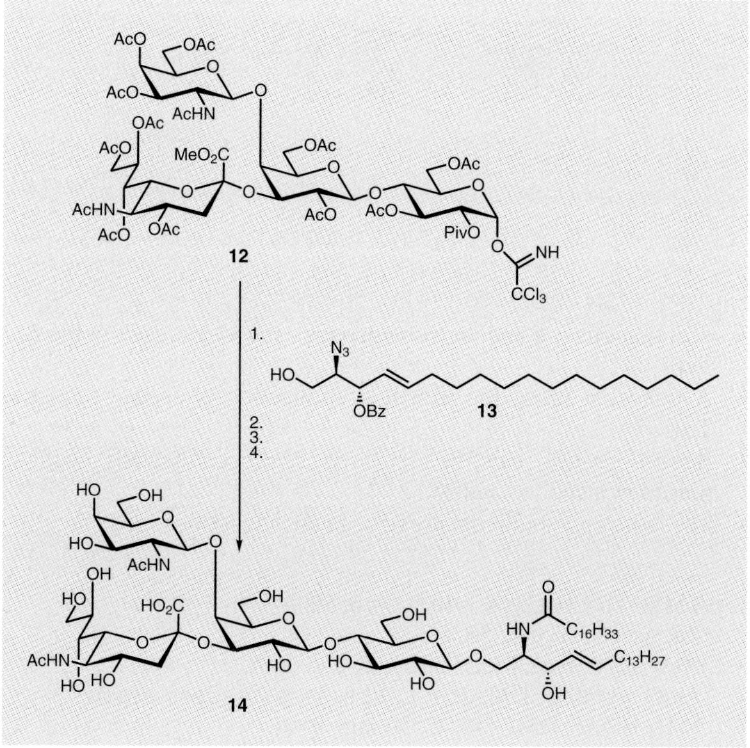

Hints

- The azidosphingosine (**13**) is connected with the oligosaccharide.
- The azide is transformed to an amino functionality.
- Coupling with the correct fatty acid occurs next.
- Cleavage of all protecting groups is the last step.

1. BF$_3$·OEt$_2$, CH$_2$Cl$_2$, –40 °C → r. t., 2 h, 89 %	*Solution*
2. H$_2$S/ pyridine, r. t., 3 d	
3. C$_{17}$H$_{35}$CO$_2$H, WSC, r. t., 18 h, 71 %	
4. NaOMe, MeOH, r. t., 24 h, quant.	

First azidosphingosine (**13**) is connected to the oligosaccharide moiety. Boron trifluoride etherate in dichloromethane at temperatures of –40 °C to room temperature has been proved to be very suitable with regard to yield and diastereoselectivity of the glycosylation. With neighboring group participation of 2-*O*- (or 2-*N*-) protective groups the 1,2-*trans*-products were obtained. The pivaloate group at C-2 prevents the formation of the α-glycoside (**67**) and leads to **69** with a β-glycosidic bond:

Discussion

The reason for this is steric hindrance of the acyloxonium carbon atom for a nucleophilic attack. Inversion of configuration at the anomeric center would be preferred with the nonparticipating 2-*O*-benzyl protective group.

Azidosphingosine **13** was received in different synthetic pathways.[32] The synthesis by *Schmidt*[33] is mentioned here.

D-galactose **15** is transformed to 4,6-*O*-benzylidene ketal **70**. Subsequent sodium periodate treatment at pH 7.6 provides 2,4-*O*-benzylidene-D-(+)-threose (**71**) in high yield. *Wittig* reaction of this compound with hexadecanylidene triphenylphosphorane (**72**) in presence of lithium bromide affords the *trans*-eicosenetriol derivative **73**. Followed by trifluorosulfonylmethane activation of the hydroxy group the azide functionality is introduced (**74**). Acidic cleavage of the protecting group with subsequent selective benzoylation of the secondary alcohol leads to **13**.

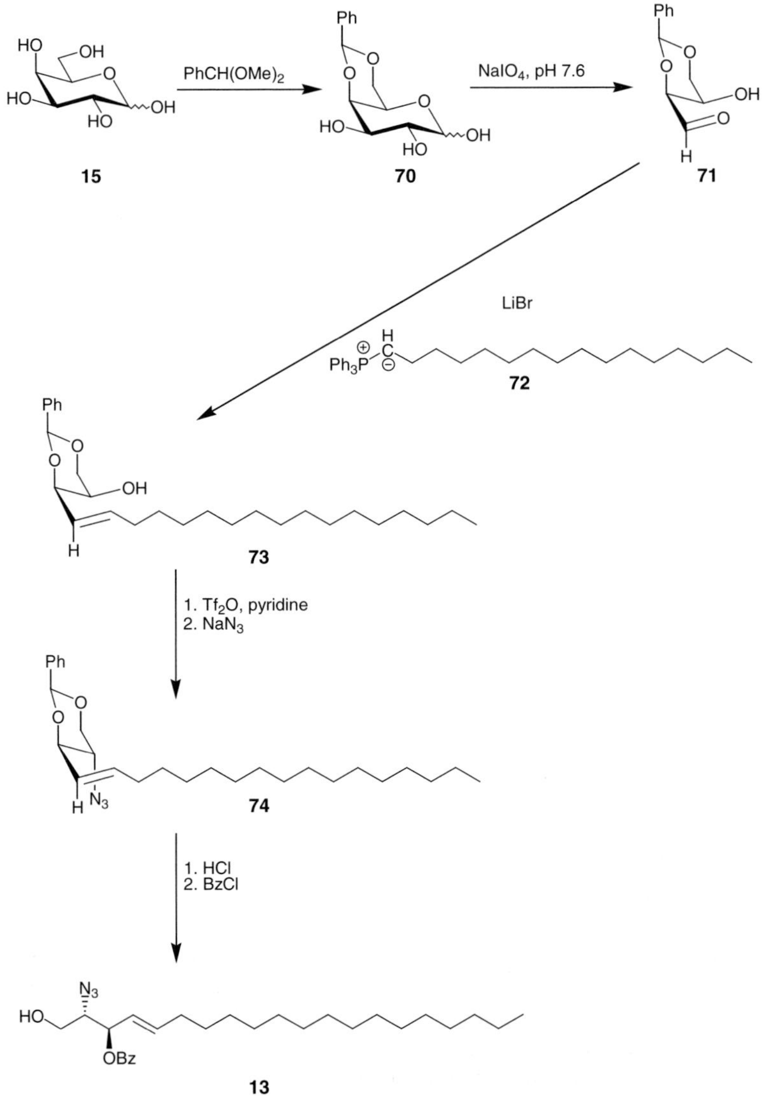

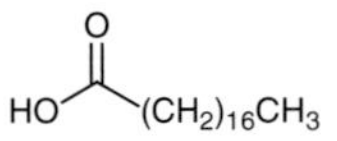

After glycosylation with sphingosine the azide group is reduced to an amine. Hydrogen sulfide in pyridine / water hereby acts as reducing reagent. Next stearic acid (**75**) has to be connected to the amine – a peptide bond is formed.

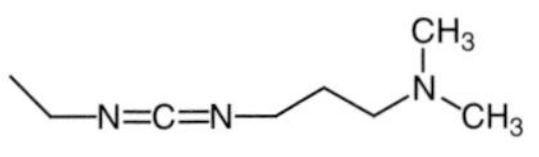

1-Ethyl-3-(3-dimethylaminopropyl)carbodiimide (**76**) is a water-soluble carbodiimide (WSC), which has been used as a condensing reagent in peptide synthesis. The major advantage of WSC-based couplings is an easy removal of both the excess reagent and the corresponding urea derivative by washing the reaction mixture with dilute acid or water. WSC is also known as EDC, EDAC, ECDI or ethyl-CDI in papers. The WSC-based coupling reaction scheme is the following:

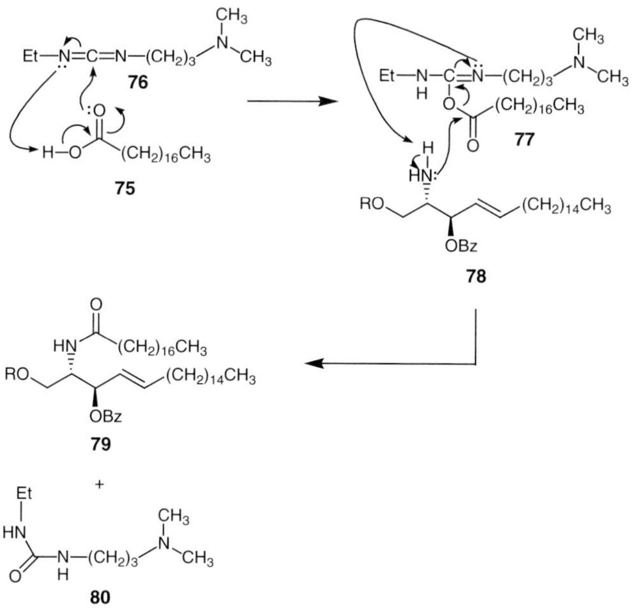

After the coupling of **76** with **75** the nitrogen of the sphingosine couples with the intermediate **77** to yield **79**.

The last step is the cleavage of all protecting groups under *Zemplén* conditions, using NaOMe in MeOH.[34] After deprotection, GM2 (**14**) is obtained.

15.4 Conclusion

The total synthesis of GM2 presented here is an introductory exercise in carbohydrate chemistry. During the synthesis the importance of a suitable protecting group strategy to achieve the desired stereochemistry has been pointed out. There are reactions specific for carbohydrates and many other transformations, which can be utilized in "non-carbohydrate-synthesis", too. However there is no ideal way in carbohydrate chemistry. Each synthesis has been optimized for the respective sugar.

In this case *R. R. Schmidt* et al. have presented another GM2 total synthesis, which, according to them, is more efficient and allows the synthesis of large amounts of GM2.[2]

15.5 References

1. K. O. Lloyd, K. Furukawa, *Glycoconj. J.* **1998**, *15*, 627-636.
2. J. C. Castro-Palomino, G. Ritter, S. R. Fortunato, S. Reinhardt, L. J. Old, R. R. Schmidt, *Angew. Chem.* **1997**, *109*, 2081-2085; *Angew. Chem. Int. Ed. Engl.* **1997**, *36*, 1998-2001.
3. R. R. Schmidt, *Angew. Chem.* **1986**, *98*, 213-236; *Angew. Chem. Int. Ed. Engl.* **1986**, *25*, 212-235.
4. R. R. Schmidt, W. Kinzy, *Liebigs Ann. Chem.* **1985**, 1537-1545.
5. P. C. Kocienski, *Protecting Groups*, Thieme, Stuttgart, **1994**.
6. a) T. K. Lindhorst, *Essentials of Carbohydrate Chemistry and Biochemistry*, Wiley-VCH, Weinheim, **2000**; b) Y. Chapleur, *Carbohydrate Mimetics*, Wiley-VCH, Weinheim, **1997**.
7. W. Dullenkopf, J. C. Castro-Palomino, L. Manzoni, R. R. Schmidt, *Carbohydrate Research* **1996**, *296*, 135-147.
8. R. Behrend, P. Roth, *Ann.* **1904**, *331*, 359-382.
9. B. Neises, W. Steglich, *Angew. Chem.* **1978**, *90*, 556-557; *Angew. Chem. Int. Ed. Engl.* **1978**, 17, 522-524.
10. G. Excoffier, D. Gagnaire, J.-P. Utille, *Carbohydr. Res.* **1975**, *39*, 368-373.
11. K. Toshima, K. Tatsuta, *Chem. Rev.* **1993**, *93*, 1503-1531.
12. R. R. Schmidt, *Pure & Appl. Chem.* **1989**, *61*, 1257-1270.
13. E. Juaristi, G. Cuevas, *Tetrahedron* **1992**, *48*, 5019-5087.
14. R. U. Lemieux, A. R. Morgan, *Can. J. Chem.* **1965**, *43*, 2205-2213.
15. M. A. Fabian, C. L. Perrin, M. L. Sinnott, *J. Am. Chem. Soc.* **1994**, *116*, 8398-8399.
16. M. Hudlicky, *J. Org. Chem.* **1980**, *45*, 5377-5378. T. H. Black, *Aldrichimica Acta* **1983**, *16*, 3.
17. S. M. Hecht, J. W. Kozarich, *Tetrahedron Lett.* **1973**, 1397-1400.
18. R. Kuhn, P. Lutz, D. L. Mac Donald, *Chem. Ber.* **1966**, *99*, 611-617.
19. T. J. Martin, R. R. Schmidt, *Tetrahedron Lett.* **1992**, *33*, 6123-6126.
20. G. Magnusson, K. Jansson, S. Ahlfors, T. Frejd, J. Kihlberg, *J. Org. Chem.* **1988**, *53*, 5629-5647.
21. H. Wetter, K. Oertle, *Tetrahedron Lett.* **1985**, *26*, 5515-5518.
22. J. S. Brimacombe, *Methods Carbohydr. Chem.* **1972**, *6*, 372.
23. J. M. Lassaletta, R. R. Schmidt, *Synlett* **1995**, 925-927.
24. H. Kunz, W. Sager, D. Schanzenbach, M. Decker, *Liebigs Ann. Chem.* **1991**, 649-654.
25. W. Koenigs, E. Knorr, *Chem. Ber.* **1901**, *61*, 1257-1270.
26. R. R. Schmidt, J. Michel, *Tetrahedron Lett.* **1984**, *25*, 821-824.
27. A. Lubineau, A. Malleron, *Tetrahedron Lett.* **1985**, *26*, 1713-1716.
28. F. J. Urban, B. S. Moore, R. Breitenbach, *Tetrahedron Lett.* **1990**, *31*, 4421-4424.
29. J. M. Lassaletta, K. Carlsson, P. J. Garegg, R. R. Schmidt, *J. Org. Chem.* **1996**, *61*, 6873-6880.
30. R. R. Schmidt, M. Behrendt, A. Toepfer, *Synlett Letters* **1990**, 694-696.
31. H. Kondo, S. Aoki, Y. Ichikawa, R. L. Halcomb, H. Ritzen, C. Wong, *J. Org. Chem.* **1994**, *59*, 864-877.
32. P. Tkaczuk, E. R. Thornton, *J. Org. Chem.* **1981**, *46*, 4393-4398.
33. R. R. Schmidt, P. Zimmermann, *Tetrahedron Lett.* **1986**, *27*, 481-484.
34. G. Zemplén, *Ber. Dtsch. Chem. Ges.* **1927**, *60*, 1555-1564.

16

H-Type II Tetrasaccharide Glycal
(Danishefsky 1995)

16.1 Introduction

This time you will be engaged in the synthesis of blood group determinants. Besides the described glycoconjugates there are carbohydrate antigens of the A, B, H (O) and *Lewis* families on cell surfaces too. Because of their biological significance, these glycoconjugates are used to stimulate antibody production.
You can divide the blood group determinants into two basic categories, type I and type II. Type I is characterized by a backbone comprised of a galactose 1→3β linked to *N*-acetyl-glucosamine while type II contains, a 1→4β linkage between the same building blocks.
The position and extent of α-fucosylation of these backbone structures gives rise to the *Lewis*-type and H-type specifications. Presence of an α-monofucosyl branch, solely at the C-2c-OH in the galactose moiety of the backbone, constitutes the H-type specificity (types I and II). Further permutation of the H-types by substitution of α-linked galactose or α-linked *N*-acetyl-galactosamine at its C-3c-hydroxy group provides the molecular basis of the familiar serological blood group classifications A, B and O.[1]
In this chapter our synthetic target **4** is the tetrasaccharide of the blood group classification O.[2]
S. J. Danishefsky et al. used the "glycal assembly strategy" for the synthesis of blood group determinants and H-Type oligosaccharides and took advantage of the "glycal epoxide method" to build β-selective glycosyl bonds. Halogen sulfonamide was utilized to initiate the *N*-acetyl-glucosamine building block.

16.2 Overview

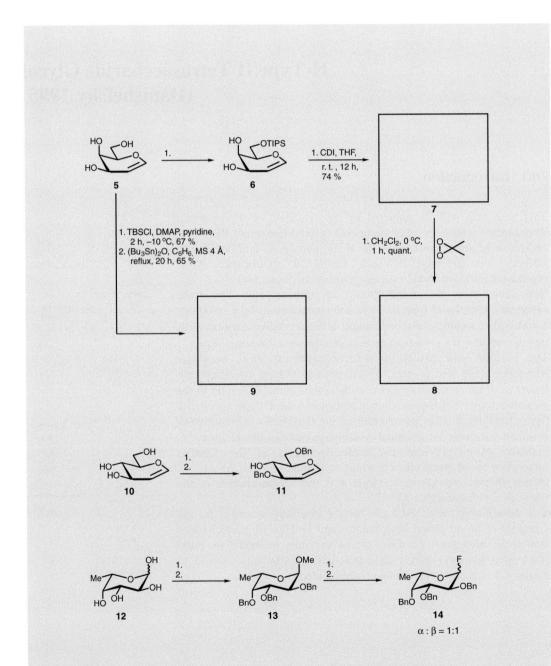

16 H-Type II Tetrasaccharide Glycal

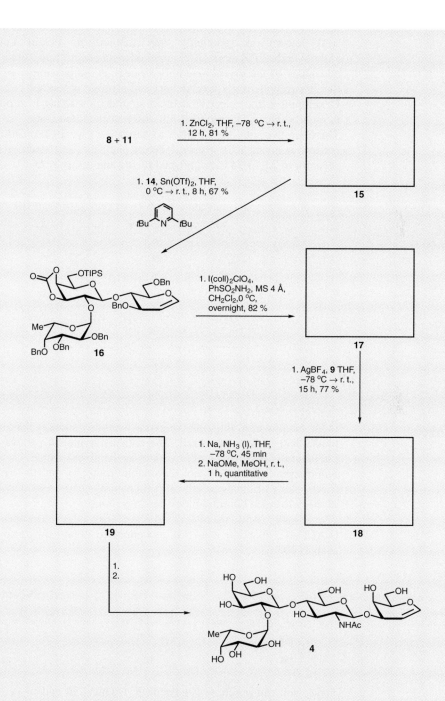

A few words...

... on the glycal assembly strategy. In the last few years this method has been primarily connected with the work of *Danishefsky* who is well known in practicing this method to obtain complex carbohydrates. Pioneers of this method are *Lemieux* and *Thiem*.[3] The advantage of the glycal assembly strategy is the possibility of using the glycals both as glycosyl donors and glycosyl acceptors.

Developing some selective activating strategies, today it is possible to modify the glycals rapidly with different functionalities.

There are two vital possibilities to activate a glycal (**21**). Either you activate a glycal to a glycosyl donor (**22**); this one reacts with an acceptor to **23**, or **24** is established *in situ* and reacts to **23**.

5 D-galactal

10 D-glucal

20 L-fucal

In the beginning of utilizing the glycal assembly strategy the building blocks had to be synthesized. The simplest synthetic pathway is to use naturally occurring carbohydrates as the glycal source.

Sometimes, if there is no reasonable route to get the required functionality of the target glycal from natural sugars, a total synthesis of the demanded building block has to be done.

The *Lewis* acid catalyzed diene-aldehyde cyclocondensation (LACDAC) reaction is a fast procedure to obtain dihydropyrones such as **28**.[4]

Today you can purchase D-galactal **5**, D-glucal **10** and D- or L-fucal **20** e. g. peracetylated or with unprotected hydroxy groups.

16.3 Synthesis

Problem

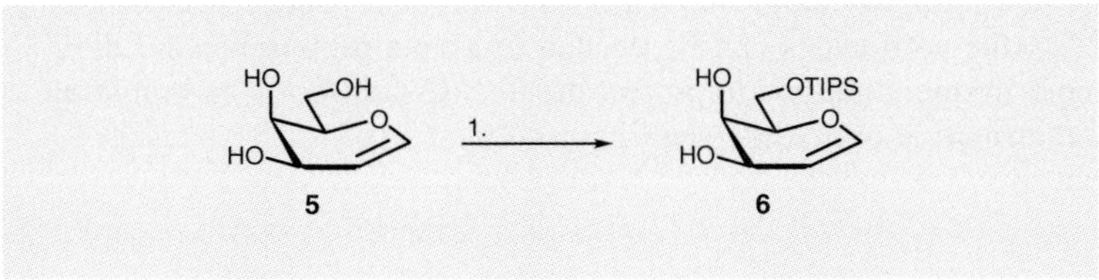

[Structure 5: galactal with HO, OH, HO groups] → 1. → [Structure 6: galactal with HO, OTIPS, HO groups]

Hints
- The primary hydroxy group has to be selectively protected.
- A silyl protecting group is introduced.

1. TIPSCl, imidazole, THF, r. t., 48 h, 90 %

Solution

The triisopropylsilyl (TIPS) group is introduced under the same conditions as TBS groups.[5] Instead of imidazole DMAP can be used, too. Under these conditions only the primary alcohol functionality is selectively protected as TIPS ether.

Discussion

[Structure 6] 1. CDI, THF, r.t., 12 h, 74 % → 7

Problem

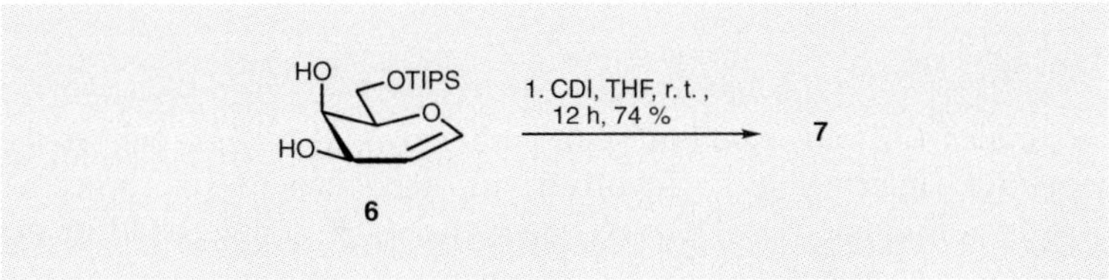

- Two possibilities: Protecting secondary hydroxy groups or transforming the double bond?
- CDI is a derivative of H_2CO_3.

Hints

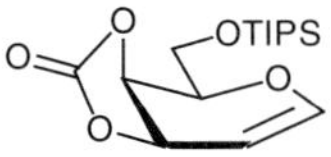

[Structure 7: cyclic carbonate with OTIPS]

Solution

N,N'-carbonyldiimidazole (CDI) **29** is used to protect the 3- and 4-OH group of the galactal **6** as a cyclic carbonate.[6] The first report of a cyclic carbonate was in 1883 by *Nemirowsky*.[7] He treated ethylene glycol with phosgene (**30**). Phosgene is known as a highly toxic

Discussion

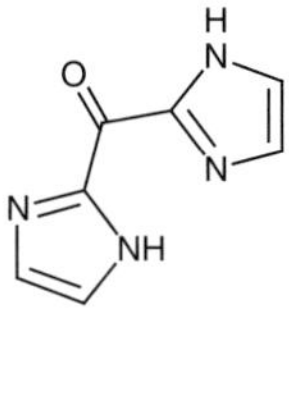

29 CDI

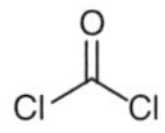

30 phosgene

reagent; thus CDI is an alternative leading to the same result. Cyclic carbonates are very stable to acidic hydrolysis and more stable to basic conditions than esters. As with the benzylidene acetals (see p. 272), the carbonates can be opened to give a monoprotected diol.[8] A more frequent method to protect the C-3/ C-4 diol of a carbohydrate is the formation of acetals (see Chapter 7).

Problem

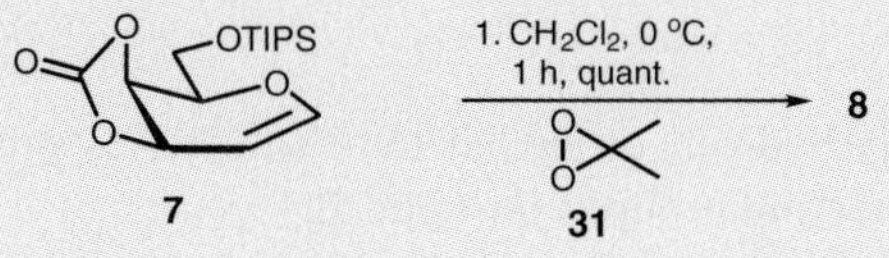

Hints

- What kind of functionality is generated?
- Which anomer is favored?

Solution

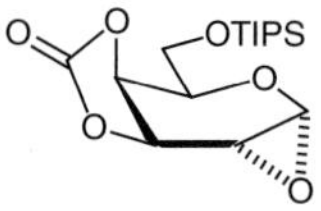

Discussion

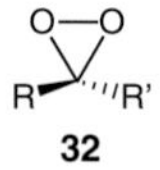

32

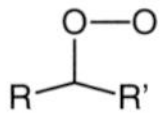

33

Dioxiranes (**32**) are isomeric with carbonyl oxides (**33**), one of the peroxidic intermediates involved in the ozonolysis process.[9] Dimethyldioxirane (DMDO) (**31**) epoxidizes the double bond of the protected galactal favoring the α-epoxide **8** with a selectivity of 20 : 1.[10,11] You can use **31** likewise to transform an aldehyde into a carboxylic acid.

The high α-selectivity is based on the absence of neighbor group participation. Protection of sugar **5** with acetyl groups and using the same epoxidation protocol leads to a 1:1 α : β mixture of the epoxides.

Problem

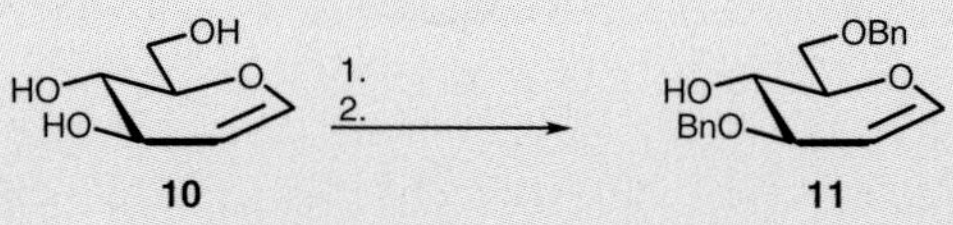

Hints

- It is not necessary to protect and deprotect the OH group at C-4.
- The 4-OH has to be "deactivated" until the benzylation is performed.
- Elements with free electron pairs can act as ligands on transition metals.

Solution

1. $(Bu_3Sn)_2O$, C_6H_6, reflux, 20 h
2. BnBr, TBABr, reflux, 17 h

65 % (over two steps)

Discussion

Trialkyltin alkoxide forms a coordination bond with a neighboring oxygen atom at the tin atom as in **34** and **35** because of its high nucleophilicity. Therefore intramolecular coordination could be expected to occur when equivalent amounts of stannylating reagent $(Bu_3Sn)_2O$ are employed for the reaction with **10**.

D-glucal (**10**) and di-(tributyltin) oxide in dry benzene are refluxed using a *Dean-Stark* trap. Under these conditions stannyl complex **36** of the D-glucal is formed. *Ogawa* was the first to report on the stereoselective stannylation.[12]

Subsequent reaction of **36** with tetrabutylammonium bromide (TBABr) and benzyl bromide gives the regioselectively benzylated product **37**.

After the nucleophilic attack of the bromide anion on the Sn atoms of the tributylstannyl ether **36** you get oxyanions at C-3 and C-6 as the reactive species. Now selective benzylation can proceed and the alcohol at C-4 is only benzylated as a by-product in low yield.

To avoid stannyl reagents, there is the possibility to synthesize the same building block in one more step. First step is the formation of the benzylidene acetal **38** at the positions C-4 and C-6 of the glucal **10**. The acetal formation at C-4 / C-6 is favored because you can get a six-membered instead of a five-membered ring at the position C-3 / C-4. To form **38** benzaldehyde dimethyl acetal **39** can be used with catalytic amounts of *para*-toluenesulfonic acid.

16 H-Type II Tetrasaccharide Glycal

A following benzylation of the alcohol at C-3 yields **40**. Conversion of **40** into **43** with the primary alcohol functionality protected is realized with sodium cyanoborohydride (NaBH₃CN). Reduction with diisobutylaluminum hydride (DIBAH) **42** furnishes **44** leaving the C-6 alcohol unprotected.

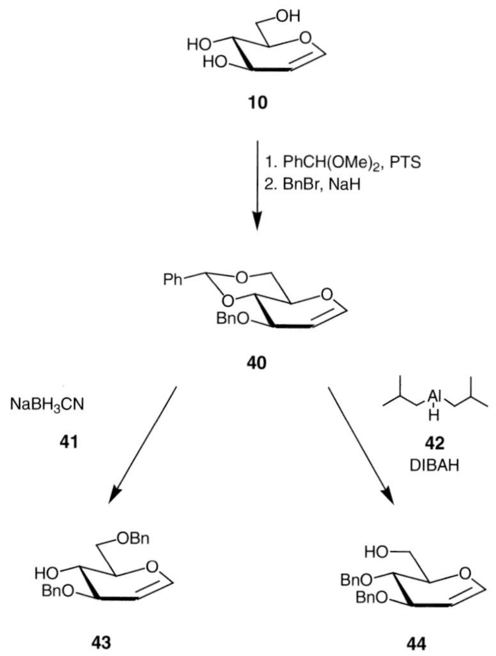

Problem

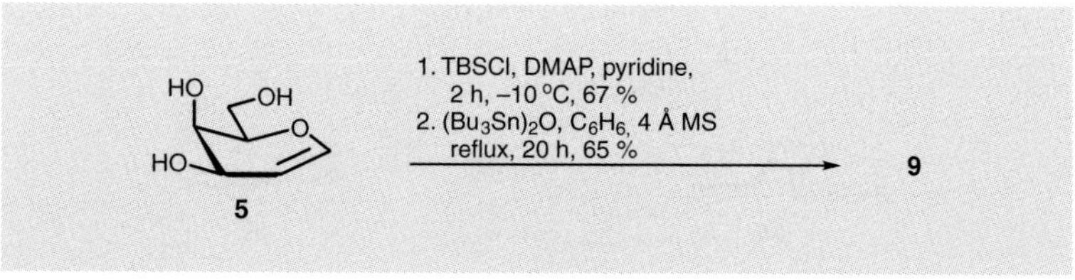

Hints

- The reaction performed in the first step should already be known.
- How do stannyl reagents react in the presence of diols?

Solution

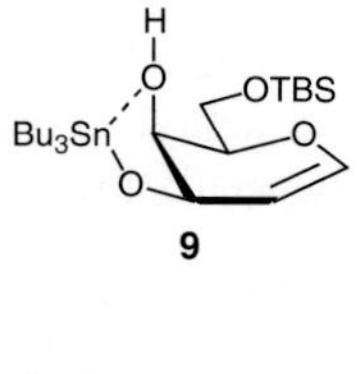

Discussion

Again, at first the primary alcohol is protected with a silyl protecting group. Next the galactal **5** is converted into **9** with di-(tributyltin) oxide. The same mechanism as in the previous problem takes place in this step yielding **9**.

Problem

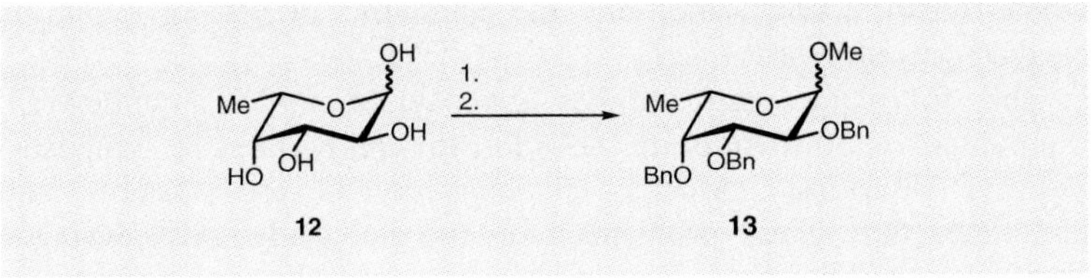

[Structure 12 → Structure 13, with steps 1. and 2.]

Hints

- The most reactive OH group is protected first.
- Which methods are known to form benzyl ethers?

Solution

1. MeOH, HCl, reflux, 24 h, 92 %
2. NaH, BnBr, DMF, 0 °C, 0.5 h; r. t., 1 h, 96 %

Discussion

Starting with *hemi*-acetal **12** the anomeric center has to be protected. Under acidic conditions in methanol the C-1 methyl ether is obtained as a mixture of anomers. Therefore the α-anomer is favored because of the anomeric effect (see p. 251). Separation of the two anomers by crystallization is possible. This reaction is connected to the research of E. Fischer (*Fischer-Helferich* method).[13]
Then the other hydroxy groups are benzylated. If (Bu$_3$Sn)$_2$O and BnBr are used for benzylation the 4-OH unprotected sugar **45** is formed, because of a stannyl complex which acts as in the case of **39**.

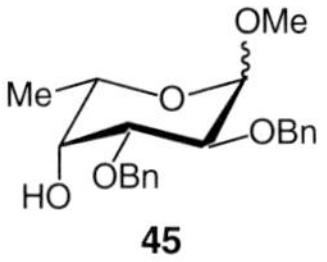

[Structure 45]

Problem

[Structure 13 → Structure 14, $\alpha : \beta = 1:1$, with steps 1. and 2.]

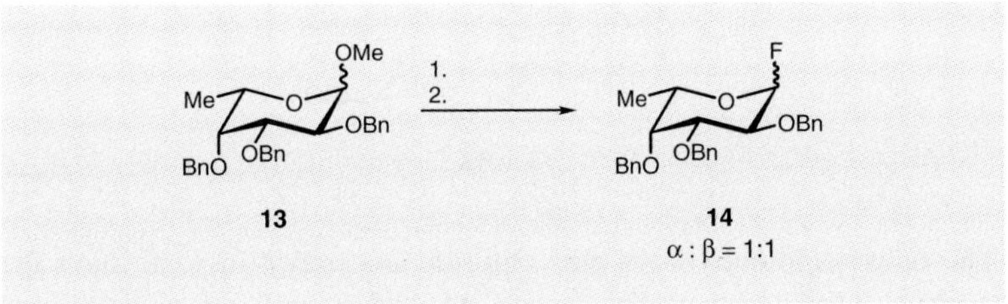

Hints
- The methyl ether is cleaved.
- The introduction of fluorine is non-stereoselective.

Solution
1. HOAc, HCl, 100 °C, 3 h, quant.
2. DAST, THF, –30 °C → r. t., 20 min, quant.

Discussion

Under harsh acidic conditions the anomeric protecting group is cleaved quantitatively.

In the following step another glycosyl donor is synthesized. Diethylaminosulfurtrifluoride (DAST) (**46**) can be used to synthesize glycosyl fluorides.[14] Mostly, glycosyl donors are unstable and sometimes they cause problems with the stereoselectivity during a glycosylation. Therefore it was proposed to use glycosyl fluorides as glycosyl donors, because of the strong C-F bond. With these compounds more stereoselective couplings are possible – if the fluoride donor can be activated. However, there are only a few catalysts for fluoride donor activation known which operate stereoselectively. This is a restriction to this methodology.

Problem

Hints
- Which glycoside is obtained in this step?

Solution

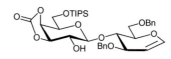

15

Discussion

The *Lewis* acid anhydrous zinc chloride as catalyst in THF gives the desired β-linkage between **8** and **11**. After epoxide opening and nucleophilic attack of **11** compound **15** can function as glycosyl acceptor at C-2b in the next step. Both high stereoselectivity and yield are attributed to the absence of participating groups.[15]

16 H-Type II Tetrasaccharide Glycal

Next the trisaccharide **16** has to be synthesized. For this purpose tin (II) chloride is used to get selectively a new β-linkage between **15** and **14**.

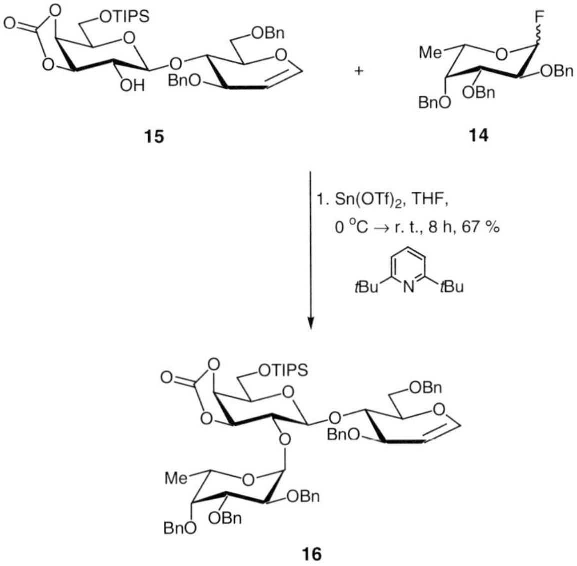

Problem

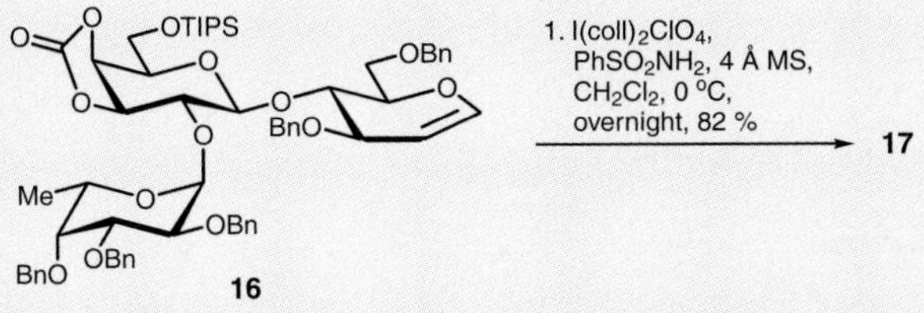

- Another glycosyl donor is formed.
- I(coll)$_2$ **48** is an electrophile.

Hints

Solution

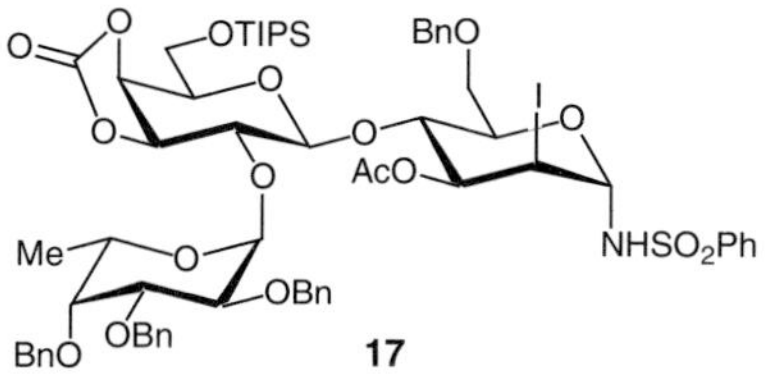

Discussion

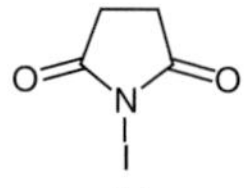

Lemieux and *Thiem* developed the iodine glycosylation.[16] You need an I$^+$-reagent e. g. *N*-iodosuccinimide (1-iodo-2,5-pyrrolidinedione, NIS) (**47**) or I(coll)$_2$ (iodonium di-*sym*-collidine perchlorate) (**48**). The mechanism of this step with the bromine reagent of **48** was evaluated by *Brown*.[17]

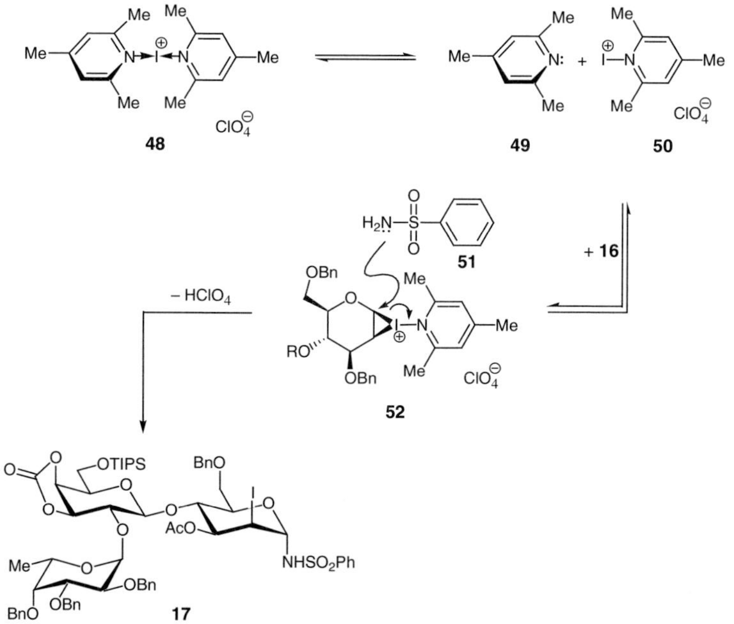

The first step is a reversible dissociation of **48** into free collidine (**49**) and a reactive intermediate, coll-I$^+$ (**50**). The electrophilic collidine-iodonium cation forms with glycal **16** the cyclic compound **52**. The perchlorate anion of **48** is not nucleophilic enough to attack the sustained intermediate **52**, but **52** is captured by nucleophilic attack of benzenesulfonamide (**51**). This attack proceeds from the back and leads to *trans* product **17**.

The C-N bond formation takes place exclusively at the anomeric center because of stereoelectronical effects of the *endo* cyclic oxygen.

Problem

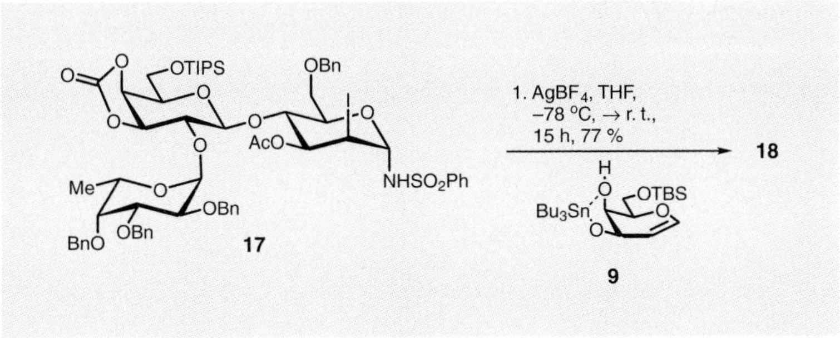

[Scheme: compound 17 + compound 9, 1. AgBF$_4$, THF, −78 °C, → r.t., 15 h, 77 % → 18]

Hints

- AgBF$_4$ acts as catalyst for glycosylations.
- What happens to the nitrogen at C-1?

Solution

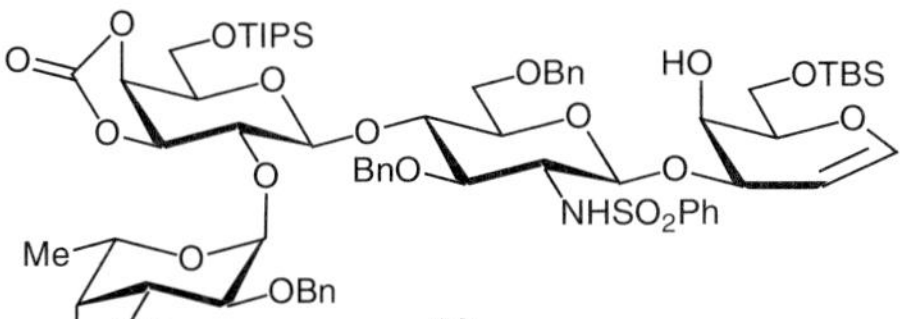

[Structure of compound 18]

Discussion

Iodine is precipitated by the silver cation as AgI. The resulting cation is stabilized by the nitrogen of the sulfonamide group at C-1 and compound **53** is formed as intermediate.

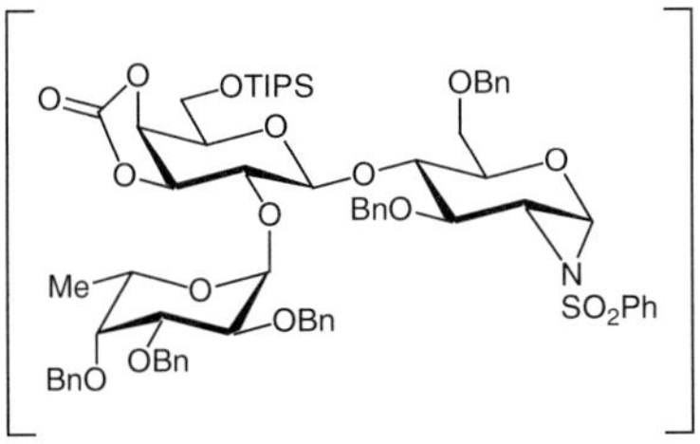

[Structure of intermediate 53 in brackets]

The attempts to isolate the intermediate 1,2-*N*-sulfonylaziridine **53** were unsuccessful. Nucleophilic attack of the acceptor **9** at C-1 opens the aziridine; the nitrogen rearranges to the C-2 position and yields the desired tetrasaccharide **18**.

16 H-Type II Tetrasaccharide Glycal

Problem

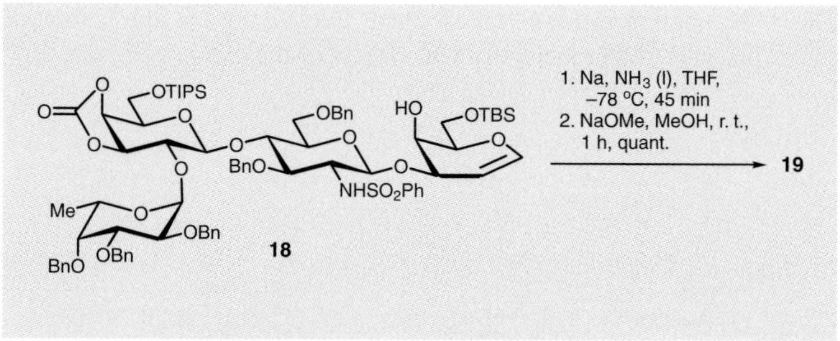

[Structure of compound **18** with reagents: 1. Na, NH₃ (l), THF, −78 °C, 45 min; 2. NaOMe, MeOH, r. t., 1 h, quant. → **19**]

Hints

- The first step is a radical reaction.
- Sodium methoxide in methanol is used for cleavage of the remaining protecting groups.

Solution

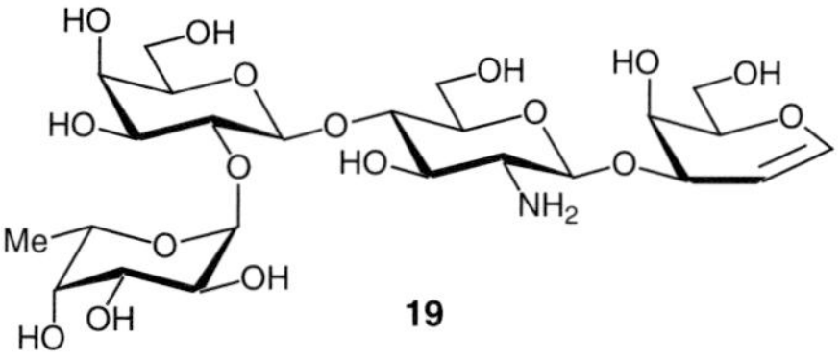

[Structure of compound **19**]

Discussion

Sodium in liquid ammonia removes the protecting groups except the carbonate to yield **54**.

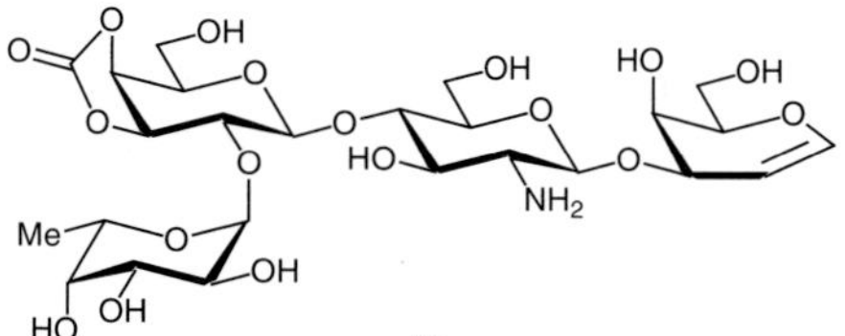

[Structure of compound **54**]

This is a radical reaction similar to the *Birch* reduction (see p. 148). In general benzyl groups are removed hydrogenolytically (see p. 260), but under these conditions elimination of the remaining double bond would occur. In the second step the carbonate is cleaved and the amine is deprotected to get the unprotected oligosaccharide **19**.

16 H-Type II Tetrasaccharide Glycal

Problem

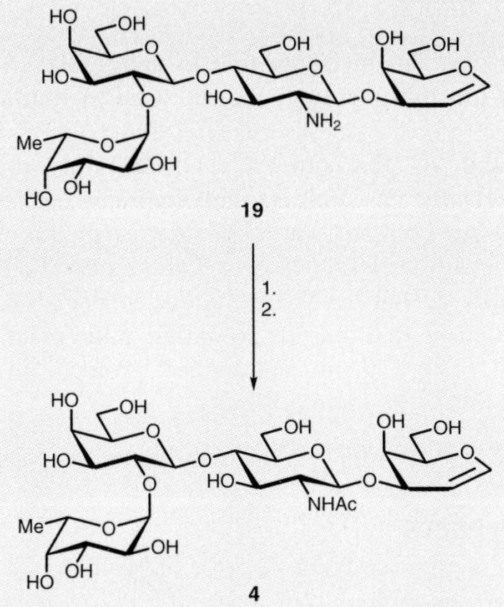

Hints
- How are acetyl groups introduced?
- In the end the oligosaccharide has to be deprotected (except the NHAc group).

Solution

1. Ac$_2$O, DMAP, r. t., 24 h, quant.
2. NaOMe, MeOH, r. t., 15 h, quant.

In the first step the molecule is peracetylated with acetic anhydride and catalytic amounts of DMAP to yield **55**.

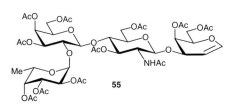

Within this step the amino functionality is transferred to the *N*-acetylamine. Subsequent deprotection under *Zemplén*[18] conditions with sodium methoxide in methanol leads to **4**, the oligosaccharide belonging to blood group type H (O) of the blood group determinant type II. The remaining glycal functionality can be used to connect the oligosaccharide with a spacer and this spacer to a protein. This is necessary for immunological and biological purposes.

16.4 Conclusion

Carbohydrates play an important role in nature. This chapter shows that the glycal assembly strategy can be used to synthesize complex carbohydrates.

Chapters 15 and 16 give only a small insight into carbohydrate chemistry. Lately the research on carbohydrates e.g. on solid phase has increased. For complex carbohydrates as in this case the solid phase method is still useless, however. On the other hand, if you want to get simple polysaccharides this technique is very powerful.

The main problem is that one has to optimize the reaction conditions for each sugar. There is no standardied protocol in carbohydrate chemistry, and this is the main difficulty.

16.5 References

1. G. Stamatoyannopolous, *The Molecular Basis of Blood Diseases*, W. B. Saunders Inc., Philadelphia, 2nd Ed. **1994**.
2. S. J. Danishefsky, V. Behar, J. T. Randolph, K. O. Lloyd, *J. Am. Chem. Soc.* **1995**, *117*, 5701-5711.
3. a) R. U. Lemieux, A. R. Morgan, *Can. J. Chem.* **1965**, *43*, 2190-2198; b) J. Thiem, H. Karl, J. Schwentner, *Synthesis* **1978**, 696-698.
4. S. J. Danishefsky, *Chemtracts. Org. Chem.* **1989**, 2, 273.
5. F. Bennett, D. W. Knight, G. Fenton, *J. Chem. Soc., Perkin Trans. I* **1991**, 1543-1547.
6. M. T. Bilodeau, T. K. Park, S. Hu, J. T. Randolph, S. J. Danishefsky, P. O. Livingston, S. J. Zhang, *J. Am. Chem. Soc.* **1995**, *117*, 7840-7841.
7. J. Nemirowsky, *J. Prakt. Chem.* **1883**, *28*, 439-440.
8. K. C. Nicolaou, C. F. Claiborne, K. Paulvannan, M. H. D. Postema, R. K. Guy, *Chem. Eur. J.* **1997**, *3*, 399-409.
9. R. W. Murray, R. Jeyaraman, *J. Org. Chem.* **1985**, 2847-2853.
10. a) M. T. Bilodeau, S. J. Danishefsky, *Angew. Chem.* **1996**, *108*, 1380-1419; *Angew. Chem. Int. Ed. Engl.* **1996**, *35*, 1482-1522; b) S. J. Danishefsky, D. M. Gordon, *J. Carbohydr. Res.* **1990**, *111*, 361-366.
11. R. W. Murray, *Chem. Rev.* **1989**, *89*, 1187-1201.
12. T. Ogawa, M. Matsui, *Tetrahedron* **1981**, *37*, 2363-2369.
13. a) E. Fischer, L. Beensch, *Ber. Dtsch. Chem. Ges.* **1894**, *27*, 2478-2486; b) E. Fischer, *Ber. Dtsch. Chem. Ges.* **1895**, *28*, 1145-1167.
14. a) W. Rosenbrook, D. A. Riley, P. A. Lavley, *Tetrahedron Lett.* **1985**, *36*, 3-4;
    b) K. C. Nicolaou, J. L. Randall, G. T. Furst, *J. Am. Chem. Soc.* **1985**, *107*, 5556-5558.
15. R. L. Halcomb, S. J. Danishefsky, *J. Am. Chem. Soc.* **1989**, *111*, 6661-6666.
16. J. Thiem, W. Klaffke, *J. Org. Chem.* **1989**, *54*, 2006-2009.
17. A. A. Neverov, R. S. Brown, *J. Org. Chem.* **1998**, *63*, 5977-5982.
18. G. Zemplén, *Ber. Dtsch. Chem. Ges.* **1927**, *60*, 1555-1564.

Abbreviations

9-BBN	9-borabicyclo[3.3.1]nonane
Ac	acetyl
AIBN	2,2'-azobisisobutyronitrile
AIDS	acquired immunodefiency syndrome
Ar	aryl
ATP	adenosine triphosphate
AZT	azidothymidine
BINAL-H	2,2'-dihydroxy-1,1'-binaphthylaluminum hydride
Bn	benzyl
BOC	*tert*-butyloxycarbonyl
BOP	benzotriazol-1-yl-oxytris(dimethylamino)phosphonium hexafluorophosphate
Bt	benzotriazole
Bz	benzoyl
ca.	circa
CAN	ceric (IV)-ammonium nitrate
cat.	catalytic
CBZ	benzoyl carbonyl
CDI	*N,N'*-carbodiimidazole
cGMP	cyclic guanosine monophosphate
CSA	camphorsulphonic acid
DAST	diethylaminosulfurtrifluoride
dba	dibenzylideneacetone
DBU	1,8-diazabicyclo[5.4.0]undec-7-ene
DCC	1,3-dicyclohexylcarbodiimide
DDQ	2,3-dichloro-5,6-dicyano-1,4-benzoquinone
DET	diethyl tartrate
DHQ	dihydroquinine
DHQD	dihydroquinidine

DIAD	diisopropyl azodicarboxylate
DIBAH	diisobutylaluminum hydride
dig	digonal
DIPT	diisopropyl tartrate
DMAP	4-dimethylaminopyridine
DMDO	dimethyldioxirane
DME	dimethoxyethane
DMF	*N*,*N*-dimethylformamide
DMPM	3,4-dimethoxybenzyl
DMS	dimethylsulfide
DMSO	dimethylsulfoxide
DNA	desoxyribonucleic acid
EC	effective concentration
EDC	*N*-(3-dimethylaminopropyl)-*N'*-ethylcarbodiimide hydrochloride
ee	enantiomeric excess
Et	ethyl
FDA	Food & Drug Administration
GMP	guanosine monophosphate
HIV	human immunodeficiency virus
HMPA	hexamethylphosphorus triamide
HOBT	1-hydroxybenzotriazole
HOMO	highest occupied molecular orbital
IBX	1-hydroxy-1,2-benziodoxol-3(1*H*)-one 1-oxide
IC	inhibition concentration
im	imidazole
Ipc	isopinocampheyl
KHMDS	potassium hexamethyldisilazide
LACDAC	*Lewis* acid catalyzed diene-aldehyde cyclocondensation
LC	lethal concentration
LDA	lithium diisopropylamide

LUMO	lowest unoccupied molecular orbital
m	*meta*
MCPBA	*meta*-chloroperbenzoic acid
Me	methyl
MOM	methoxymethyl
MS	molecular sieves
Ms	methanesulfonyl
MTBE	methyl-*tert*-butyl ether
NaHMDS	sodium hexamethyldisilazide
*n*Bu	*n*butyl
NIS	*N*-iodosuccinimide
NMO	*N*-methylmorpholine-*N*-oxide
o	*ortho*
p	*para*
PCC	pyridinium chlorochromate
Pd/C	palladium on activated charcoal
PDC	pyridinium dichromate
PDE	phosphonodiesterase
PEG	polyethylene glycol
Ph	phenyl
PHAL	phthalazine
Piv	pivaloyl
PMB	*para*-methoxybenzyl
PPTS	pyridinium *para*-toluenesulfonate
PTS	*para*-toluenesulfonic acid
py	pyridine
PYDZ	pyridazine
PYR	diphenylpyrimidine
r. t.	room temperature
RCM	ring closing metathesis

ROM	ring opening metathesis
SDA	scopadulcic acid A
SDB	scopadulcic acid B
TAS-F	tris-(dimethylamino)-sulfur-(trimethylsilyl)-difluoride
TBABr	tetra-*n*-butylammonium bromide
TBAF	tetra-*n*-butylammonium fluoride
TBAT	tetra-*n*-butylammonium triphenyltrifluorosilicate
TBCO	2,4,4,6-tetrabromocyclohexadienone
TBS	*tert*-butyldimethylsilyl
*t*Bu	*tert*-butyl
TDS	thexyldimethylsilyl
TEMPO	2,2,6,6-tetramethyl-1-piperidinyloxy (free radical)
Teoc	2-(trimethylsilyl)ethoxycarbonyl
TES	triethylsilyl
tet	tetragonal
Tf	triflate, trifluoromethanesulfonyl
TFA	trifluoracetic acid
THF	tetrahydrofuran
TIPS	triisopropylsilyl
TMEDA	N,N,N',N'-tetramethylethylendiamine
TMS	trimethylsilyl
TPAP	tetra-*n*-propylammonium perruthenate
TPP	5,10,15,20-tetraphenyl-21H,23H-porphine
TPPTS	triphenylphosphine-3,3',3"-trisulfonic acid trisodium salt
TPS	*tert*-butyldiphenylsilyl
trig	trigonal
WSC	water soluble carbodiimidazole (see EDC)
X	halogen atom

Index

$(Me_3Si)_4Si$	134
acetals	4, 64, 109, 223, 253
-hydrolysis	64, 200, 256
acetogenins	137
acetyl	250
acetylene	20
acyclic diene metathesis polymerization	145
acyloxonium salt	201
addition	
-1,2	39, 61
-1,4	61, 160, 222
-1,4 conjugated	180
-*syn*	65, 220
$AgBF_4$	277
AIBN	182
alcohol	
-allylic	22, 160, 197
-homoallylic	42
-propargylic	219
aldol reaction	30, 224
-asymmetric	142, 161
alkylation	131
alkyllithium aggregate	76
alkynes	77, 91, 169
alkynylation	92
-enantioselective	81
allene	155
allyl addition	18
allylation	
-asymmetric	44
allylic alcohol	23, 160, 197
-oxidation	23, 187
allylMgBr	65
(*S*)-Alpine Borane	212
Alpine Borane	91, 212
Alzheimer's disease	177
amide synthesis	74, 189
amides	46, 74, 132, 161, 189, 237, 242
ammonia	238
anomer	
-α/β	250
-kinetical	250
-thermodynamical	250
anomeric center (selective deprotection)	250
anomeric effect	251
anomerization	255
anti-anti-stereotriad	18
antibiotics	15, 157, 193
anticancer drugs	35
antifungal activity	157, 193
anti-syn diastereomer	18
antitumor agent	121
Appel reaction	9, 44, 149, 213
Arbuzov reaction	23, 196
asteriscanolide	1
azido functionality	203
azidosphingosine	260
azidothymidine	71
bafilomycin A_1	15
Baldwin rules	183
9-BBN	160, 213
benzenesulfonamide	276
benzylether	148, 186, 254, 260, 278
benzylidene ketal	272
benzyloxycarbonyl (CBZ)	108
BF_3 (g)	66
$BF_3 \cdot OEt_2$	30
$BH_3 \cdot DMS$	65
bidentate chelate	46
(*S*)-BINAL-H	213
Birch reduction	58, 148, 184, 278
bis(triphenylphosphine)palladium dichloride	92
blood group classification	265
blood group determinants	265
BOC	88, 108, 203, 206
bond energies Si-C/Si-F	184
BOP	205
boron enolates	161
Bürgi-Dunitz angle	148
Burgess reagent	49
Bu_3SnH	40, 182
camptothecin	121
CAN	272
cancer therapy	121
carbamates	82, 132
carbamoylation	205
carbinols	68
carboalumination	21
carbodiimide	94
carbometalation	21
carbon disulfide	141
carbon tetrabromide	150
carbonyl oxide	270
carboxyl group activation	
-anhydride	140, 172

-BOB	205
-carbodiimide	94
-carbonyldiimidazole	240
-chloride	74, 140, 189
caranyl borane	43
Castro's reagent	205
catecholborane	19, 173
CBZ	108
CDI	269
ceramide	245
cerium ammonium nitrate (CAN)	26
cerium chloride	160
chain extension	151
Charette cyclopropanation	46
chemotherapy	121
chiral pool	43
chiral Ru complex	107
chitin synthetase	193
chromium(II) chloride	24
chromium(VI) oxide	229
cinchona alkaloid	62
cis-elimination	61
cisoid 1,3-diene	69
cisoid conformation	97
cis selective reduction	170
cis-selectivity	38
Claisen ester condensation	116
Claisen-Ireland rearrangement	163
Claisen rearrangement	163
colchicine	35
collidine	276
condensation	110
1,4 conjugated addition	180
Cope rearrangement	216
copper(I)-catalysis	92, 180
Corey-Fuchs reaction	20, 170
$Cp_2Zr(H)Cl$	38, 170
Cram chelate	148
cross-coupling reaction	172
crotyl alcohol	127
crotylation	18
curacin A	35
cyanocuprates	165
cyclic carbonate	269
cyclic semi-acetal	273
cyclizations	
-5-*exo*-dig	134, 182
-5-exo-*trig*	182
cycloadditions	
-1,2	69
-[2+2]	1, 146
-[4+2]	69, 96, 110, 187
cyclopropanation	45
D-(+)-galactose	248
D-(+)-lactose	245
DAST	274
DBU	96, 251
DCC	132
DDQ	20, 81
decarboxylation	105, 111
dehydratization	77
dehydro-sulfenylation	61
desoxygenation	40
Dess-Martin periodinane	12, 29, 67, 160
desulfurization	6
D-galactal	268
D-glucal	268
$(DHQD)_2PHAL$	130
$(DHQD)_2PYDZ$	62
$(DHQD)_2PYR$	128
DIAD	49
diazomethane	107, 204, 252
DIBAH	22, 40, 108, 144, 160, 197, 272
Diels-Alder reaction	96, 110
1,5-diene	216
dienophile	69
(−)-diethyl D-tartrate	197, 202
dihydroquinidine	62
dihydrocorynantheine	101
dihydropyrones	268
dihydroxylation	
-asymmetric	62, 128
-OsO_4	24
diisopropoxytitanium diazide	198, 203
β-diketones	234
3,4-dimethoxybenzyl (DMPM)	26
dimethyldioxirane	270
diols	24
dioxirane	270
dioxoborolane	46
dissolving metal	58
di-*sym*-collidine perchlorate	276
di-*tert*-butyldicarbonate	88, 108
di-(tributyltin)oxide	271
DMAP	47, 94, 108, 250
DMPM	26
DMSO	47, 196
$DMSO/CO_2Cl_2$	see *Swern*
domino radical cyclization	182
domino reaction	5, 110, 116, 134, 220

domino-*Heck* reaction	220	-as glycosyl acceptor	268
domino-*Knoevenagel*-hetero-		-as glycosyl donor	268
Diels-Alder reaction	110, 116	1,2-glycols	24
DOWEX 50W	199	glycols cleavage	62, 199, 204
dye-sensitizer	68	1,2-*trans*-glycoside	250
ECDI (see WSC)	263	glycosyl acceptor	248
EDAC (see WSC)	263	glycosyl donor	
EDC	47, 263	-donor phosphite	252
E-enolate	164, 214	-semi-acetal	248
efavirenz	71	-imidate	260
electrophilic aromatic substitution	236, 238	glycosylation	
endocyclic double bond	66	-boron trifluoride	261
endoperoxide	69	-stannous triflate	257
ene reaction	11	-TMSOTf	259
enol triflate	6	GM2	245
α,β-enone	60	*Grigg's* condition	128
α-epoxide	270	*Grignard* reagent	9, 77, 148, 180
epoxide opening	198	*Grubbs* catalyst	10, 145
epoxidation	197, 223, 202	β-H-elimination	41, 220
E-silyl ketene acetal	163	H_2	63, 93, 113, 132, 237
ester	165, 172	H_2, Pd/C	113, 132
-α,β-unsaturated	160	halogen-metal exchange	4, 124
-formation	94	*Heck* reaction	128
-hydrolysis	205	HF-acetonitrile	154
-methyl	252	HF-pyridine	49, 150
ethyl-CDI see WSC	263	$Hg(OAc)_2$	152
ethyleneammonium diacetate	111	high pressure	64
(*S*)-2-ethylpiperidine	88	himbacine	85
Et_3SiH	126	himbeline	85
Evans aldol addition	142, 161	hirsutine	101
Evans auxiliary	141, 162	histidine	241
Evans-Mislow rearrangement	166	HOBT	47
facial selectivity	116, 227	homoallylic alcohol	42
Felkin selectivity	18	*Horner-Wadsworth-Emmons*	23, 25, 196
Felkin-Anh model	31, 149	*Hünig's* base	69, 142, 225, 252
Finkelstein reaction	45	hydrazine	234
Fischer-Helferich activation	273	hydrazine acetate	250
fluoride donor	274	hydrazone	234
free electrons	58	1,2-hydride migration	169
Friedel-Crafts catalysts	200	hydroboration	65, 213
γ,δ-unsaturated aldehyde	164	-(*S*)-Alpine Borane	212
ganglioside	245	-rhodium(I) complexes	19
gauche effect	147	hydrogen sulfide	262
geminal vinyl dibromide	170	hydrogenation	64, 93, 113, 132, 191, 200
γ-hydroxybutenolide	69	-asymmetric	107
Gilbert-Seyferth phosphonate	25, 169	hydrogenolysis	191, 254
Gilman-cuprate	61, 181	hydrometalation	38
glycal assembly strategy	265, 268	hydrozirconation	38
glycal epoxide method	265	hyperconjugation	131
glycals		β-hyperconjugation	66

I(coll)$_2$	276
IBX	160
ICl	126
ICN	126
imide	190
iminium ion	105, 190
indole	108
indole alkaloid	101
insecticidal activity	138
insertion	41, 220
intersystem crossing	68
iodination	67
iodine glycosylation	276
[(−)IPC]$_2$B-allyl	42
ipso substitution	130
irinotecan	121
isomerization	216
isopinocampheyl-borane	43
isopropylidene ketal	253
Jones reagent	46, 189
β-keto carboxylic acid	104
ketone	
-α,β-unsaturated	58
-prochiral	81
KHMDS	6, 40
kinetic control	40
KMnO$_4$	106
Knochel cuprate	181
Knoevenagel condensation	110
Koenigs-Knorr reaction	257
KO*t*Bu	242
LACDAC reaction	268
laurallene	137
Lewis acids	
-BF$_3$·OEt$_2$	30
-SnCl$_2$	30
-TiCl$_4$	30, 142
-TMSI	130
L-fucal	268
LiAlH$_4$	9, 40, 144, 190
LiBH$_4$	144, 161
LiCl	7, 47
LiClO$_4$	188
Lindlar catalyst	93
lithium diisopropylamide	61
Luche reduction	160
2,6-lutidine	144
Lycopodium alkaloids	177
macrocyclization	28
manganese dioxide	23, 38, 106, 160, 171
Markovnikov's rule	152
MCPBA	60, 166, 181, 197, 223
Meerwein's reagent	235
Meldrum's acid	110
mesylate	9
meta sense	97
metathesis reaction	145
methoxymethylenation	151
methylamine	190
methyl *tert*-butyl ether (MTBE)	74
Michael reaction	5, 160, 180, 222
Michael system	5, 38
mismatched reaction	18
Mitsunobu reaction	49
MnO$_2$	23, 38, 106, 160, 171
Mukaiyama reaction	30
multicomponent reaction	110
myxalamide A	157
N,N'-carbonyldiimidazole	240, 270
N,O-acetal	126
NaBH$_4$	46, 62, 160
N-acetyl-D-(+)-galactosamine	245
N-acetyl-galactosamine	265
N-acetyl-glucosamine	265
N-acetyl-neuraminic-acid	245
NaIO$_4$	45, 62
neighboring group effect	202, 249, 261
N-iodosuccinimide (NIS)	126, 219, 276
nitration	236
nitrile effect	258
nitronium ions	236
N-methylation	99
N-methylmorpholine-*N*-oxide	24, 129, 168, 222
N-methylpiperazine	239
norephedrine	78
Normant-cuprate	61, 180
N-pyrrodinyl norephedrine	78
nucleophilic addition	65, 74, 77
nucleophilic substitution	67, 78, 124, 237, 239
β-nucleosides	200
olefin metathesis	10, 145
-mechanism	146
organocopper reagents	61
organometallic compound	124
organozirconium	38
ortho-lithiation	75, 89, 125
OsO$_4$	24, 128
oxalyl chloride	see *Swern*

oxaphosphetane	218	*p*-nitrophenyl chloroformate	205
oxazaborolidine	68, 213	polymer supported reagents	200
oxazolidinones	140, 161	polyoxin J	193
oxazolidine-2-thione	140	porphyrin	68
oxazoline	49	potassium permanganate	106
oxidation		PPTS	66
-allylic alcohols	23, 187	prochiral ketone	81
-*Dess-Martin*	12, 29, 67, 160	propargylic alcohol	219
-IBX	160	protodesilylation	186
-*Jones*	46, 189	pseudoequatorial position	162
-MnO_2	23, 38, 106, 160, 171	pyrazoles	234
-PCC	185, 229	pyridazine	62
-PDC	185	pyridinium chlorochromate (PCC)	185, 229
-photooxidation	11, 68	pyridinium *para*-toluenesulfonate	64
-RuO_4	13, 198, 204	pyrimidine bases	193
-*Swern*	20, 47, 148, 149, 160, 196	quinoline alkaloid	121
-TPAP	13, 160, 168, 222	radical annulation	134
oxidative addition	92, 173	radical domino reaction	134
oxidative cleavage	63	radical reaction	182, 278
oxocene	145	*Raney* nickel	6, 93, 98
π-deficient	236	RCM	10, 145
π-excessive	236	rearrangements	
palladium catalysis	40, 92, 173	-[2,3]-sigmatropic	166
palladium on charcoal	113, 132, 191, 200, 237	-[3,3]-sigmatropic	163, 216
		-1,2-rearrangement	65
(+)-paniculatine	177	-*Claisen*	163
PCC	185, 229	-*Claisen-Ireland*	163
$Pd(OAc)_2$	173	-*Cope*	216
peptide coupling reagent	205	-*Wagner-Meerwein*	105
peptide linkage	193	reduction	
peptidyl nucleosides	193	-azide	262
periodic acid	199	-DIBAH	22, 40, 108, 144, 160, 197, 272
PHAL	130		
phase transfer catalyst	235	-double bond	64, 98
phosgene	82, 269	-enantioselective	91, 149, 185, 212
phosphonate esters	196	-EtSiH	127
phosphonates	23	-H_2	63, 93, 132, 113, 191, 200
phosphonium salts	48, 153	-$LiAlH_4$	9, 40, 144, 190
phosphorous ylide	48	-$LiBH_4$	144, 161
photochemical oxidation	11, 68	-Li/NH_3	58, 143, 184, 278
photolysis	134	-$NaBH_4$	46, 62, 144 160
photooxidation	11, 68	-1,2-reduction	160
photosensitizer	11	-1,4-reduction	160
Pictet-Spengler reaction	105	-*L*-Selectride	149, 185
(−)-α-pinene	43, 213	-triple bond	219
piperidine	89, 190	reductive amination	114
piperidine alkaloids	85	reductive elimination	7, 40, 92, 173
pivaloyl	256	reductive iodination	219
pivaloyl chloride	74, 140, 172	*re* face	42, 143
PMB	20	regioselectivity	62, 69, 96, 220

reverse anomeric effect	251	*Stille* reaction	7, 27, 40
ring-closing metathesis	10, 145	substrate control	142
ring-opening metathesis polymerization	145	sulfenylation-dehydrosulfenylation	60
Rose Bengal	68	sulfide	61
RuCl$_3$	13, 45, 199, 204	sulfonate	49
RuO$_4$	13, 204	sulfonic acid	238
saponification	28, 104, 172, 135	sulfoxide	4, 61, 166
Schlosser conditions	124	*Suzuki* reaction	27, 173
Schrock catalyst	145	*Swern* oxidation	20, 22, 47, 148, 160, 187, 196
Schwartz reagent	38	*syn* addition	65, 220
scopadulcic acid A	209	*syn* elimination	220
L-Selectride	149, 185	*Takai* reaction	24, 90
selenium dioxide	187	tandem radical cyclization	182
L-serine methyl ester	47	L-tartaric acid	88
sesquiterpene	1	TAS-F	31
SET	184	TBABr	271
Seyferth-Gilbert reagent	25	TBAF	23, 44, 67, 154, 170, 184, 256
Sharpless dihydroxylation	62, 129	TBAT	31
-(DHQD)$_2$PHAL	130	TBCO	155
-(DHQD)$_2$PYDZ	62	TBS	25, 49, 65, 143, 154, 163, 167, 170, 184, 213, 251, 269, 272
-(DHQD)$_2$PYR	128	TBSOTf	144
Sharpless epoxidation	197, 201	TDS	253
si-face	42, 142	TEMPO	95
sildenafil	231	TMSOTf	250
silyl enol ether	30, 163	Troc	249
silyl group migration	254	*tert*-butyloxy urethane	88
silyl protecting groups		*tert*-butyl-diphenylsilyl ether	64
-TBS	25, 49, 65, 143, 154, 163, 167, 170, 184, 213, 251, 269, 272	*tert*-butylhydroperoxide	202
-TDS	253	tetrabutylammonium iodide	235
-TES	25	tetrahydro-β-carboline	106
-TIPS	38, 154, 167, 269	TFA	29, 99, 115, 186, 256
-TMS	25, 30, 186	thallium hydroxide	27
-TPS	38, 45, 64, 66, 213	thermodynamically controlled	215
Simmons-Smith reaction	46	thermolysis	61
singlet oxygen	11, 68	thioether	166
SnCl$_4$	30	thionyl chloride	237, 239
sodium cyanoborohydride	99, 272	thymine	193
sodium periodate	199	TiCl$_4$	30, 142
solvated metal cations	58	tin(II) chloride	275
solvent effect	258	TIPS	154, 167, 269
solvomercuration	152	TIPSOTf	167
Sonogashira coupling	91	titanium(IV) tetraisopropoxide	202
spiroindolenine intermediate	105	TMEDA	76, 143
β stabilization	131	TMSCl	180
stannanes	7, 170	TMS ether	30
stannous triflate	258	TMSI	131
stannylation reagent	271	TMSLi	61
stereoelectronic effects	114	TMSOTf	200
stereoselective allylation	42		

tosyl chloride	165
TPAP	13, 160, 168, 222
TPPTS	173
TPS	38, 45, 64, 66
trans-decalin	59
transfer hydrogenation	107
transition states	
-boat-like	114, 213, 216
-chair-like	42, 114
-*endo-E-anti*	97
-*endo-Z-syn*	97
-*exo-E-anti*	97
-*exo-Z-syn*	97
-four-membered	146, 153, 169, 219
transmetalation	39, 79, 180
trialkyltin alkoxide	271
tributylstannyl ether	271
tributyltin hydride	182
trichloroacetimidate	250
trichloroacetimidate method	250
trifluoroacetic acid	29, 99, 115, 186, 256
trimethoxonium tetrafluoroborate	235
trimethylaluminum	21
ultrasound	112
α,β-unsaturated aldehyde	39, 168
α,β-unsaturated ester	160
α,β-unsaturated ketone	58
urethanes	204
VIAGRA™	231
vicinal diols	62
Vilsmeyer intermediate	238
vinyl stannanes	171
Wagner-Meerwein rearrangement	105
Weinreb amide	144, 214
Wilkinson's catalyst	64
Williamson ether synthesis	140
Wittig reaction	22, 48, 62, 90, 151, 152, 168, 171, 196, 218
-*E/Z*-selectivity	152
WSC	263
ylides	48, 90, 152, 163, 168, 171, 196, 218
Zemplén conditions	263, 279
Z-enolate	142, 161, 214
Ziegler reaction	124
Zimmerman-Traxler transition state	42
zinc chloride	274
zirconium	38
$Zn(CH_2I)_2$	46
$ZnEt_2$	79